THE STRUCTURE

WORKS OF MAHENDRA RAJ

THE STRUCTURE

WORKS OF MAHENDRA RAJ

Vandini Mehta
Rohit Raj Mehndiratta
Ariel Huber

PARK BOOKS

Editors:
Vandini Mehta
Rohit Raj Mehndiratta
Ariel Huber

Book Design: Rohit Raj Mehndiratta, Nidhi Sah
Project Text: Vandini Mehta, Rohit Raj Mehndiratta
Text Editors: Neeraja Sankaran, Megha Singh
Copy Editors: Megha Singh, Aditi Ghosh
Image Curators: Rohit Raj Mehndiratta, Vandini Mehta, Ariel Huber, Nidhi Sah
Photo Editor: Ariel Huber
Printing: Thomson Press India Ltd

ISBN 978-3-03860-025-1

Published by
Park Books AG
Niederdorfstrasse 54
8001 Zurich
Switzerland

www.park-books.com

Front and back cover: View of the Hindon River Mills roof, Ghaziabad.

Page 1: Exterior view of the Municipal Stadium, Ahmedabad.

Page 2: Folded-plate frame structure, Tagore Memorial Hall, Ahmedabad.

Page 4, 140: Mahendra Raj at the Hindon River Mills site.

Page 5: Shukla and Mahendra Raj at their residence in Kailash Colony, New Delhi.

Pages 6-7: Exterior view of the Municipal Stadium, Ahmedabad.

Page 8: (Partial) Drawing no. 31.DH.37 B: Reinforcement Detail of Girder P.G.2: 6 August 1967. Project: Akbar Hotel, New Delhi.

Page 36: Raj at a construction site.

Page 236: Raj with RC Jain, BB Choudhary and Dr PC Jain.

Page 340: Raj receiving the Chairman's award from the former Prime Minister Deve Gowda, 1996.

To my wife Shukla, without whom all this could not have been possible.

– Mahendra Raj

GATE
10

Contents

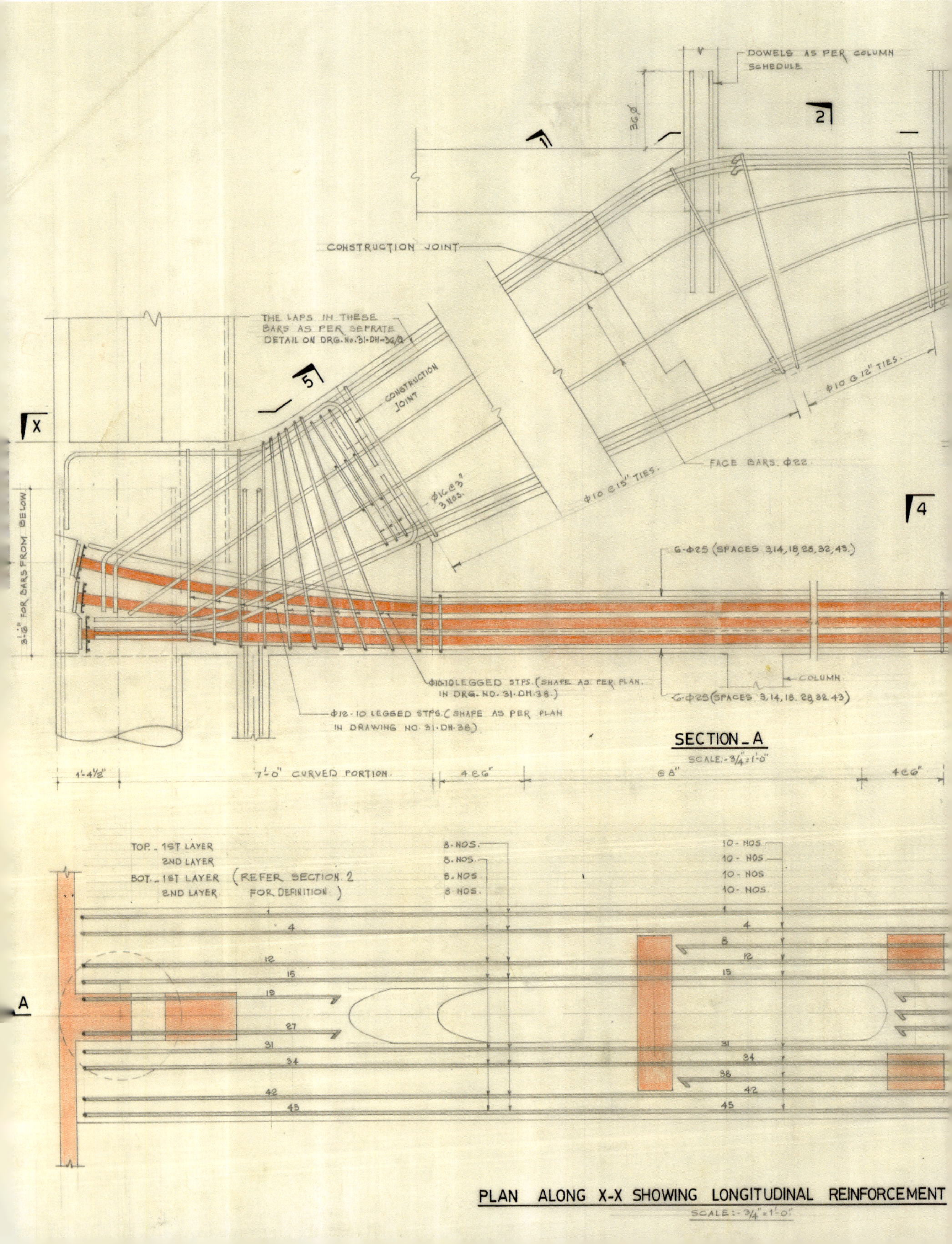

DOWELS AS PER COLUMN SCHEDULE
CONSTRUCTION JOINT
THE LAPS IN THESE BARS AS PER SEPRATE DETAIL ON DRG. No. 31-DH-36/2
CONSTRUCTION JOINT
FACE BARS. Φ22
Φ10 @ 12" TIES.
Φ10 @ 15" TIES.
Φ16 @ 3" 3 NOS.
3'-6" FOR BARS FROM BELOW
6-Φ25 (SPACES 3,14,18,28,32,43.)
Φ10-10 LEGGED STPS. (SHAPE AS PER PLAN IN DRG. NO. 31-DH-38.)
Φ12-10 LEGGED STPS. (SHAPE AS PER PLAN IN DRAWING NO. 31-DH-38.)
COLUMN
6-Φ25 (SPACES 3,14,18,28,32,43)
SECTION_A
SCALE:- 3/4"=1'-0"
1'-4½"
7'-0" CURVED PORTION
4 @ 6"
@ 8"
4 @ 6"
TOP. _ 1ST LAYER
2ND LAYER
BOT. _ 1ST LAYER
2ND LAYER
(REFER SECTION. 2 FOR DEFINITION.)
8-NOS.
8-NOS.
8-NOS
8 NOS.
10-NOS.
10-NOS
10-NOS
10-NOS.
PLAN ALONG X-X SHOWING LONGITUDINAL REINFORCEMENT
SCALE:- 3/4"=1'-0"

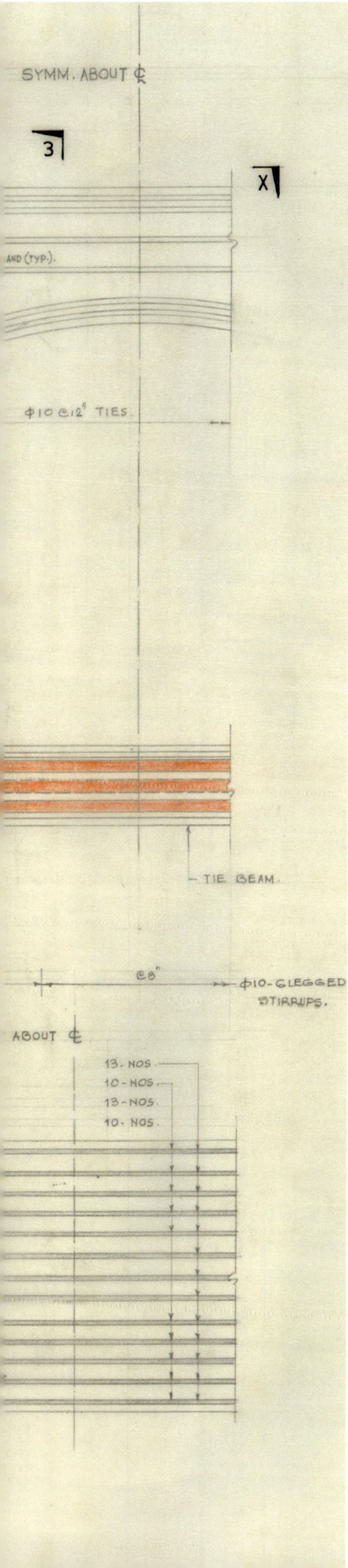
SYMM. ABOUT ℄
3
X
AND (TYP.).
Φ10 @12" TIES.
TIE BEAM.
@8"
Φ10-6 LEGGED STIRRUPS.
ABOUT ℄
13-NOS.
10-NOS.
13-NOS.
10-NOS.

Editors' Note

Putting together a book on Mahendra Raj has been an intimidating and monumental task. We must confess right at the outset that even though this book is a bit of a family production, working on a book on one's father/-in-law from the beginning was a place of pride. Pride in the sheer enormity of his achievements through the iconic works of engineering he has envisioned over the last half century. His oeuvre heralds a glorious time of nation-building in post-Independence India. Along with such a grand practice is his astute sensibility and precision as an engineer coupled with his unusual humility.

There was also a sense of discomfort with the proximity of this very biased place we stand in, and the vulnerabilities of perception that a close relationship brings. To offset that we decided that the book needed to bring his voice along with contributions of others as we sought to relate the story of a man, of his practice through images, words and drawings.

This book brings together 28 of Mahendra Raj's most important projects from over 250 projects to choose from. Each project that we have chosen shows a visionary and pioneering engineering solution, where **The Structure** makes a bold statement and is integral to the architectural expression. Some of Raj's buildings are unique, one of their kind, in India and the world, for instance, the Hall of Nations and Industries with their large-span space frame in cast-in-situ concrete, the Hindon River Mills, with bow-string arches in concrete spanning 48m. His practice also boasts of heroic structures built in the '60s and '70s such as the large folded-plate cantilevered structure in concrete for the Municipal Stadium, Ahmedabad, the folded-plate frame for Tagore Hall, the 26-storey high-rise of Usha Kiran, the transfer girders that helped realise the architects' needs for the Great Insurance Building and Akbar Hotel–all of them being first of their kind in India. The '80s saw further innovative and pioneering structures to help architects realise their intended forms such as the zig-zag columns stabilised by corridors as structural elements (NCDC Office), and the use of Vierendeel trusses (STC Offices) to create cantilevered large-span cascading forms.

Almost all these projects have invented and evolved analysis and design procedures with limited or no access to computers and more importantly, have a base in ingenious engineering intuition and judgement. Structural elements like transfer girders and sheer walls, procedures like post-tensioning and staging and details in every joint and node, in every taper and twist that shaped different elements, allude to unique structural logic.

Many of Raj's important works also belong to a time where inventive use of concrete as a building material was popular the world over. However, countries such as India, with lower levels of mechanisation and cheaper construction and labour, allowed experimentations that Raj exploited to create artefacts that are uniquely indigenous but also remarkably global in nature.

A key component of the book in understanding Raj's work comes through images. Archival and contemporary photographs indexing the power and elegance of different structures as they were constructed and as a final form make evident Raj's enormous contribution to the fields of architecture and engineering. The archival images reveal the technologies of construction in different decades of Raj's practice. They also show the context–the scale of modernisation of the society. These buildings were iconic objects, many with a commanding presence and setting. Then a rigorous task of photographing the selected projects was taken on as they exist today, first to break away from history and second, to also capture the structures as they have withstood the changing times and context. These photographs also bring the projects built over 55 years into a common contemporaneous narrative.

With greater understanding of Raj's work the recent photographs also capture the nuances of the engineering design, the fine structural details that make many of Raj's buildings elegant and enduring.

We also chose to use structural drawings as the principal means of presenting the buildings. In her unflinching support to her husband, Shukla Raj always joked that for every drawing architects made, the structural engineer made ten, but only got half the attention, if that. She may not have been exaggerating, as we discovered through the process of making this book! Most of the major projects we have presented here have over nearly a hundred handmade final drawings (not including the development drawings or almost full sets of alternative proposals made). The economy of resources and materials was not taken lightly even at the drawing level and typically every inch of the drawing sheet was carefully composed and packed with information. These are tremendous documents that we felt deserved to be presented as the primary record of the building and as relics of historic value.

An interesting element we discovered in the process of studying and choosing the structural drawings was of the drawings elaborating the construction sequences. These are a very crucial part of Raj's work for three reasons. First, the construction of the large-span structures was a carefully staged process and meant analysing the different structures that emerged due to changing forces and shifting loads at different stages of construction. Second, the construction of these complex and large buildings, with no previous precedence and with the rudimentary and manual construction technology of the early '60s to '80s, required instilling courage and confidence in the contractors, which was done through detailed communication. Third, the notes helped analyse and resolve issues such as economy of time, material and construction resources, and reduce the use of scaffolding and other materials while ensuring the required strength for the setting concrete.

Innovation from scarcity and a discipline of control echoes in Raj's work at various levels and are manifest in his drawings. It is also the point where economy and sensibility of detailing and an intuitive understanding of structural behaviour can only be attributed to the ways of a genius. From the smallest element, connection, detail and shaping of members, to the larger structural systems, there is a continuity of idea, simplicity of resolution of forces and an expressive flow of stresses. The artfully controlled use of reinforcing steel, a limited resource at the time, is presented through detailed visualisations of its multifarious formations and tensioning. These drawings reflect a quality of Raj's work that aptly resonates the words of Pier Luigi Nervi, *'The pattern of steel should always have an aesthetic quality and give the impression of being a nervous system capable of bringing life to a dead mass of concrete.'*

This book also shows some of the calculation sheets of different projects as historical documents, and records of the copious, detailed manual calculations done for each project and their several alternatives in those days. The sheets are used in the book not so much for explaining the project but to express a rigour and imagination of the engineer's mind and method that, at an esoteric level, are almost Klee esque figures of moments and forces.

Many people have been involved in the making of this book over the last few years. Mrinal Rajgopal helped with the enormous task of archiving the drawings and calculation sheets, which Arthur Lemaire continued and Noureen Nazir completed with her industrious work on the final document. The consistent support from Chaitali Sinha, Nirmala Devi, Parkash Chand Verma, Manish Bhardwaj and all of the MRC and Studio VanRO staff is much appreciated. We also appreciate the initial discussions with Christian Brunner, the early book design efforts of Sudeep Chaudhuri and the initial editorial inputs from Samprati Pani. We wish to give our thanks to the editorial team, particularly Neeraja Sankaran and Megha Singh, who laboured with us in the final weeks of this project, making the journey of revisions and editions pleasant. Nidhi Sah, thank you for your professionalism and hard work that helped realise our vision for the book. We are indebted to Abba Tor, Mark Jarzombek, Jaimini Mehta, Neelkanth Chhaya, Hans Ulrich Obrist and Sanjay Prakash for their contributions; to Rajeev Moudgil, Sunit Paul, Monika Correa, Rahul Mehrotra, Kathy Fehl, Devika Daulet-Singh, Pavan and Pankaj Mahatta, Bärbel Högner, Stephane Paumier, Kaiwan Mehta, KT Ravindran, Vikramaditya Prakash, Durganand Balsavar, Aakar Foundation, Charles Correa Foundation, Vastu Shilpa Foundation and the Doshi family for supporting the book in various ways. Our deepest gratitude for the sustained blessings and impatient encouragement from BV Doshi, and now late Charles Correa who we deeply regret could not see the final product.

Finally, we wish to express our gratitude to our families for their many years of support–specifically Uma, for believing in us and pushing us along, Dheera for indulging in us and Ratna mashi for reading some of the contributions. Last and most importantly, to the man himself, Mahendra Raj, for his trust in us, for his immense patience and for his readiness to explain large and little structural concepts to us repeatedly, which has been an unparalleled learning experience.

Double arches and hangers of the Hindon River Mills, Ghaziabad.

Preface

Mark Jarzombek

It is one of the most ironic aspects of our understanding of modern architecture that even though Le Corbusier and others celebrated the engineer's aesthetic, the actual engineers, the flesh and blood people, who produced that aesthetic, remain largely unknown. Some, like Iannis Xenakis and Vladimir Bodiansky, who worked on projects under Corbusier, or August Komendant, who worked with Louis Kahn, or John Skilling, who worked under Minoru Yamasaki, have attained a type of cult status among those who know something about modern architecture. But for the most part, engineers are anonymous figures. Who knows the name Stephen Giblett, for example, whose job it was to make the undulating brick facade of Frank Gehry's Dr Chau Chak Wing Building? Part of the problem is in the nature of the beast. Engineers are collaborators, they work in teams. Engineering is not a culture of egos. But that still doesn't quite explain why we–as moderns–celebrate the profession that has had such a profound impact on our lives and imaginations and yet can hardly come up with individual names.

On the surface Mahendra Raj falls into that predictable profile. His career was defined by the great period of India's post-Independence modernisation, which saw a vast construction boom that changed the look and feel of a young nation. Its ambition was marked by its buildings, bridges, dams, factories, airports, hospitals and banks. And though the tone of India's economy and government changed over the decades, sometimes favouring large-scale public infrastructures, at other times the private sector, it was Raj's unique gift to always bring to the conversation the magic of engineering. Whether working in the background for well-known architects, or serving a client directly as the designer, his astonishing contribution to the history of architecture in India has no real parallel anywhere in the world. This means that here in India we can perhaps start to rewrite the history of modern architecture by acknowledging the role that figures like Raj have played. This book is certainly an important and positive step in that direction.

Previous pages: Interior view of the transfer girder, New Great Insurance Company, Bombay.

Opposite: Triangulated plates and the post-tensioned joints of the Indoor Stadium at Srinagar.

Page 20-21: Interior view of the Hall of Nations, New Delhi.

Foreword

Abba Tor

I had the good fortune of first meeting Mahendra Raj in early 1957. We were both young engineers, in our mid-thirties, recently hired by Ammann & Whitney–a prestigious consulting engineering firm with its head office in New York City. Othmar Ammann was the original chief engineer of the Port Authority of New York and New Jersey and the designer of the George Washington Bridge. Charles Whitney was the father of American reinforced concrete design. They teamed up in 1946, after Ammann's retirement from the Port Authority, to found Ammann & Whitney.

Mr Ammann, Swiss born and educated, was a formal, soft-spoken gentleman who addressed everyone with the honorific 'Mister'. Charles Whitney, 'Charlie' to everyone, was the ebullient Midwesterner from Milwaukee. His mottos, 'Remember! Concrete cracks!' and 'Only lousy engineers don't have failures,' have stayed with me throughout my career.

Under their leadership, Ammann & Whitney rose to become one of USA's leading structural engineering firms in the very first decade of its establishment. The firm was responsible for many innovations. Its Special Structures Division, where Mahendra and I worked, was informally run, serving as an incubator for new ideas as well as developing the talent of its carefully chosen young engineers.

The firm was international both in terms of project locations and the origins of its personnel. Eager young engineers from the US, Switzerland, Scandinavia, Holland, Greece, China, Pakistan, Israel and India all worked together, each contributing from the special vantage point and strengths of his or her culture to a harmonious exchange of ideas.

The partners in charge were good mentors and supportive leaders. They led us 'riding easily between the reins,' with no raising of voices, no recriminations, no drama.

Boyd Anderson, Andy to us, was the partner-in-charge of the Special Structures Division. He was an inspired and inspiring engineer. He made us look at structure as a breathing, living, dynamic, moving and sometimes even mischievous, creature.We all got rid of the notion of 'statics' and its corollary 'make it stand up.' Our real job was to 'tame' the structure–allow it to move within limits without affecting its serviceability in the process.

Project meetings with Andy, around the small conference table in his office, were a pleasure and a privilege for us. Sometimes we did not even have to look at sketches or drawings; it was more of an exchange of ideas–a chess game without the board. If anyone ventured to say something which made no sense, Andy would not correct it head on. Rather, he would calmly say, 'If I understand you correctly…,' followed by the corrected statement. No arguments, judgement or drama. If one did not get the hint, one would most probably not be invited to the next meeting of the inner circle.

It was in this heady, nurturing and challenging environment that Mahendra and I found ourselves. We were encouraged to spread our wings, try different approaches, to learn by trial and error. This free professional and personal atmosphere was very conducive to forming professional and personal friendships. We both blossomed in this environment. Mahendra's quiet, deliberate and thoughtful manner, his engineering sense and people skills were appreciated by all–his peers as well as our superiors in the firm.

This was the period of innovative large-span concrete folded plates and/or concrete shell structures in the US.

Following in the footsteps of the great European engineers Nervi, Torroja, Esquillant and the Mexican, Candela, Ammann & Whitney came up with innovative designs for the Trans World Airlines hangar and the terminal at the Idlewild (now John F. Kennedy) Airport.

When I first encountered Mahendra, he was working on TWA's major hangar at Idlewild as its design engineer. The roof structure of this hangar consisted of cable-held large-span concrete folded plates. It was the first of its kind.

His next project was the Assembly Hall for the University of Illinois on which we worked in collaboration. This facility had 17,000 fixed seats, covered by a 122m-diameter concrete folded-plate dome (10cm thick) resting on a circumferential concrete ring. From above it looked like a giant orange squeezer. In order to avoid external buttresses, the ring was post-tensioned and the natural reaction forces of the dome were directed inwards along the sloping seating and vomitories down to the foundation ring. Mahendra designed the dome, while I took care of the lower functional container, that is, the usable spaces and foundations.

In the course of our professional collaboration our personal relationship developed and grew over time. I appreciated his calm pipe-smoker's demeanour, his judgement and sensitivity in all matters, professional and personal. Mahendra struggled with the decision between continuing at Ammann & Whitney and returning to India. He finally decided to return home. We continued our friendship by mail and reunited professionally in the early Seventies when he returned to New York City on a one-year sabbatical.

By then, I was a principal in my own firm, Pfisterer, Tor & Associates. We collaborated on some US-based projects and invented the now common practice of outsourcing: after his return to India, I used to send him (via air mail, there being no fax or internet back then) design development project documents for which his firm completed the detailed design and produced construction drawings. This arrangement was possible only for select projects that were relatively immune to daily or even weekly changes by the architect.

Mahendra and I last collaborated in 1990 on the Government of India Mission in New York City–a high-rise, 'needle' building housing offices and staff. Charles Correa, the architect, and Mahendra's firm in India developed the early schemes in New Delhi. The architects Bond/Ryder and I developed the final design and construction documents and followed the construction phase here.

Dealing with the combined bureaucracies of the Government of India and New York City further cemented our friendship. We survived and came out of the battles only scratched, almost intact.

Over the intervening years, I kept abreast of and admired Mahendra's many innovative projects in India, some of which I was able to see during my last visit to the country in 2007. What I saw in his structures was the continuation of a growth process that started in those exciting years of induction at Ammann & Whitney.

My Initiation into Structural Engineering

Mahendra Raj

I belong to a lower-middle class family. My grandfather dealt in wholesale fruits and vegetables in my native place Jhang Maghiana, now in Pakistan. My father did not want to pursue his father's line of work but wanted to get educated and take a job. He studied for a diploma in civil engineering and afterwards, joined the Punjab Public Works Department (PWD), Building and Roads branch, as an overseer. Quite early in his career he acquired the reputation for being a hardworking, knowledgeable, dedicated and honest officer. As a result, soon he got promoted to the level of Assistant Engineer and later to that of Executive Engineer, the position from which he retired in 1933.

We were a family of eight, five brothers and three sisters. The difference between our ages was substantial. The eldest in the family was a sister and the youngest a brother; the difference between their ages was over twenty years. My father was determined to educate all his children properly and was eager that all five sons become civil engineers. But this was an impossible task in such a large family with his meagre salary. He made it very clear to all his children that as soon as one of them got a job, he would have to educate the next brother and others in the family. As soon as the next one got a job he would, in turn, have to pitch in.

My eldest brother did not want to be a civil engineer; he became an automobile engineer instead, and got a job in Lahore, the then capital of Punjab, India. My next brother got through a competitive examination and got admission into one of the leading colleges of engineering in Roorkee and graduated as a civil engineer. He was supported by our eldest brother through his education. Meanwhile my third brother also managed to pass the admission test, joined Roorkee and graduated as a civil engineer. His education expenses were taken care of by the second brother.

Next was my turn. This was in 1943 and the Second World War was on. The engineering college at Roorkee was educating and training army engineers and was closed to civilians. So I joined the Punjab College of Engineering and Technology, Lahore, and graduated as a civil engineer in 1946. My education was supported by my third brother.

My youngest brother refused to become an engineer and wanted to be a writer in Hindi. He studied in Lahore, Allahabad and Banaras, got a doctorate and pursued his career as a teacher and a writer. It was my turn now to support him through his education. He retired as a Professor of Hindi from Punjab University, Chandigarh, and is a very well-known and respected Hindi author with a number of published books to his credit.

Meanwhile our sisters were also educated, supported by one brother or the other. That is how our father lifted us all up from a lower-middle to a middle class status.

Focusing on my own education, because this book essentially covers my career as a civil and structural engineer, I will offer some more details about it and my introduction to structural engineering.

I was born in 1924 in Gujranwala where my father was posted at that time. My earliest memories go back to the time when in my pre-school years I used to accompany my father on his tours. I distinctly remember one such trip in which he had to travel through a desert area on camelback with his little son perched behind and firmly holding on to him. It was a very enjoyable ride, a memory which I still relish at times.

I remember attending schools in Gurgaon and Hissar, where my father was posted. He ultimately retired from Hissar in 1933 when I was just nine years old. I moved with him to our native place Jhang Maghiana and was educated in a school located midway between the twin cities of Jhang and Maghiana. It was called 'Adhiwal School' meaning 'Midway School.' I studied there until 1937.

In 1937 I moved to Lahore from Jhang Maghiana. By that time my eldest brother who had become an automobile engineer was doing very well. He had a good job and a large house. He invited his younger brothers and sisters, who were still studying in schools and colleges, to come to Lahore, stay with him and get educated in known institutes with high standards of education.

In Lahore, I joined the Central Middle School as a student in the ninth class. I passed tenth class in 1939 and gained admission to Government College, Lahore, which was considered to be the best and most privileged educational institution of that time (it still has that reputation). I completed a Faculty of Science (F.Sc.) in 1941 and graduated with a BA Honours degree in Mathematics, my strongest subject, in 1943.

To gain admission to the Punjab College of Engineering and Technology was not that easy at the time. Seats were reserved on a religious basis among Hindus, Sikhs, Muslims and others, and further divided between agricultural and non-agricultural sections. For non-agricultural Hindus–the category I belonged to–there were only two seats. That is why even though the minimum educational qualification required for admission was FSc, it was impossible for non-agricultural Hindus to gain admission with this degree. It was only with BA (Hons.) and a high rank in the Government College that I could gain admission to the engineering college.

From 1943 to 1946 I studied at the engineering college. Every year there was a qualifying examination for promotion to the next year. Along with two of my other class fellows, I shared the top three ranks every year. In the three years of our engineering education, each of the three of us in turn stood first with the other two in second and third place.

In 1946 I graduated with honours as a civil engineer. The same year I was selected as an Assistant Engineer in the Punjab Public Works Department in the Buildings and Roads branch, the organisation in which my father had worked earlier. After an initial period of training in Lyallpur I was attached to the Soil Stabilisation Department of PWD in Lahore in early 1947 to learn the art of using stabilised soil in buildings and roads. This was the period when the non-violent struggle to gain independence under the leadership of Mahatma Gandhi was in the final stages of success. It was still not clear whether we would gain independence as one country or if we would be divided. At the same time fundamentalists of both religions were busy inciting Hindus and Muslims resulting in riots in Lahore and other parts of Punjab.

Finally it was clear that the country would be divided into India and Pakistan. But there were speculations on the line of division between the two countries.

Mahendra Raj with his classmates from the Punjab College of Engineering and Technology, Lahore, on a survey camp.

The Shimla-Tibet road and mule track.

There were constant rumours, unconfirmed though, of where Lahore would be assigned: one day it would be rumoured that it would be in India and the next day that it would be in Pakistan. All these speculations intensified the riots and the exodus of Hindus and Sikhs out of Lahore.

While the riots and exodus were going on, the government machinery was busy ascertaining from its various employees whether they would work in India or Pakistan after Independence.

I opted to work in India and soon after was posted to Shimla, which had been chosen as the temporary capital of that part of Punjab assigned to India. This choice was due to the fact that Shimla, being the summer capital of British India, already had the complete infrastructural facilities required for a capital city. On August 11, I took a train from Lahore to Shimla, one of the last two or three trains which had left Lahore and were not stopped en route for the Hindu and Sikh travellers to be massacred. I reached Shimla safely on the morning of August 12.

Earlier, one afternoon in the last week of July 1947, I had gone to the Lahore railway station on my bicycle, without telling anyone in the family, to book a railway wagon for transporting my household goods to Shimla, which I was entitled to in my new position. During this time, although I hadn't realised it, there had been some riots and cases of arson and stabbing at the railway station. Meanwhile my family at home had learnt of the disturbances and when they found out upon missing me that I had gone to the railway station, all by myself on a bicycle, they became very anxious. When I, oblivious of all this tension, returned home after completing my task, I received a big slap on my face from my eldest brother–the only time in my life that he got so angry with me. It was a slap of concern, anxiety, affection and all the stress which every individual in Punjab was going through on account of the partition of the country.

Coming back to my story, when I arrived in Shimla I was informed that I had been given charge of the Shimla-Tibet Road; jeep-able for about ten to fifteen kilometres out of Shimla, this road was just a mule track after that. But it was a very important and heavily-used trade route between Shimla and Tibet. During the rainy and winter seasons there were innumerable landslides which would block this track. It was my task to maintain it and keep it open for trade in all seasons.

Keeping this mule track of an approximate length of 200 kilometres open to trade was a very interesting assignment. The only means of communication between me, sitting in Shimla, and my maintenance staff posted along this track, was through hand-written messages, which were brought to Shimla by the maintenance staff members through a relay system on foot. Five or six days after its dispatch I would get a hand-written message from more than 100 kilometres away that on such and such night it rained very heavily and landslides had blocked the track, and that in anticipation of my sanction, labour had been employed to clear it. I never had any option but to approve the actions of my maintenance staff without any physical verification.

Survey and construction work along the Shimla-Tibet road.

After one year I was given the task of maintaining all the government property in Shimla as well as the Shimla-Kalka road. This assignment was fairly challenging and brought me close to higher officials of PWD and other departments.

It was during this assignment that I would meet senior officers of my department quite frequently while inspecting and maintaining their residences. Perhaps some of them noticed potential in me, which shaped my future career in the department. This assignment, which lasted from 1948–49, took place during a period of unusual activity in our department. Immediately after Partition in 1947, PWD was given the task of providing shelter to all the displaced persons who had moved to the Indian part of Punjab from Pakistan. A separate Rehabilitation Cell was set up with Mr PL Verma as the Chief Engineer. The task assigned to this cell was to build small housing colonies in various existing towns for displaced persons of middle class social status who could afford to pay for these houses; mud houses for the lower-middle class; and tented colonies for the very poor. It was an onerous task on account of shortage of funds and non-availability of construction materials. Very imaginative and ingenious techniques of construction such as stabilised soil and cheap treated wood, mud, bricks tiles were used quite effectively to build all these facilities.

Meanwhile, the search for a new capital city for the Indian part of Punjab had also begun. This search activity is amply documented and is well known. After investigating different existing towns of Punjab, the decision was made to build a new capital city in Chandigarh. Le Corbusier, the internationally renowned French architect, was chosen to prepare a master plan of the city and design four major buildings in its Capitol Complex: the High Court, Secretariat, Assembly and Governor's Residence. PL Verma, Chief Engineer of the rehabilitation project, was chosen to head this endeavour as well.

As I have said earlier, while in charge of all the government residences in Shimla my senior officers had taken notice of me and now they decided to try me out in a higher position. As future developments in my career would show, this attention and fast promotion worked in my favour to begin with, but later worked against me.

Verma created a new position of 'Technical Personal Assistant' to the Chief Engineer and appointed me to this post. It changed my life completely. Since it was a new position the scope of service of this assignment was not well defined. I was allotted a small room near the Chief Engineer's room in the Secretariat building in Shimla. Some stray files would come to my room but I was not doing anything worthwhile. However, I was supposed to accompany the Chief Engineer when he went out on tours to different cities where rehabilitation work was under progress and keep notes of our inspections and meetings. On our return to Shimla I was supposed to present these accounts to the Chief Engineer. Considering that the rehabilitation work in the state was under the charge of different Senior Superintendent Engineers, they did not much appreciate a junior officer like me, of the rank of an Assistant Engineer, questioning them and seeking details of the progress of work. In due course the senior officers developed a strange resentment towards me and I was transferred from the post, which was deemed unsuccessful and so abolished.

By this time there were further developments on the building of Chandigarh and I was posted as the first 'Resident Engineer – Chandigarh.' My initial task was to collect scarce materials like coal and steel and create a stockpile for the city's future construction. I was also required to assist in the search for a new site for a captive cement plant for Chandigarh along with identifying a source from where limestone for the plant could be mined. At first we investigated the possibility of setting up a new plant, but after investigations this idea was dropped. Instead it was decided to enhance the capacity of an existing plant in the nearby city of Surajpur. I worked in this position for almost one year and eventually got tired of the monotonous, unproductive work. In late 1949 I requested for a transfer.

For the next two or three years I worked in a couple of routine positions in two or three different cities in Indian Punjab. During this period the department held a competitive examination for three coveted positions of Class I officers. I entered the competition but got the fourth rank and missed being selected for these positions. By this time I was no longer the blue-eyed boy of the department. All such officers are normally posted in design cells. I got posted in the design cell of Chandigarh in late 1951/early 1952. Land for Chandigarh had been acquired by now and construction activity at the site had started.

When Le Corbusier was selected to design Chandigarh he had requested the Indian government to post three senior foreign architects of his choice to oversee the architectural and planning work in the city. An architectural department of Indian architects was set up to work under these three senior foreign architects who were, namely, Maxwell Fry and Jane Drew, a husband-and-wife team, and Pierre Jeanerette, a cousin of Corbusier. At the same time, Corbusier had also requested the government to set up a similar cell of structural engineers under a couple of senior French engineers. Before Chandigarh there used to be only one senior architect and one senior town planner attached to each state, while all the planning design work was undertaken by engineers. As such Corbusier's second request was denied. The fallout of this refusal to Corbusier had to be faced by all the structural engineers who worked on the Chandigarh project later on. I was one of them.

When I was appointed as Assistant Engineer Designs in late 1951 or early 1952, I was assigned to work under the Executive Engineer, Gulzar Singh, who had been one year senior to me in college but had been fortunate enough to go to Canada and acquire a Master's degree in Structures. Gulzar Singh was my friend as well as my boss. He gave me the American codes and design literature, introducing me to the art of structural design. In my four or five years in the PWD until then, I had not designed so much as a lintel! The first building assigned to the two of us was the structural design of the High Court, architecturally designed by Le Corbusier. Thus began my career as a structural engineer, which I stuck to for the rest of my life.

I started on this project with a clean slate. Every day was a day of learning something new, which I enjoyed considerably. It was a delight to study the American codes and design aids of Portland Cement Association, given to me by Gulzar Singh, and apply this knowledge to the design and detail of one part of the High Court or the other and prepare construction drawings.

Even though Gulzar Singh had done his Master's in Canada, this was also the first building that he was designing.

He would have meetings with seniors in the engineering and architectural departments and decide on a structural system while I would carry out the design, establish the sizes of members, provide the requisite steel and get structural drawings prepared. At that time the design process of every element of the building was a great adventure for me.

Since both of us were doing this type of architectural building for the first time it was hard for us, in the beginning, even to completely understand the architectural drawings prepared in Paris in Corbusier's office.

The Indian architect who helped us considerably in this task was Jeet Malhotra. He was very good at his job and cooperated with us in developing further the drawings he received from Paris. As luck would have it, these architectural drawings in Paris were also being prepared by an Indian, BV Doshi, who at that time was working for Corbusier.

Sizes of all members were established and fixed on the basis of Le Corbusier's modular dimensions, which he had developed. It was always interesting for us to argue with Jeet Malhotra to agree on slightly bigger dimensions of a member than those based on the Corbusier module. We would say, 'Jeet we need to make this beam 60cm deep and 23cm wide.' He would look at his module and say, 'Sorry, it can be only 58cm deep and 21cm wide.' Ultimately we had to accept the dictates of the module, but it was an interesting struggle.

Corbusier had designed the roof of the High Court as a series of cantilevered conoidal shells. These shells were difficult for us to design structurally and we also had trouble making them work functionally.

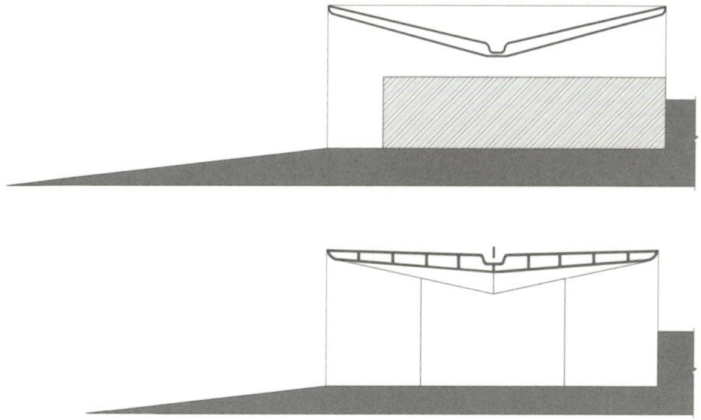

(Redrawn) Sketches showing the evolution of the roof parasol of the Chandigarh High Court, from the cantilevered conoidal shell structure designed by Corbusier (above), to the one suggested and finally built by the engineering team (below).

The roof parasol visible at one end of the High Court. An archival photo.

Gulzar Singh and I proposed a balanced cantilever roof supported on two columns in each bay from which the conoidal shell could be hung. This resulted in an upwards shift of a huge sculpturally-shaped gutter which Corbusier had expressed on each side elevation of the High Court building. It was a big struggle for our seniors to get Corbusier's approval for this change but eventually he agreed to it, and that is how it was built. Even so, as the one who had to size it and put reinforcement in it, I found its design to be a big challenge.

Concrete surfaces in Corbusier's buildings were supposed to be untouched and had to stay as they came out of the shutters. Even honeycombed surfaces and marks of construction joints were not supposed to be touched and repaired. If any part of the concrete surface with heavy honeycombing was touched up, even by mistake, and it came to the notice of Jeanerette, he would shout and scream at the engineers and workers involved saying, 'You, gentleman, have made a palace look like a hut after touching it up. Please do not do it again.'

Once the design of the High Court was completed in early 1953, Gulzar Singh wanted to move over to construction. By this time I had become a reasonably expert structural engineer, having been exposed to the design of so many different elements earlier. Moreover, I had acquired a real taste for structural design and had taken to it as a fish takes to water. So I got promoted as the Executive Engineer and took charge from Gulzar Singh. My first assignments in this post were to complete the remaining part of the High Court and to design the Secretariat building. I now had two assistant engineers, whom I trained to be structural engineers, working under me. They ran the numbers on some of the designs, all of which were checked by me. I still designed other parts of the building.

Cobusier had designed the Secretariat with six blocks, separated from each other with expansion joints. Blocks number 1, 2 and 3 are similar in design and accommodate all the administrative staff. These three blocks have a system of sun breakers in the front which are similar to each other and not very difficult to design.

Above: Archival photo of the High Court soon after construction.

Opposite, top and middle: Construction photos of the Chandigarh High Court.

Opposite, bottom: Photo taken at a construction site in Chandigarh.

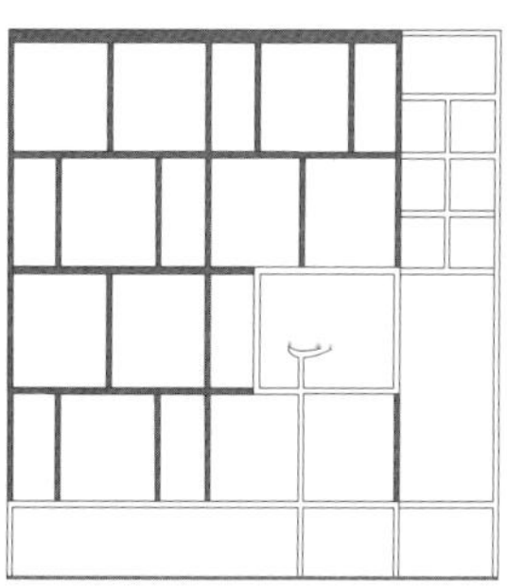

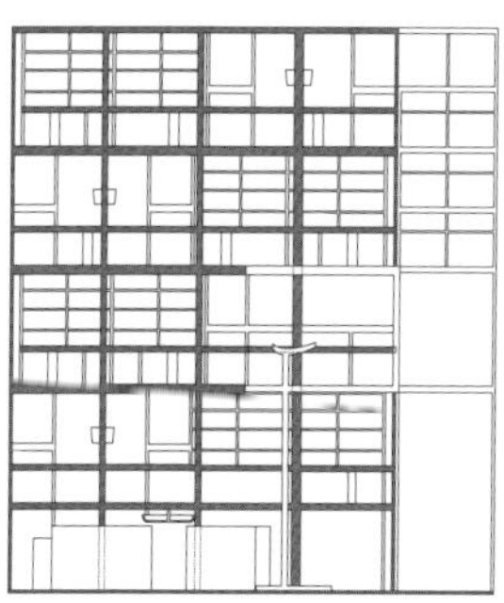

(Redrawn sketches) The design evolution of the sun breakers in Block 4 of the Secretariat building of Chandigarh, from the jumping columns originally proposed by Le Corbusier to the final design that was built.

Block number 4 seats the ministers and the governor, and has a system of sun breakers that is different from the previous three.

It was not possible to design the facade of the Secretariat as imagined by Corbusier for he had designed it with jumping columns that were not directly one below the other. If the facade had to be designed as drawn, some of the horizontal elements which supported the columns above and transferred load to staggered columns needed to be increased in size and strengthened. This alignment of columns did not change the facade much but when I explained all this to my seniors, they were reluctant to make the request to Corbusier themselves. Instead they took me directly to him and left it to me to explain to him why this change was required.

The underside of the conoidal shell roof of the High Court.

The facade of the Secretariat building showing the variations in the sun breakers in Block 4.

This was the first time I actually met face to face with Le Corbusier. To begin with he was furious and told Verma, who had taken me to him, 'Mr Verma, I had told you at the beginning of the project to have some French engineers and you refused. Look what is happening. First you made me change the concept of the roof of the High Court and now you want me to change the facade of Block 4–All this because your engineers do not know how to design!' He then addressed me directly and asked if I had an alternative suggestion. In fact, I had anticipated that question and taken a plan with me. He was fairly rude and told me to leave it with him. But a few days later I received a call that Corbusier wanted to see me. This time he was all smiles and charm. He had implemented changes more or less along the lines suggested by me and asked if it was now possible to design. I took one look at the revised facade, told him that it satisfied the structural requirements and thanked him profusely for accepting the change (see page 29).

By mid-1955 I had completed designing the Secretariat and also finished whatever had been pending on the High Court building. By then I had decided to spend the rest of my life as a structural engineer but also realised that even though I had been able to design these complicated buildings, there was much more that I needed to learn. So, nearly ten years after my graduation,

I became a student again and left Chandigarh to study for a Master's degree in Structures at the University of Minnesota in Minneapolis, USA.

Archival photo of the final stages of construction of the Secretariat building.

En route to USA I broke my journey in Marseilles to see Corbusier's building, Unité d'habitation, and also in Paris to meet the architect in his office. He was very kind and gracious on this occasion.

I was awarded my Master's degree in 1956. To my great surprise I was an all 'A's' student and became the favourite of my professor, Dr Paul Andersen. After completing the course I went to New York and took up a job with Ammann & Whitney, a very well-known firm of consulting engineers with a reputation for expertise in large-span structures. I worked there for three years from 1956 to late 1959.

Raj at his home in New York.

Left: Raj at work in the Ammann & Whitney office.

Below, left: The Assembly Hall building at the University of Illinois, Urbana-Champaign.

Below, right: The US Embassy building in Dublin.

Bottom, left: The Agricultural Trade Fair Pavilion designed by Minoru Yamasaki under construction in New Delhi.

Bottom, right: Raj (extreme right) with his team at the Pavilion site.

Mr Ammann was a Swiss bridge engineer and Mr Whitney, an American innovator with expertise in building design. As a teacher in Michigan, Whitney had propagated the use of folded plates very extensively in the US. He had also developed new techniques of shaping and designing large-span slender arches. I had very exciting three years at the firm, learning a lot about the design of large-span folded-plate as well as arched structures.

My first assignment with Ammann & Whitney was the design of a hangar for Mohawk Airlines, which was a 150-feet-long balanced cantilever in folded plates. The other main projects I worked on were a large-span arched hangar for Swiss Air and a folded-plate dome for the University of Illinois, Urbana-Champaign. Designed architecturally by Harrison & Abramovitz, this dome structure was like a saucer, with a 400-feet-diameter folded-plate seating area that lifted off the ground and a folded-plate domed roof of the same diameter. The roof and the seating were tied together with post-tensioning cables that provided the requisite horizontal stabilising force to both. I worked on the dome while another engineer friend worked on the seating. While designing the dome I found that the configuration in which the folds had been created did not give sufficient strength to the roof where required. I suggested an alternative to suit the strength requirements which was accepted by the architect and implemented.

Also during this time I worked on the design of a US Embassy building in Dublin, designed by John Johansen, where the architect twisted the exposed corridor columns.

In late 1959, a colleague and I worked on a pavilion for the US Government to be built at the Agricultural Trade Fair which was being held in New Delhi. This pavilion was designed by the architect Minoru Yamasaki, who was later to design the 100-storey-high World Trade Centre in New York City. Working with the team from Minoru Yamasaki's office we successfully completed the pavilion in time for the inauguration of the Agricultural Trade Fair by US President Eisenhower. Yamasaki too came to India to look at his pavilion and to attend the opening ceremony. During this time I had a number of interactions with Yamasaki and became somewhat close to him. As a result he offered me a position in his office as an associate on my return to USA.

I had another very interesting encounter during the early stages of this assignment. One afternoon at the construction site a tall, handsome young man walked up to me and asked, 'Are you Mahendra Raj?' 'Yes,' I said and he replied, 'I am architect Charles Correa from Bombay. I am just returning from Chandigarh where I had some work. While in Chandigarh I learnt about you as an engineer who worked on Le Corbusier buildings, the High Court and Secretariat. I also learnt that after completing this assignment you are returning to USA. I have sought you out and come to meet you to suggest that you don't go back but stay in India and work here. The country needs engineers like you.'

He continued, 'I studied in MIT in the States and have come back to work in India. Start your practice in Bombay and we will do some projects together. Don't return to USA.' I was somewhat taken aback by this interaction with a stranger, but it left quite an impression on my mind.

After the pavilion was completed, I had a number of options: I could return to USA and continue working for Ammann & Whitney or take up Minoru Yamasaki on his offer; or I could rejoin the Punjab PWD in India; or I could resign from the Punjab government and start my independent practice in Delhi or Bombay. Moreover, back in New York I had become close friends with an engineer colleague Dr RN Vakil from Ahmedabad, and the two of us often thought that whenever we went back to India we would start our own practice as consulting engineers and form a firm, 'Raj and Vakil.'

With so many options before me, I was thoroughly confused. There were pros and cons to each, which I discussed with my father and other family members as well as with some friends. My father and family members wanted me to stay back and rejoin the Punjab Government while different friends gave different, often conflicting, advice.

The final decision, of course, had to be taken by me and no one else, but it was tough.

At this time India was just waking up as a young free nation. Pandit Jawaharlal Nehru, our first Prime Minister, through his foresight and statesmanship had made India known the world over as a developing country full of potential and promise. A large number of professionals like me had gone abroad for higher studies and work experience, but motivated by the desire to serve the nation and tempted by its potential, most of them had returned or were planning to return to India, where they hoped to create a working environment and conditions similar to what they had experienced while they were abroad.

Finally I took my decision. I resigned from my job and did not return to USA. I wrote to Vakil requesting him to return so that we could launch 'Raj and Vakil' as we had talked about. He came back in early 1960, by which time I had moved to Bombay and, while staying with some friends, rented a small 250-square-foot office space from a friend. When Vakil arrived, the two of us took an apartment which we shared with another colleague who had also returned from the States and specialised in highway engineering.

Meanwhile, Charles Correa–who had left such a deep impression on me–and Doshi had started their practices, more or less at the same time, in Bombay and Ahmedabad respectively.

It was also around this time that the Bombay Presidency was split into two states. While Bombay remained the capital of Maharashtra, Ahmedabad became the capital of the newly formed Gujarat.

Bombay was already very well developed and industrialised while Ahmedabad, long a centre for the textile industry, boosted by its new status was going through a spurt of development and industrialisation. This meant that for young professionals, architects and engineers like us, it was Ahmedabad that generated more work.

The first project assigned to Raj and Vakil came from Charles Correa. He walked into our office one day and asked me, 'Raj, how would you like to do a stadium in Ahmedabad?'

I said, 'I would love to do one.' But he said, 'There is a catch. The total fee is only twenty-five thousand rupees. The client says that we have to design only one bay which will be repeated and has therefore offered twenty-five thousand for the design of one bay. I will split the fee with you half and half.' I readily agreed and we had our first project–to design one bay of a stadium to be built by the Ahmedabad Municipal Corporation in Ahmedabad.

Other projects came in from Charles Correa, BV Doshi and even from AP Kanvinde from Delhi. We hired one engineer, one draftsman and one office secretary and Raj and Vakil were in business.

Unfortunately, working and being together created some tensions and misunderstandings between the two partners and we decided to split in late 1960.

Vakil moved to Ahmedabad to start a new practice with others and I continued the practice in Bombay by myself alone.

During this time Charles, Doshi and I became good friends.

Charles had done some projects with another brilliant young engineer Shirish Patel who had done his post-graduate studies in England. The two of them had also become good friends.

One day sometime in 1964, Charles had the brilliant idea that he, Doshi, Shirish and I should become partners and start a large multidisciplinary firm. We had number of discussions but the grand alliance was never formed.

Meanwhile however, all these discussions brought Shirish and me closer and the two of us decided to get together and form a partnership under the banner 'Engineering Consultants (India)' in 1965.
For a couple of years we were quite busy. We had merged our respective offices and had taken a new, big office space where we could accommodate all our staff

members. But by1968 our workload reduced and we were in debt. I decided to return to the US via Africa and investigate possibilities for picking up some projects from these countries. Shirish was to continue looking after the office in Bombay. He shifted the office to a smaller place and I went abroad.

In Africa I went through Tanzania, Kenya, Ghana, Liberia and Senegal, and in each of these countries I tried to meet prospective clients. I had a certain measure of success and sent back some projects for conceptual design to Shirish. Unfortunately none of these prospects matured into projects that we could execute in Africa. Finally I landed in America and stayed in New York again for two years. Here I took part-time jobs, first with Paul Weidlinger and later with Abba Tor, my old colleague from Ammann & Whitney who by now had started his independent practice. I would spend part of my day scouting around for assignments which I could send back to my office in Bombay. Again, I had a certain measure of success and did manage to send some small assignments to be done in Bombay. Considering that this was a period of telephone, telex and airmail, it required quite an effort to service a project in USA while sitting in Bombay. But we managed to do a few.

By 1970 our work in the Bombay office had again picked up and we got a large project for the Interstate Bus Terminal to be done in Delhi.

Shirish asked me to return to India, which I did. Upon my return we decided that one of us should move to Delhi. I was the natural choice to go, being from that area, and Shirish stayed back in Bombay. He had his roots here.

I moved to Delhi and started a small office. I rented a place where I could have my office and residence combined. It was not difficult for me to have this arrangement because I was still a bachelor.

Later, sometime in 1971, by which time I was married to Shukla and had our first child, the partnership with Shirish also broke. We decided that I would retain the title of 'Engineering Consultants (India)' as a proprietary firm and stay in Delhi, while Shirish would have his practice as 'Shirish Patel and Associates' in Bombay.

So from 1960 to 1971, I worked for a short while under the banner of 'Raj and Vakil,' then as 'Mahendra Raj,' then as 'Engineering Consultants (India).' Sometime later in 1976, I formed 'Mahendra Raj Consultants Pvt. Ltd.,' because I had some overseas projects and foreign exchange earned on such projects would have tax benefits if the organisation was private or public limited. Ever since, I have been operating under both banners: Engineering Consultants (India) and Mahendra Raj Consultants Pvt. Ltd. It was perhaps in early 1952 that for the first time in my life I designed an element in the structure of the High Court in Chandigarh. I did not know at that time that I would keep designing structures for the rest of my life. But that is what I have been doing for the past sixty years and more. That is how destiny decided that I become a structural engineer.

A large number of engineers have helped me in this task. They are all listed at the end of the book. I will however like to name a few whose contributions have been significant. These are Krishna Hegde, TT Fatehi, Ganesh Juneja, Divekant Parikh, VV Nori, BB Choudhary, Shubhendu Pal, Mohinder Singh, CM Sardana, Ashok Nanda, Parminder Singh, J Ghose, Maya Rao, VK Dogra, Narendra Pratap, SS Mann and Aditya Sharma.

I gratefully acknowledge the contribution of my son Rohit and his wife Vandini, who have worked very hard and with utter dedication for a long time to compile this book and get it published. Their friend from Switzerland, Ariel Huber, has been of considerable help to them in this effort. He has also been responsible for a large number of photographs of different projects included in this book.

The projects in this book are those which were essentially designed by me and under my guidance, regardless of the banner under which I was operating. I hope you find them of some interest to you.

1960s

HINDUSTAN LEVER PAVILION

NEW DELHI | 1961

HLL · HLL

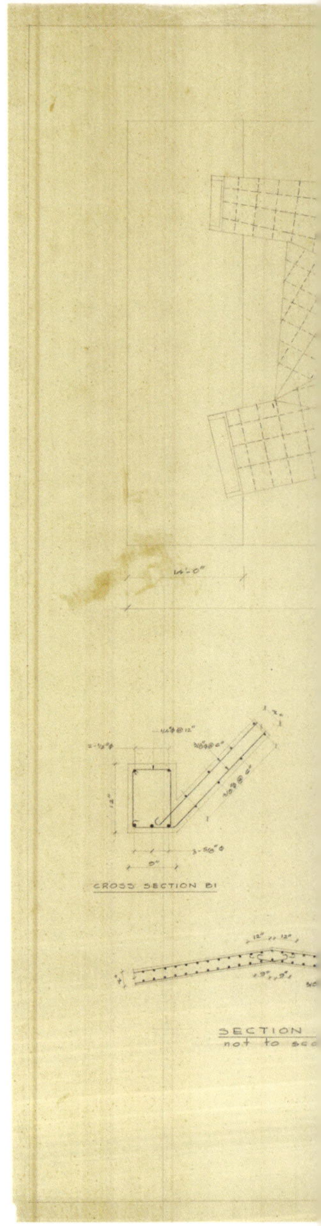

This temporary structure, designed for the 1961 Industrial Fair in Delhi, was inspired by crumpled paper.

As Mahendra Raj recalls, 'Charles (Correa) came to me and placed a crumpled paper on my desk. He said, "Raj, this is a pavilion for Hindustan Lever for the Industrial Fair in Delhi. Do you think we can build it?" We did build it, not the way he had crumpled the paper, of course, but still a crumple to satisfy some structural requirements.'

Raj suggested that every crumple have a definite geometric shape such as a triangle or quadrangle. Also the solid angle between the edges of two adjacent crumples needed to be as acute as possible. Based on these parameters, Correa realised his architectural form as a structure of intersecting irregular plates. The thickness of the plates varied from 2½ inches to a maximum of 8 inches. The thinner plates were gunited in concrete and thicker ones in shotcrete. Other than an architectural plan and approximate sections, there were no drawings. The structure was analysed, designed and detailed on the basis of a scaled model alone. Raj's structural drawings detailed a system of foundations, reinforcement, construction, methodology and thickness of different plates with constant reference to the scaled model, which finally formed the dimensional base for construction on site.

Architectural building model.

Previous pages: Photo of the completed pavilion, finished with gunited and shotcrete plates.

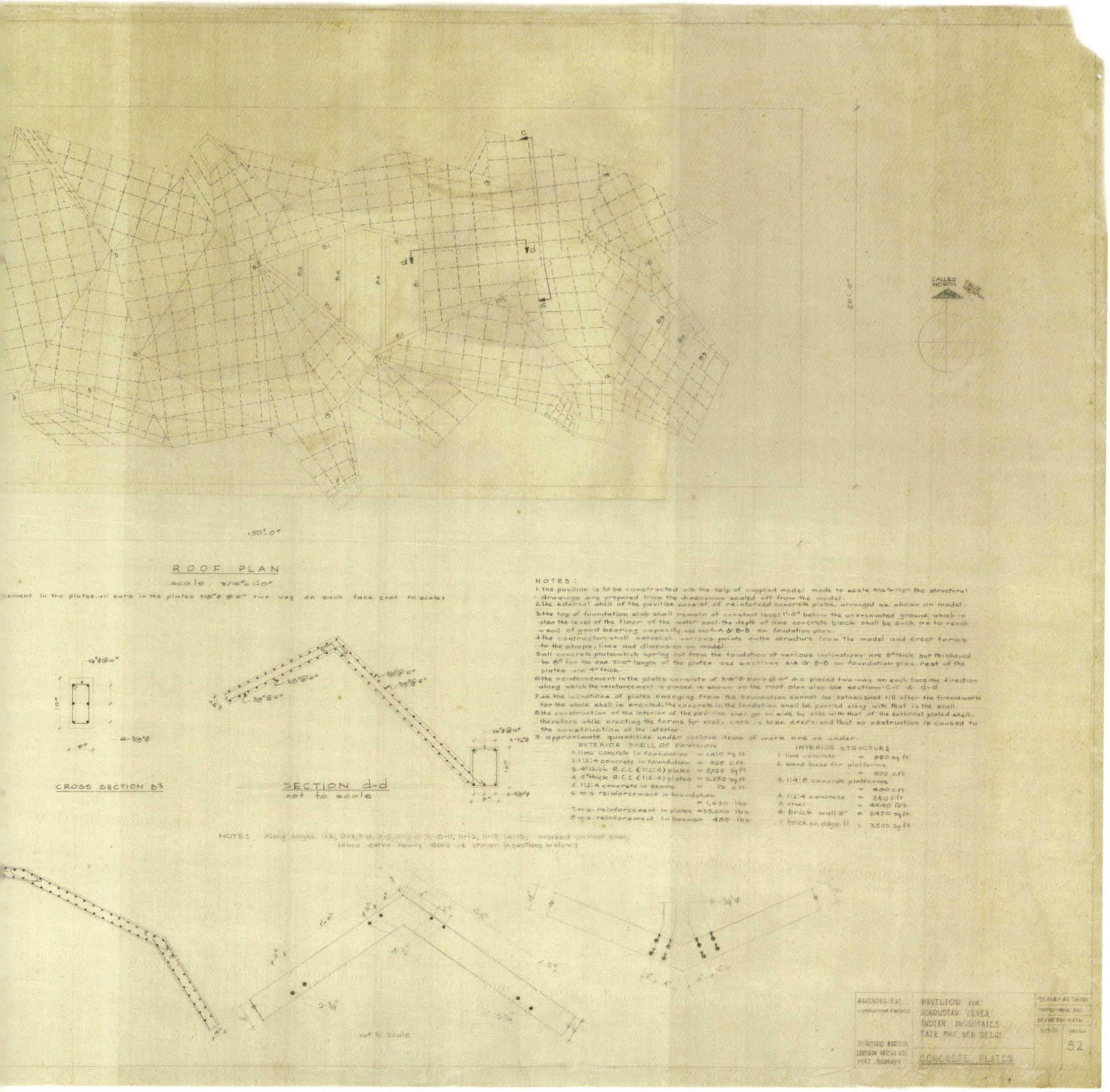

Drawing no. S2: Concrete Plates: 18 August 1961

Roof plan showing the reinforcement in plates, typical cross sections at the edges and skylight openings. The notes guide the contractor on the method of building using the model by establishing various points on the structure from the model and creating a formwork to the shape, line and dimensions on the model. All concrete plates that spring out from the foundation at various inclinations are 5" thick but thickened to 8" for the last 3'–0" length of the plates. The remaining plates are 4" thick. The reinforcement used was 3/8" diameter bars @ 6" oc placed two ways on each face. As the inclinations of the plates emerging from the foundation could not be established until after the formwork for the whole shell was erected, the concrete in the foundation could only be poured along with the shell.

View of the pavilion entrance.

HINDUSTA
EVER

Construction photographs showing the triangulated reinforcement layout of the plates and their edge conditions.

Previous pages: View of the intersecting irregular plates of the pavilion.

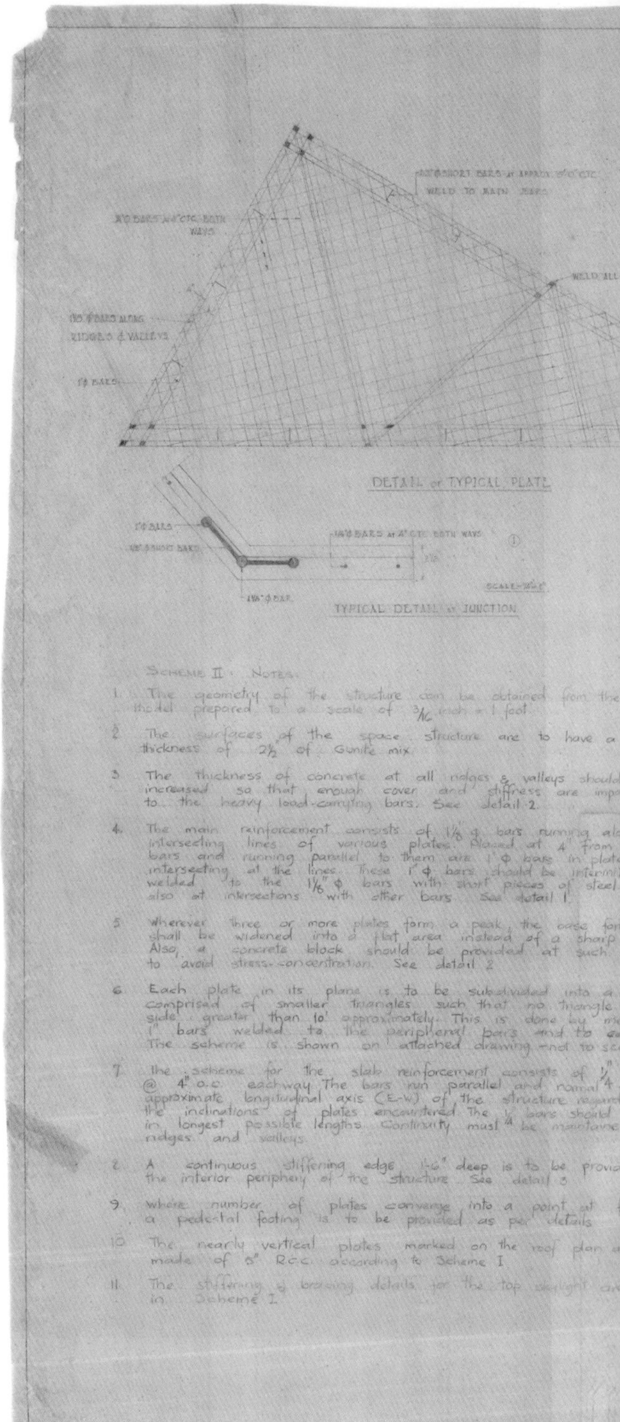

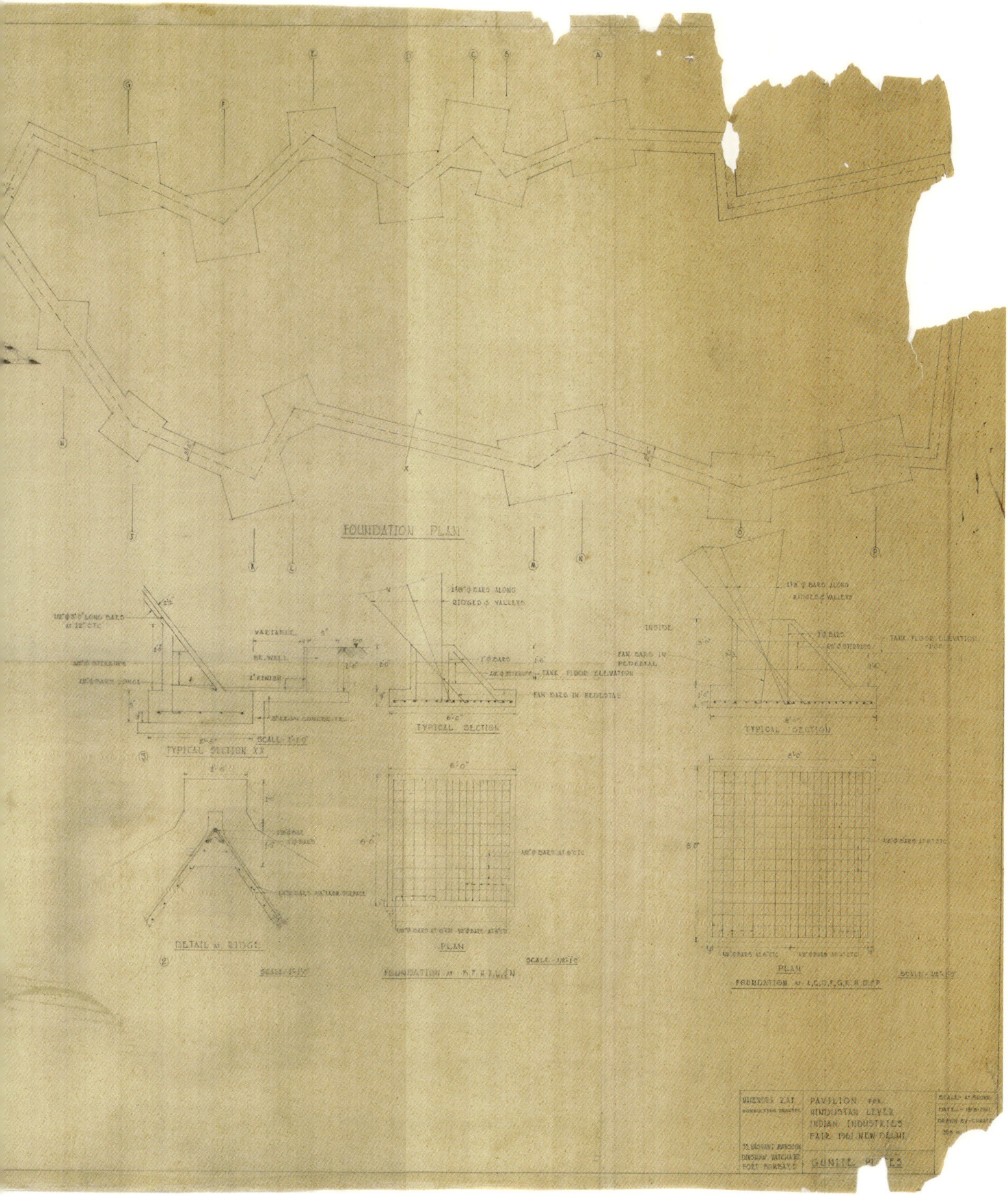

Drawing: Gunite Plates: 18 August 1961

Drawing showing typical triangulation of plates, the foundation plan and details. Each plate in its plane is subdivided into a framework of smaller triangles made of 1" reinforcement bars such that no triangle has a side greater than approximately 10'.

Interior and exterior views of the pavilion.

HINDUSTAN
LEVER
HLL·HLL·

NEW GREAT INSURANCE COMPANY
BOMBAY | 1963

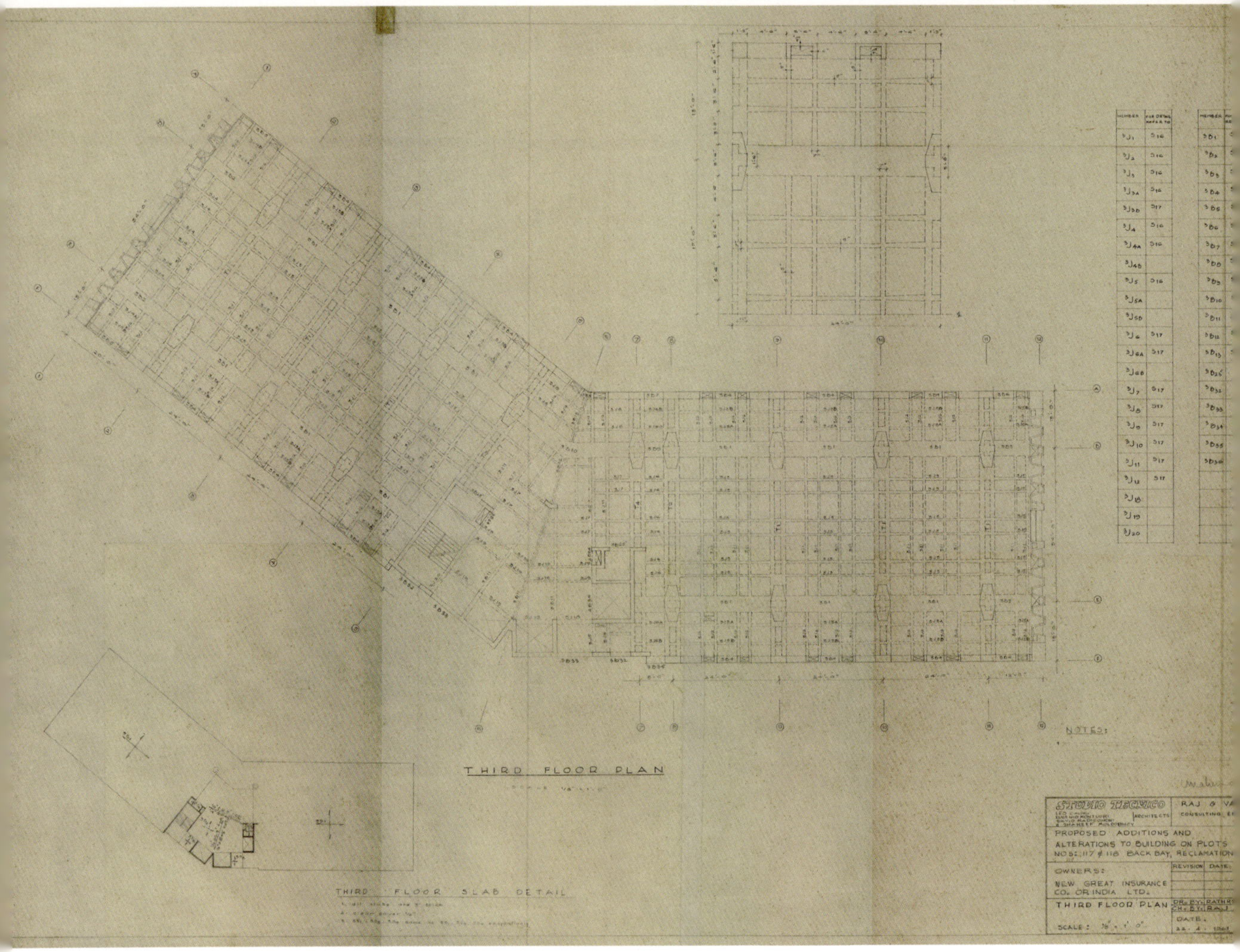

Drawing no. S13: Third Floor Plan: 22 April 1961

The New Great Insurance Building is a six-storey office block that was constructed on an existing three-storey garage located in front of Churchgate Station in Bombay. The clients wanted to utilise the revised floor area ratio that was made available by the city building codes.

Mahendra Raj explains, 'There were no old drawings and no other information, but the client wanted to make a huge six-storey structure above, keeping the existing three storeys that had a very irregular structure with columns everywhere. They wanted to carry the new structure on the existing columns without knowing their strength. This was not possible. I looked at the plan and suggested puncturing through the existing slabs and making a few additional columns supported on new piles. I made some sketches of tree-like forms and proposed a structural system. The architect Radiconcini liked the idea, and we re-planned the new upper floors based on my suggestion. At that time, this building evoked the interest and curiosity of quite a few people. A very well-reputed American magazine, *Engineering News Record*, covered it in one of its issues.'

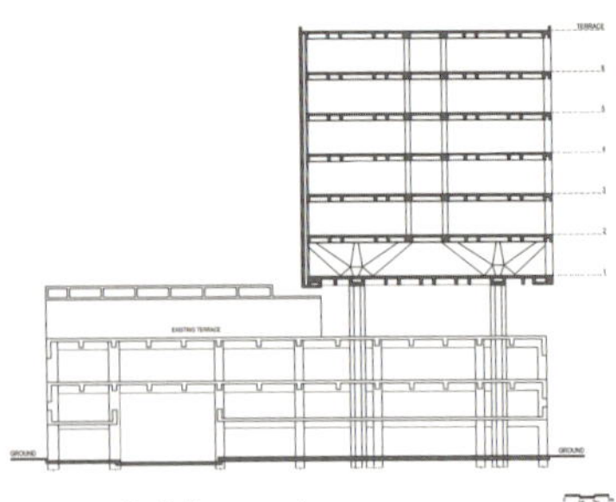

Building section.

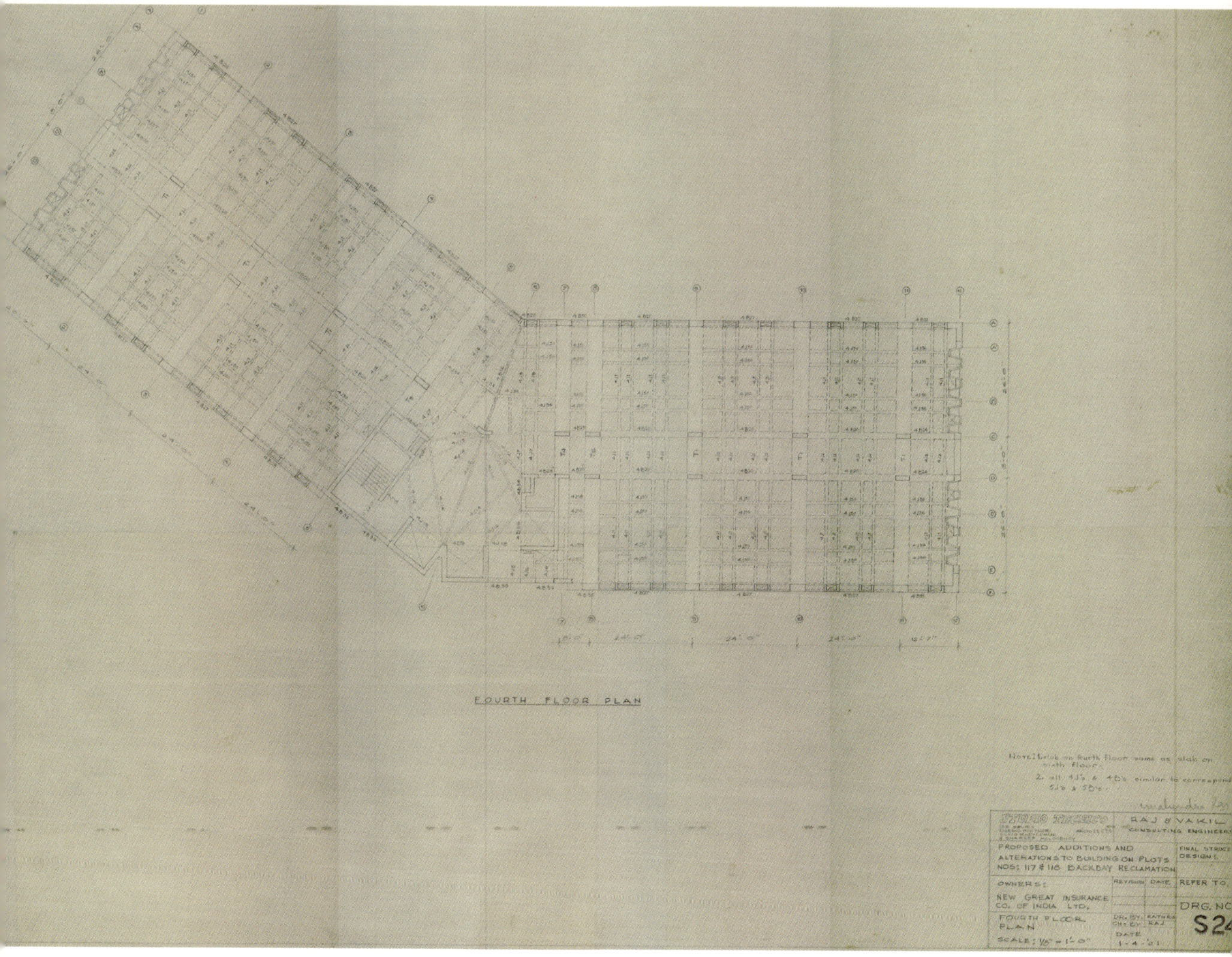

Drawing no. S24: Fourth-Sixth Floor Plan: 1 April 1961

The new office block was formed on a grid of cross-shaped columns 24 feet apart. These columns puncture the existing building and fork to carry the columns along the periphery and corridor of the office floors above. The new office floors have a two-way cross-ribbed structure of 24 x 26 feet spans, with additional mullion columns along the periphery to lighten the ribs. The form of the addition also followed the existing peripheral system of the garage, thereby forming a V junction. The junction marks the lobby area and is structurally resolved through elegant, triangulated ribbed geometry.

Structure of the lobby ceiling.

Architectural building model.

Previous pages: Tree-like girders set 24' apart transfer the loads of the new six-storey structure above.

Reinforcement model of the transfer girder.

Construction photograph showing the shuttering of the first-floor slab.

Exterior view of the new building block set on the existing garage structure.

Opposite: Interior view of the girders that support the new building above.

GANDHI MEMORIAL HALL

NEW DELHI | 1964

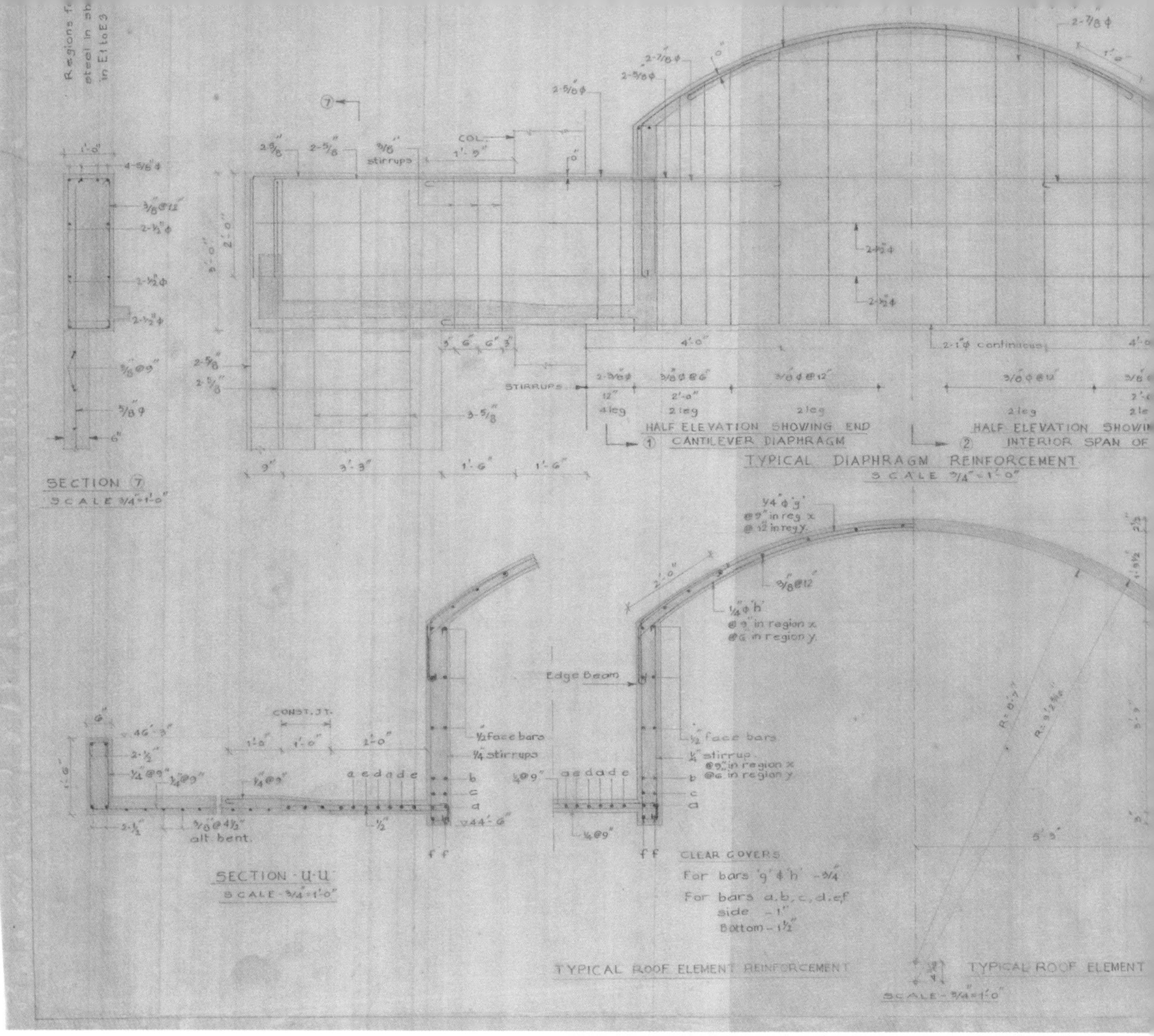

The Gandhi Memorial Hall auditorium marks the first collaboration between Mahendra Raj and architect AP Kanvinde. The concrete structure was formed by a series of slender vertical columns that support the auditorium with inclined transverse beams and longitudinal ribs. The inclined beams form a bold facia element and the 15cm-thick longitudinal ribs are curved to follow the contours of the seating. The roof is spanned by eight delicate 8cm-thick reinforced concrete cylindrical shells that span 21metres and are separated by gutters, which also act as beams. The auditorium has a seating capacity of 500.

Raj, who was based in Bombay in those days, remembers the number of proposals for changes made for this building and how the idea of the concrete shell emerged from numerous discussions between the architect and himself. Commenting on the slender and elegant form he reminisces that, '...at that time, one was always fighting to minimise thickness.'

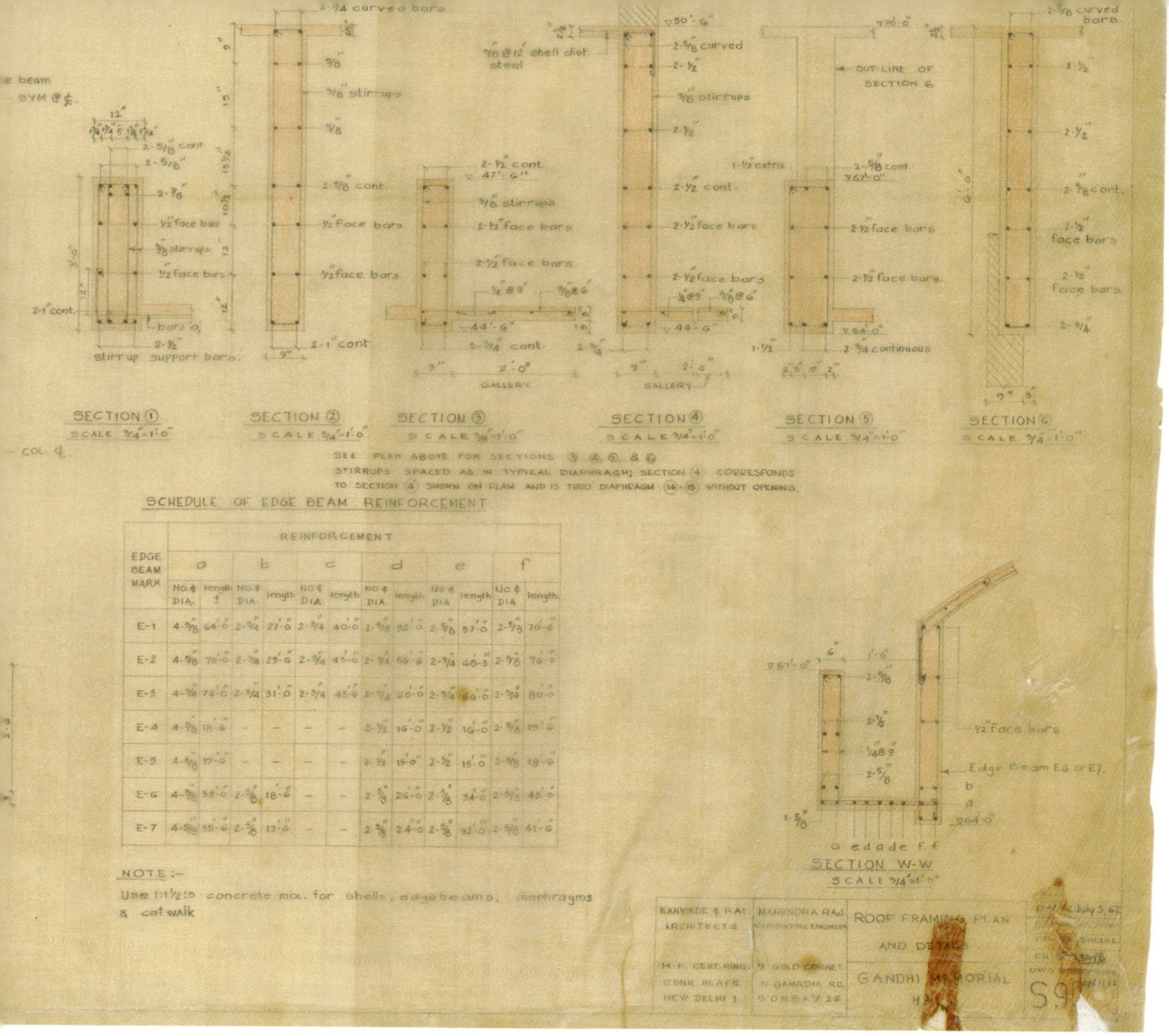

(Partial) Drawing no. S9: Roof Framing Plan and Details: 5 July 1962

Previous pages: The elegant structure of the hall before the brick was infilled.

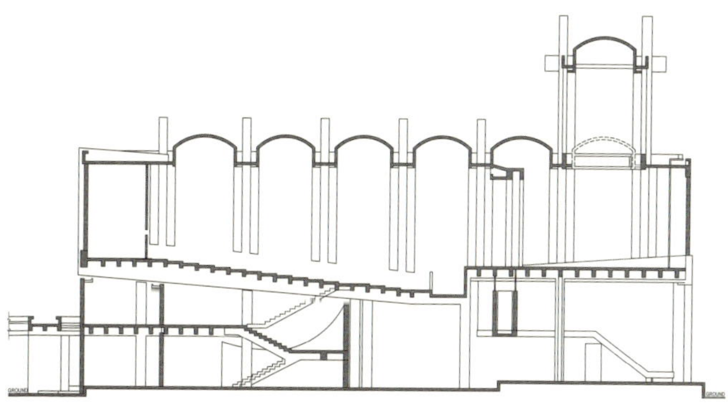

Building section.

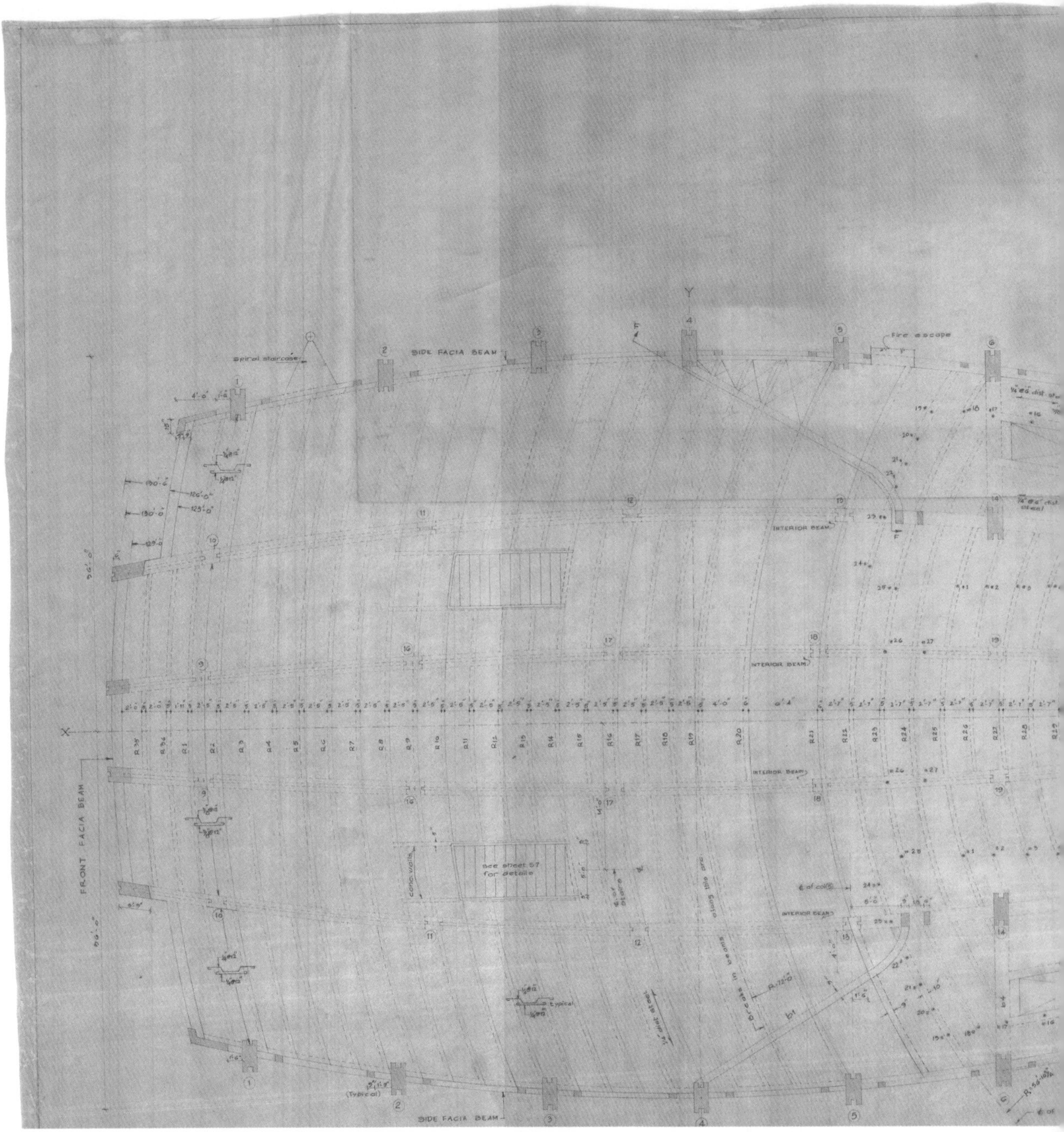

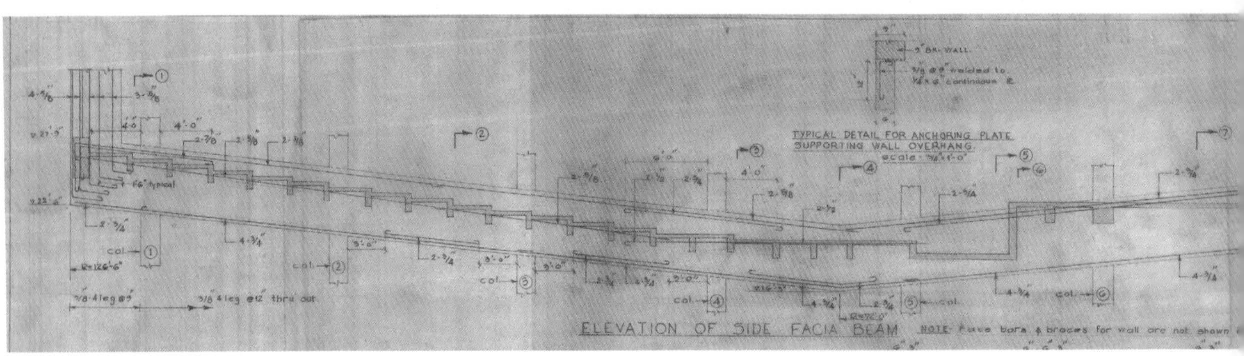
TYPICAL DETAIL FOR ANCHORING PLATE SUPPORTING WALL OVERHANG
ELEVATION OF SIDE FACIA BEAM

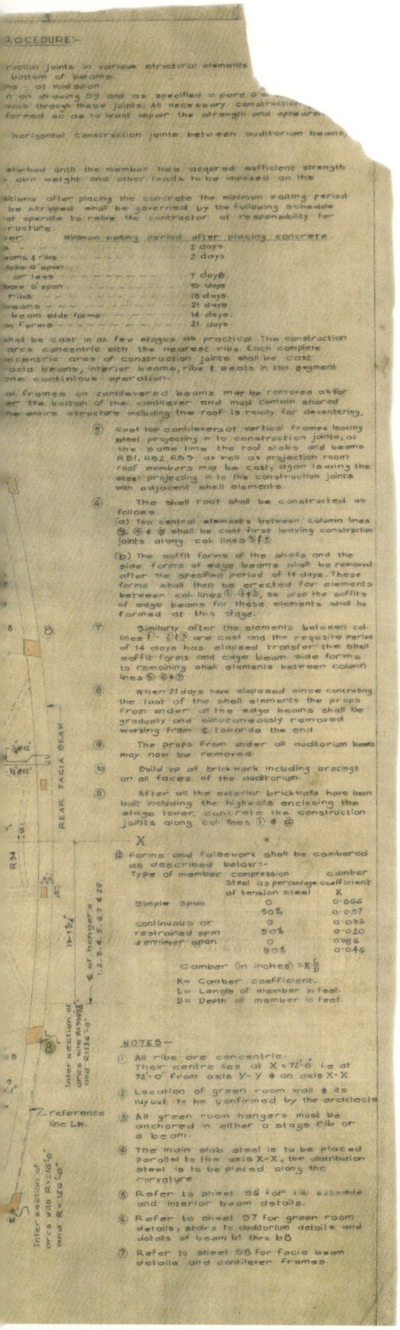

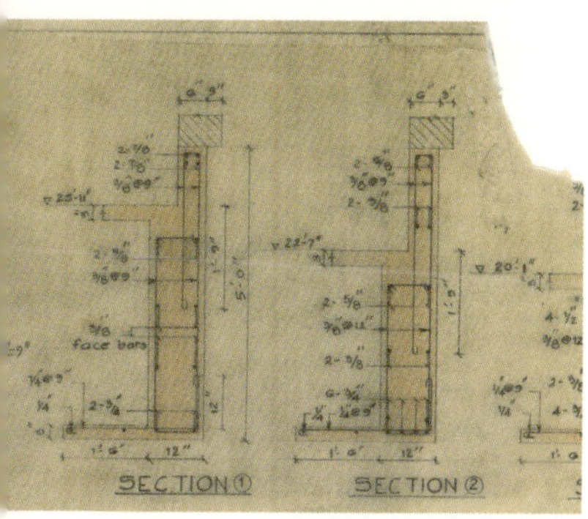

(Partial) Drawing no. S5: Auditorium Framing Plan: 24 July 1962, with detailed notes and instructions on the construction procedure, joints, casting and form-removal sequence of the shell roof.

(Partial) Drawing no. S8: Details of Facia and Roof Beams: 24 July 1962, with the elevation of the inclined side-facia beam that supports the staggered and curved ribs of the auditorium seating.

Photos at the building entry with the seating bowl structure above and the inside foyer area with its sculpted beams and staircase structure.

(Partial) Drawing: Foyer Framing Plan and Details: 6 April 1962

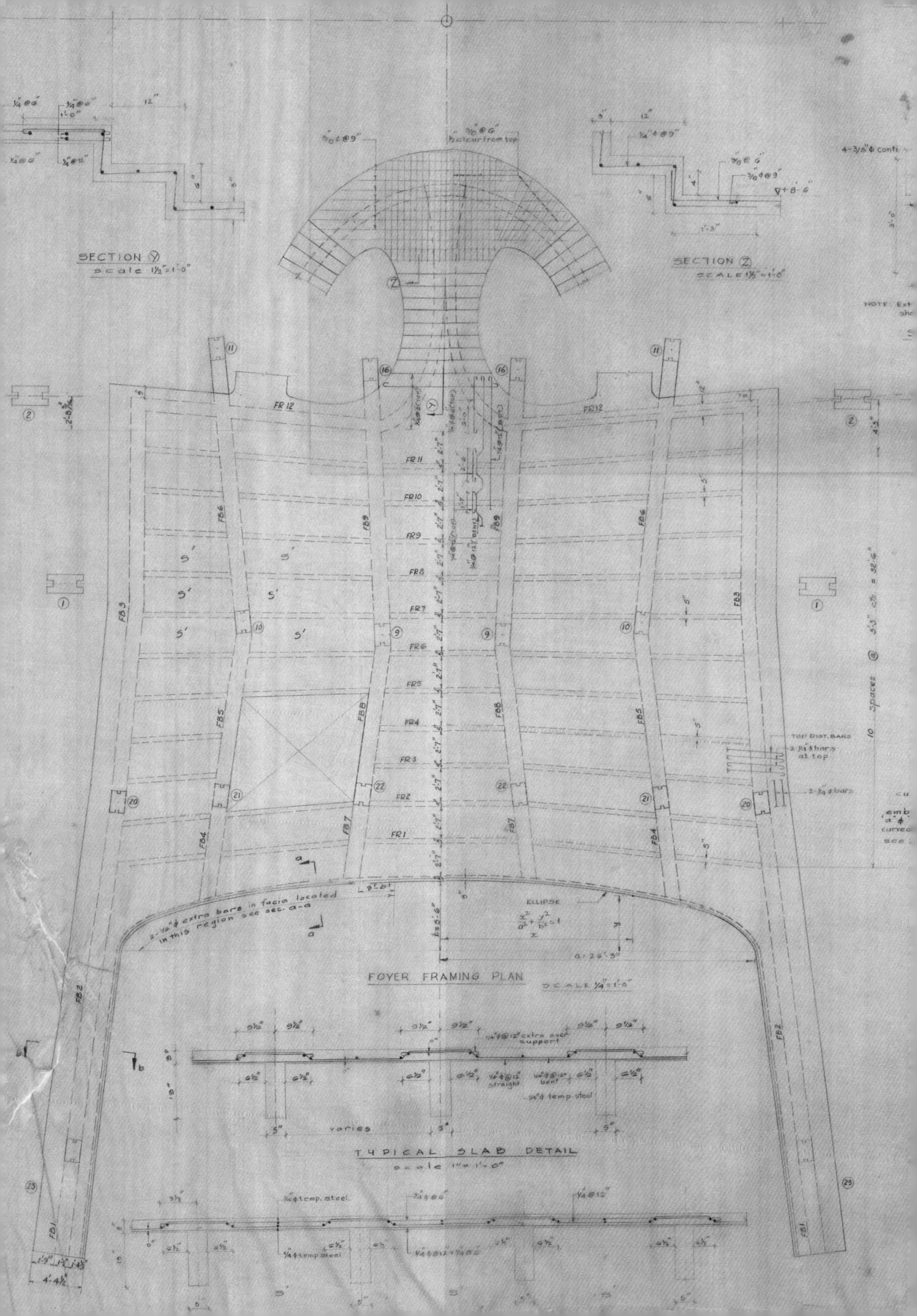
SECTION Y
scale 1½"=1'-0"
SECTION Z
SCALE 1½"=1'-0"
FOYER FRAMING PLAN
SCALE ¼"=1'-0"
ELLIPSE
$\frac{x^2}{a^2}+\frac{y^2}{b^2}=1$
a=26'-3"
b=5'-6"
2-½"ϕ extra bars in facia located in this region see sec. a-a
TYPICAL SLAB DETAIL
scale 1"=1'-0"
TOP DIST. BARS
2-¼"ϕ bars at top
2-¾"ϕ bars
10 spaces @ 3'-3" c/c = 32'-6"
¼"ϕ@12" extra over support
¼"ϕ@12" straight
¼"ϕ@12" bent
¼"ϕ temp. steel
varies
4-3/8"ϕ conti.

MUNICIPAL STADIUM

AHMEDABAD | 1965

This cricket stadium is one of the first of its kind using folded-plate cantilever frame structures in India. Particularly noteworthy is the delicacy and elegance of the structure, as the thin folded plates rise from the ground to form an integral whole with the spectator-stands and the 20m-long cantilever roof.

'To my knowledge, this structure was the first of its kind at that time–folded plate as a cantilever,' Mahendra Raj says. 'We were all very nervous. In those days there was no proof checking, so we double checked within our office.' He adds, 'Dr Vakil and I started working on the stadium with Charles, who invited his friend from MIT, Fred Taylor, to help him design it. We worked on a number of alternate designs, made models and had long discussions on the architectural, structural and constructability merits and demerits of each design. All the options we worked out were folded-plate structures. I had already worked on the design of plates and shells during my higher studies and at work, especially the Mohawk Airlines hangar, my first project in the office of Ammann & Whitney Consulting Engineers, New York, which was also a cantilever hangar, although different, as it was a balanced one.'

The structure is composed of folded-plate cantilever frames in reinforced concrete, which provide partial cover for the 35,000-person stadium. The 20 metres cantilevered roof has a tapering fold with 8–15cm thick plates anchored to the folded inclined leg of the main frame that inturn is connected to the seating frame. Both legs of the main frame and seating frame rest on a combined strip footing. The incline allows the centre-of-gravity of the superloads on the cantilevered folded-plates to pass through the combined strip-footing. This way the large cantilever moment on the frame is dissipated by the time it reaches the foundations making each part of the structure work with the other in a fine balance. At the discontinuous ends of the roof folds, cross ties are provided to control stresses in the end plates. This structure is built in mild steel, before TOR steel was introduced in India.

Views of the covered portion of the stadium and its dramatic cantilevered folded-plate roof.

Previous pages: The stadium's folded inclined legs as seen from the exterior. The gutter is integrated seamlessly within the folds that act as a diaphragm and connects the folded inclined legs.

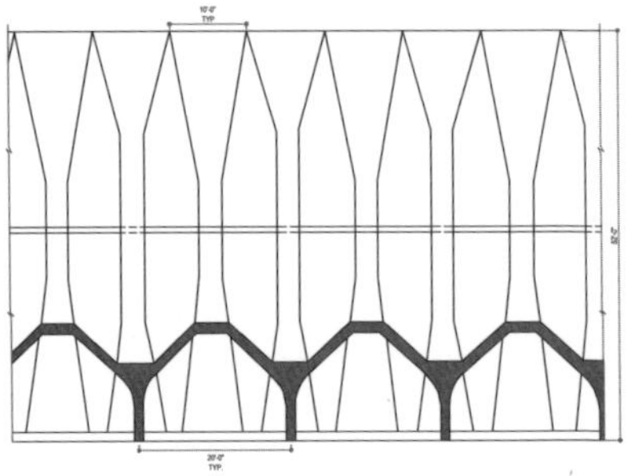

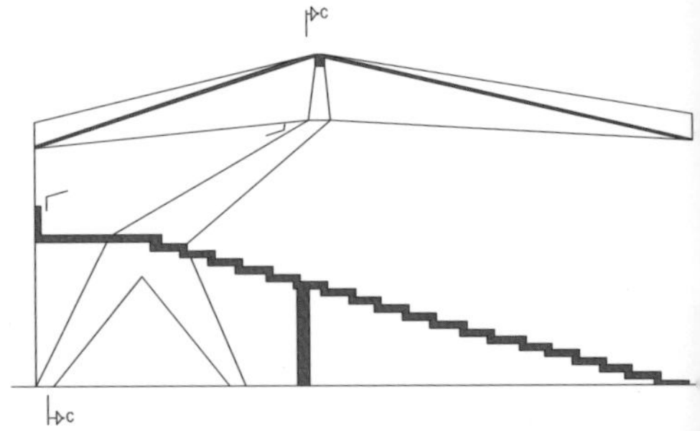

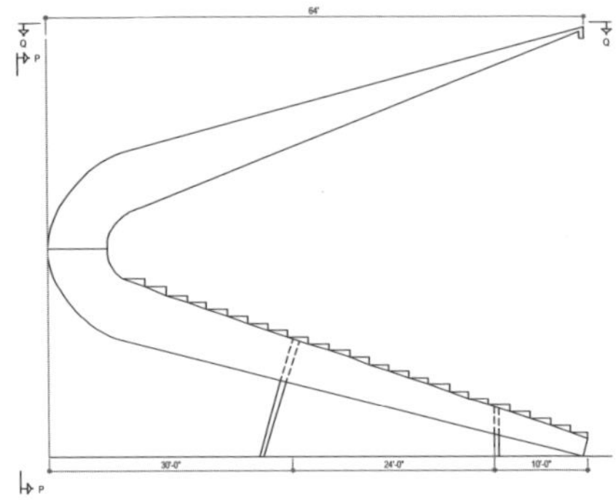

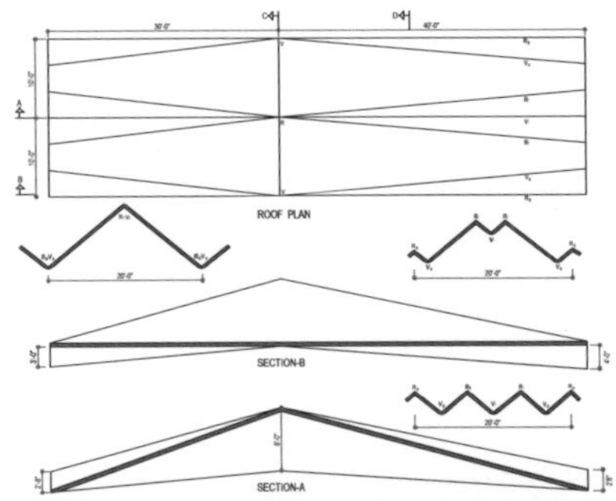

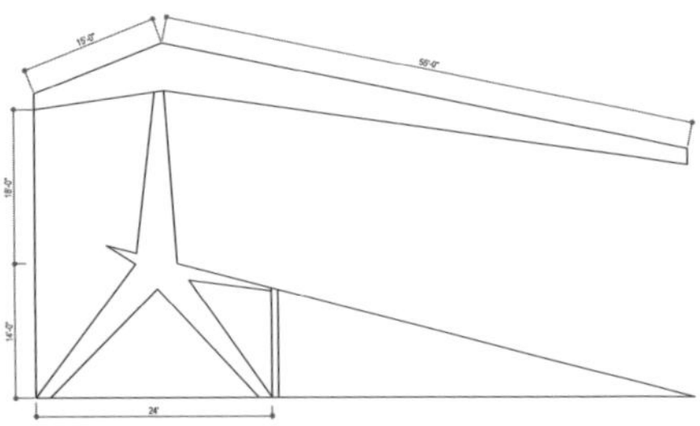

Plans, sections and models of various alternatives that were worked on before the final form and structure were determined.

Recent photos of the stadium and field from the covered stand supported on the folded-plate inclined legs.

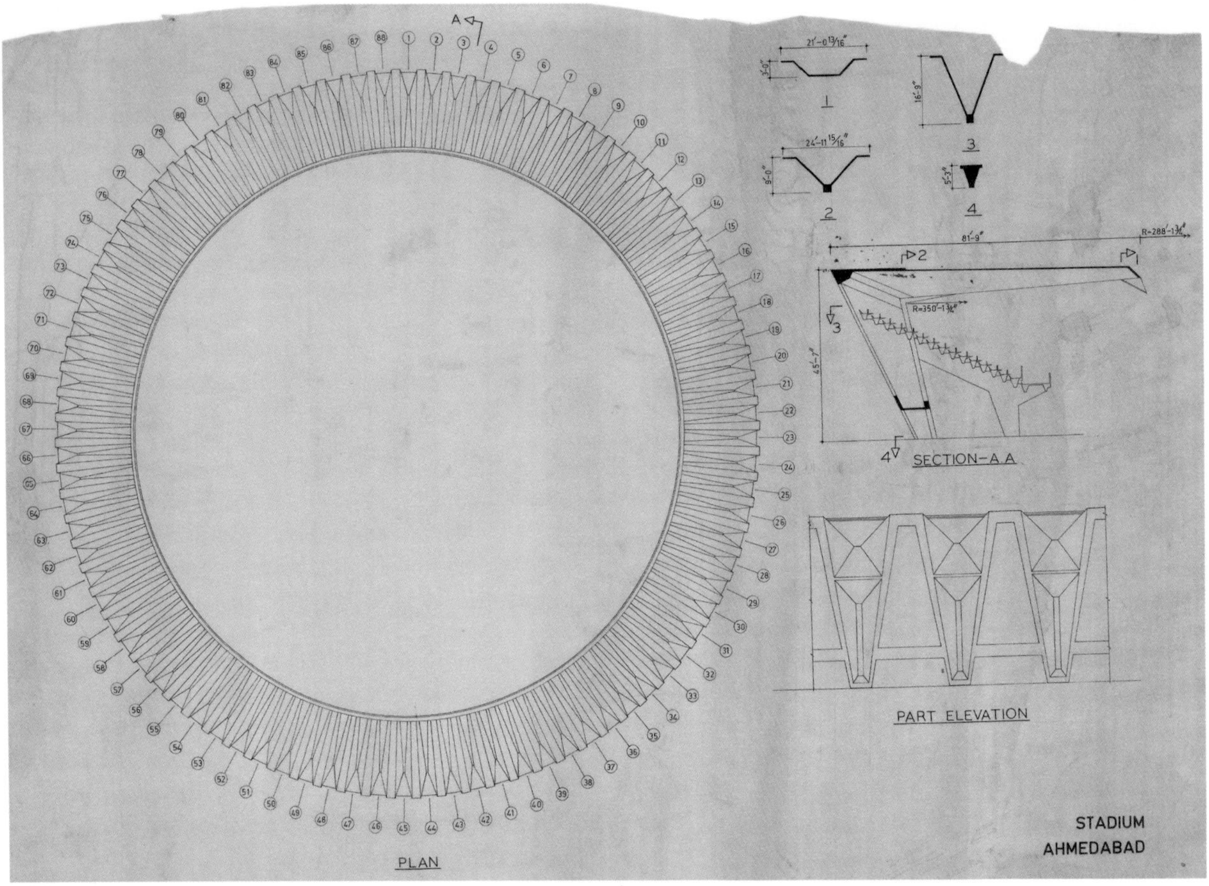

Drawing: Concept Plan, Section, Part Elevation
Structural drawing sketch showing an overall plan with the cantilever roof on the entire stadium (not built), with part elevation, section of the folded-plate frame structure and varying folds of the frame.

Model of the final proposal.

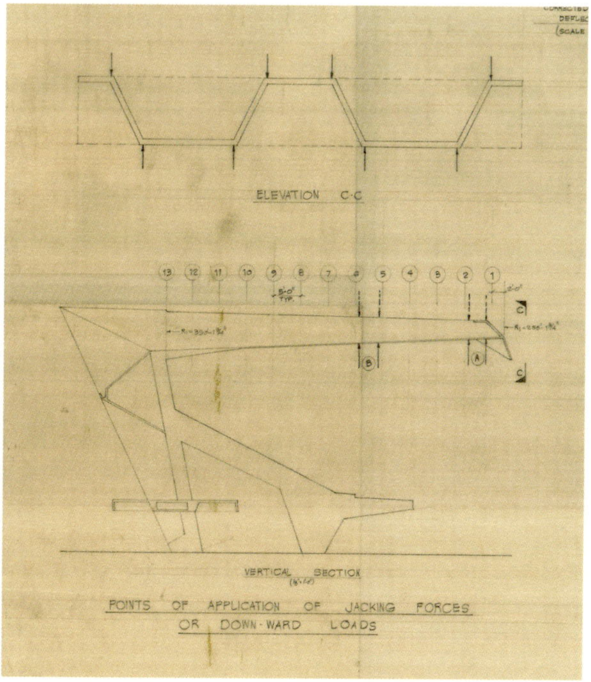

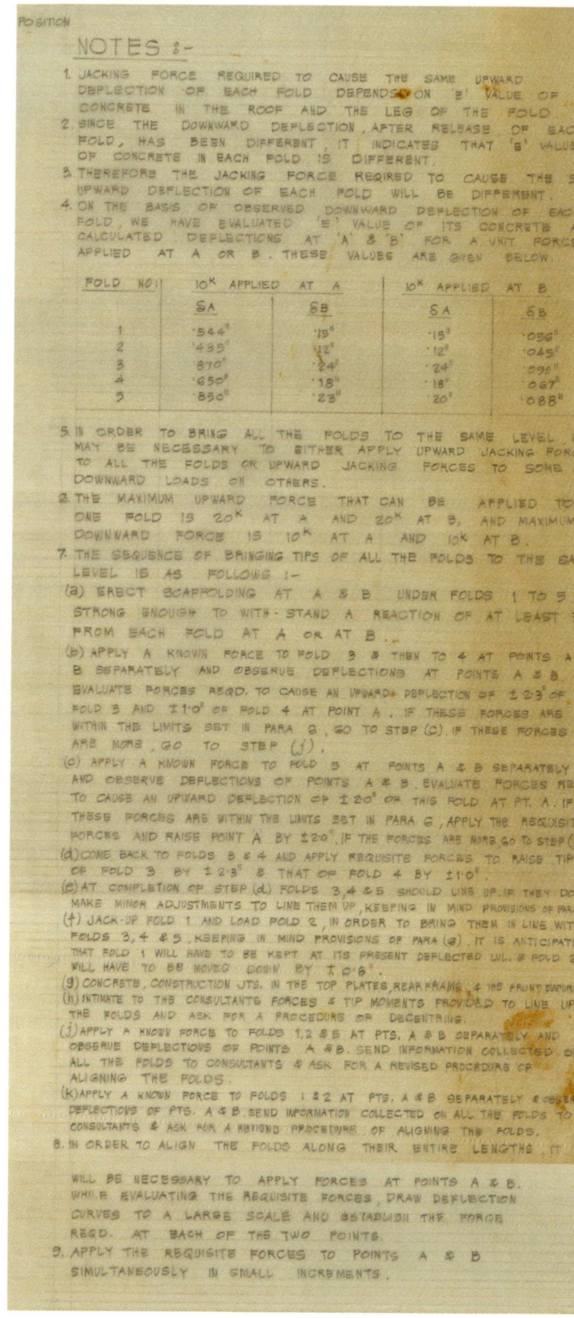

POSITION

NOTES :-

1. JACKING FORCE REQUIRED TO CAUSE THE SAME UPWARD DEFLECTION OF EACH FOLD DEPENDS ON 'E' VALUE OF CONCRETE IN THE ROOF AND THE LEG OF THE FOLD.
2. SINCE THE DOWNWARD DEFLECTION, AFTER RELEASE OF EACH FOLD, HAS BEEN DIFFERENT, IT INDICATES THAT 'E' VALUES OF CONCRETE IN EACH FOLD IS DIFFERENT.
3. THEREFORE THE JACKING FORCE REQIRED TO CAUSE THE SAME UPWARD DEFLECTION OF EACH FOLD WILL BE DIFFERENT.
4. ON THE BASIS OF OBSERVED DOWNWARD DEFLECTION OF EACH FOLD, WE HAVE EVALUATED 'E' VALUE OF ITS CONCRETE AND CALCULATED DEFLECTIONS AT 'A' & 'B' FOR A UNIT FORCE APPLIED AT A OR B. THESE VALUES ARE GIVEN BELOW.

FOLD NO.	10^K APPLIED AT A		10^K APPLIED AT B	
	δA	δB	δA	δB
1	·544"	·15"	·15"	·056"
2	·435"	·12"	·12"	·045"
3	·870"	·24"	·24"	·095"
4	·650"	·18"	·18"	·067"
5	·850"	·23"	·20"	·088"

5. IN ORDER TO BRING ALL THE FOLDS TO THE SAME LEVEL, IT MAY BE NECESSARY TO EITHER APPLY UPWARD JACKING FORCES TO ALL THE FOLDS OR UPWARD JACKING FORCES TO SOME AND DOWNWARD LOADS ON OTHERS.
6. THE MAXIMUM UPWARD FORCE THAT CAN BE APPLIED TO ONE FOLD IS 20^K AT A AND 20^K AT B, AND MAXIMUM DOWNWARD FORCE IS 10^K AT A AND 10^K AT B.
7. THE SEQUENCE OF BRINGING TIPS OF ALL THE FOLDS TO THE SAME LEVEL IS AS FOLLOWS :-
(a) ERECT SCAFFOLDING AT A & B UNDER FOLDS 1 TO 5 STRONG ENOUGH TO WITH-STAND A REACTION OF AT LEAST 30 FROM EACH FOLD AT A OR AT B.
(b) APPLY A KNOWN FORCE TO FOLD 3 & THEN TO 4 AT POINTS A & B SEPARATELY AND OBSERVE DEFLECTIONS AT POINTS A & B. EVALUATE FORCES REQD. TO CAUSE AN UPWARD DEFLECTION OF ±2·3" OF FOLD 3 AND ±1·0" OF FOLD 4 AT POINT A. IF THESE FORCES ARE WITHIN THE LIMITS SET IN PARA 6, GO TO STEP (C). IF THESE FORCES ARE MORE, GO TO STEP (J).
(c) APPLY A KNOWN FORCE TO FOLD 5 AT POINTS A & B SEPARATELY AND OBSERVE DEFLECTIONS OF POINTS A & B. EVALUATE FORCES REQD. TO CAUSE AN UPWARD DEFLECTION OF ±2·0" OF THIS FOLD AT PT. A. IF THESE FORCES ARE WITHIN THE LIMITS SET IN PARA 6, APPLY THE REQUISITE FORCES AND RAISE POINT A BY ±2·0". IF THE FORCES ARE MORE, GO TO STEP (K).
(d) COME BACK TO FOLDS 3 & 4 AND APPLY REQUISITE FORCES TO RAISE TIP OF FOLD 3 BY ±2·3" & THAT OF FOLD 4 BY ±1·0".
(e) AT COMPLETION OF STEP (d) FOLDS 3, 4 & 5 SHOULD LINE UP. IF THEY DO NOT MAKE MINOR ADJUSTMENTS TO LINE THEM UP, KEEPING IN MIND PROVISIONS OF PARA 6.
(f) JACK-UP FOLD 1 AND LOAD FOLD 2, IN ORDER TO BRING THEM IN LINE WITH FOLDS 3, 4 & 5, KEEPING IN MIND PROVISIONS OF PARA (6). IT IS ANTICIPATED THAT FOLD 1 WILL HAVE TO BE KEPT AT ITS PRESENT DEFLECTED LVL. & FOLD 2 WILL HAVE TO BE MOVED DOWN BY ±0·6".
(g) CONCRETE, CONSTRUCTION JTS. IN THE TOP PLATES, REAR FRAME, & THE FRONT DIAPHRAGM.
(h) INTIMATE TO THE CONSULTANTS FORCES & TIP MOMENTS PROVIDED TO LINE UP THE FOLDS AND ASK FOR A PROCEDURE OF DECENTRING.
(j) APPLY A KNOWN FORCE TO FOLDS 1, 2 & 5 AT PTS. A & B SEPARATELY AND OBSERVE DEFLECTIONS OF POINTS A & B. SEND INFORMATION COLLECTED ON ALL THE FOLDS TO CONSULTANTS & ASK FOR A REVISED PROCEDURE OF ALIGNING THE FOLDS.
(k) APPLY A KNOWN FORCE TO FOLDS 1 & 2 AT PTS. A & B SEPARATELY & OBSERVE DEFLECTIONS OF PTS. A & B. SEND INFORMATION COLLECTED ON ALL THE FOLDS TO CONSULTANTS & ASK FOR A REVISED PROCEDURE OF ALIGNING THE FOLDS.
8. IN ORDER TO ALIGN THE FOLDS ALONG THEIR ENTIRE LENGTHS, IT WILL BE NECESSARY TO APPLY FORCES AT POINTS A & B. WHILE EVALUATING THE REQUISITE FORCES, DRAW DEFLECTION CURVES TO A LARGE SCALE AND ESTABLISH THE FORCE REQD. AT EACH OF THE TWO POINTS.
9. APPLY THE REQUISITE FORCES TO POINTS A & B SIMULTANEOUSLY IN SMALL INCREMENTS.

(Partial) Drawing and Notes
Drawing no. ST-S 20A: Procedure for Concreting Construction Joints Between Folds: 26 August 1970

The inclined legs of the main frame and the seating frame were cast first and the cantilever roof was cast later in segments. After completing the entire sector, the variations in the deflections at the tip were equalised by either jacking up or weighing down different folds and then casting the front diaphragm connecting all the folds. The notes here show how detailed instructions on this construction procedure were transferred to the contractor who had never done a structure of this kind before.

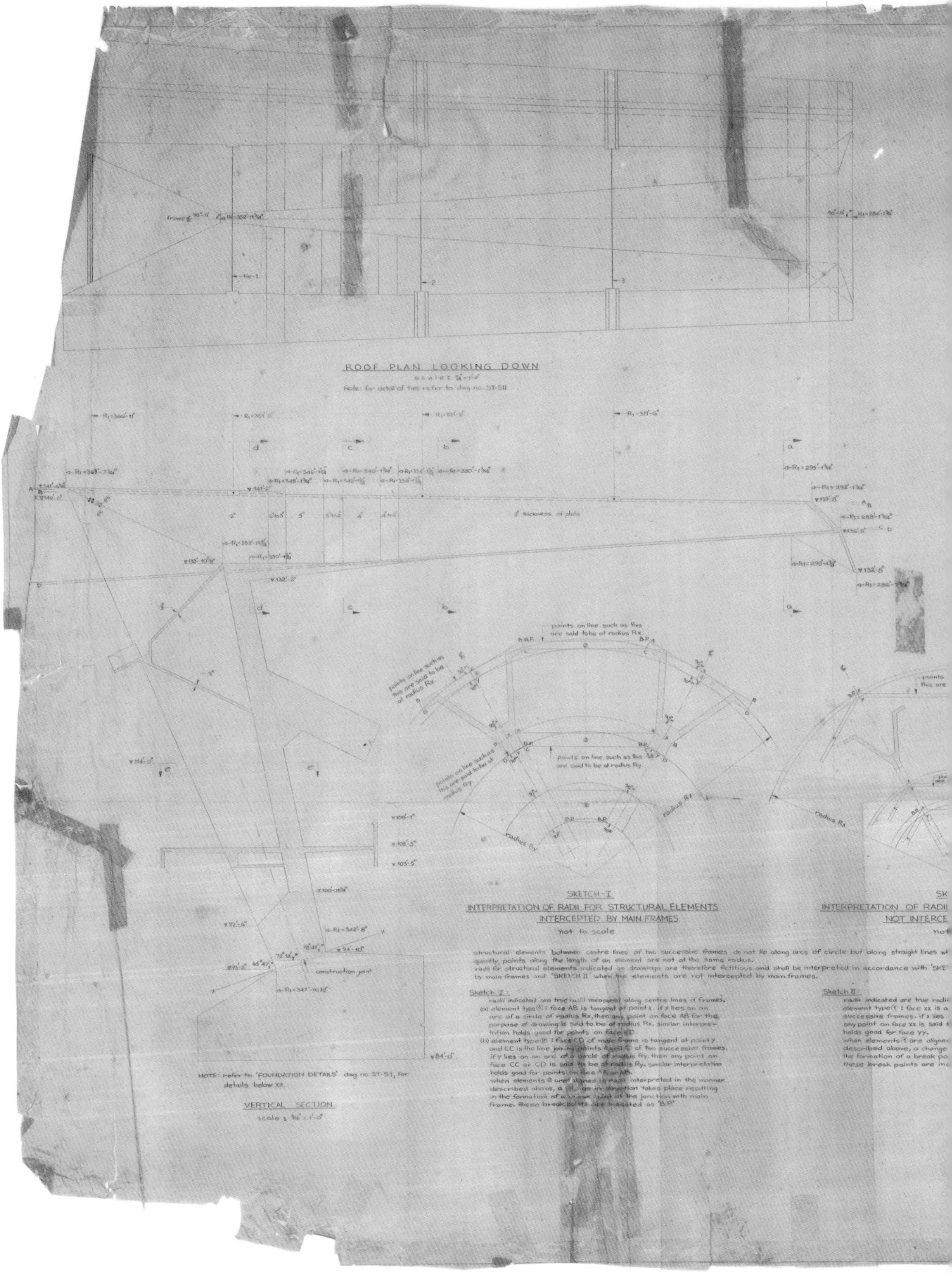
ROOF PLAN LOOKING DOWN
SKETCH-I
INTERPRETATION OF RADII FOR STRUCTURAL ELEMENTS
INTERCEPTED BY MAIN FRAMES
not to scale
construction joint
VERTICAL SECTION

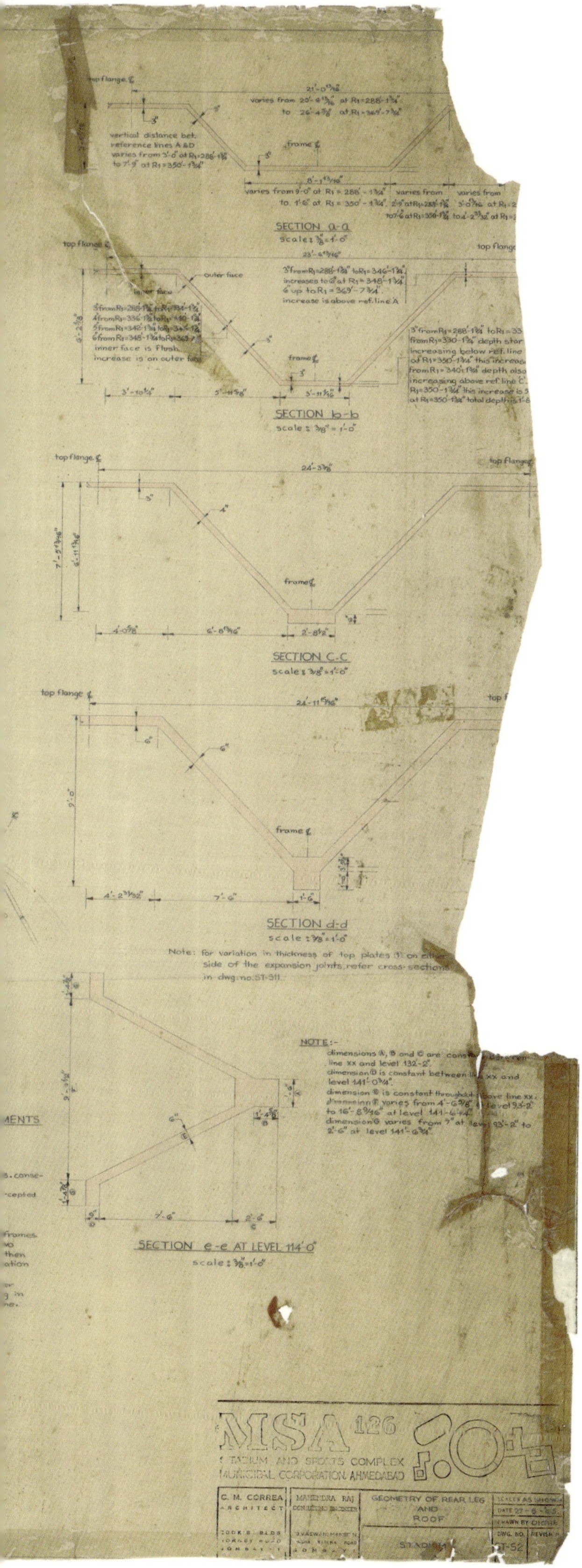

Inclined folded legs during construction.

Rear view of the Stadium.

Drawing no. ST-S 2: Geometry of Rear Leg and Roof: 7 August 1965
Drawing showing the radii of curvature for construction of the structure, its folded-plate geometry in relationship to the circular stadium plan and the geometry of the folds: the thickness of folds ranges from 3" at the front edge and increases gradually to 4", 5" and 6" (8–15cm) towards the knee joint.

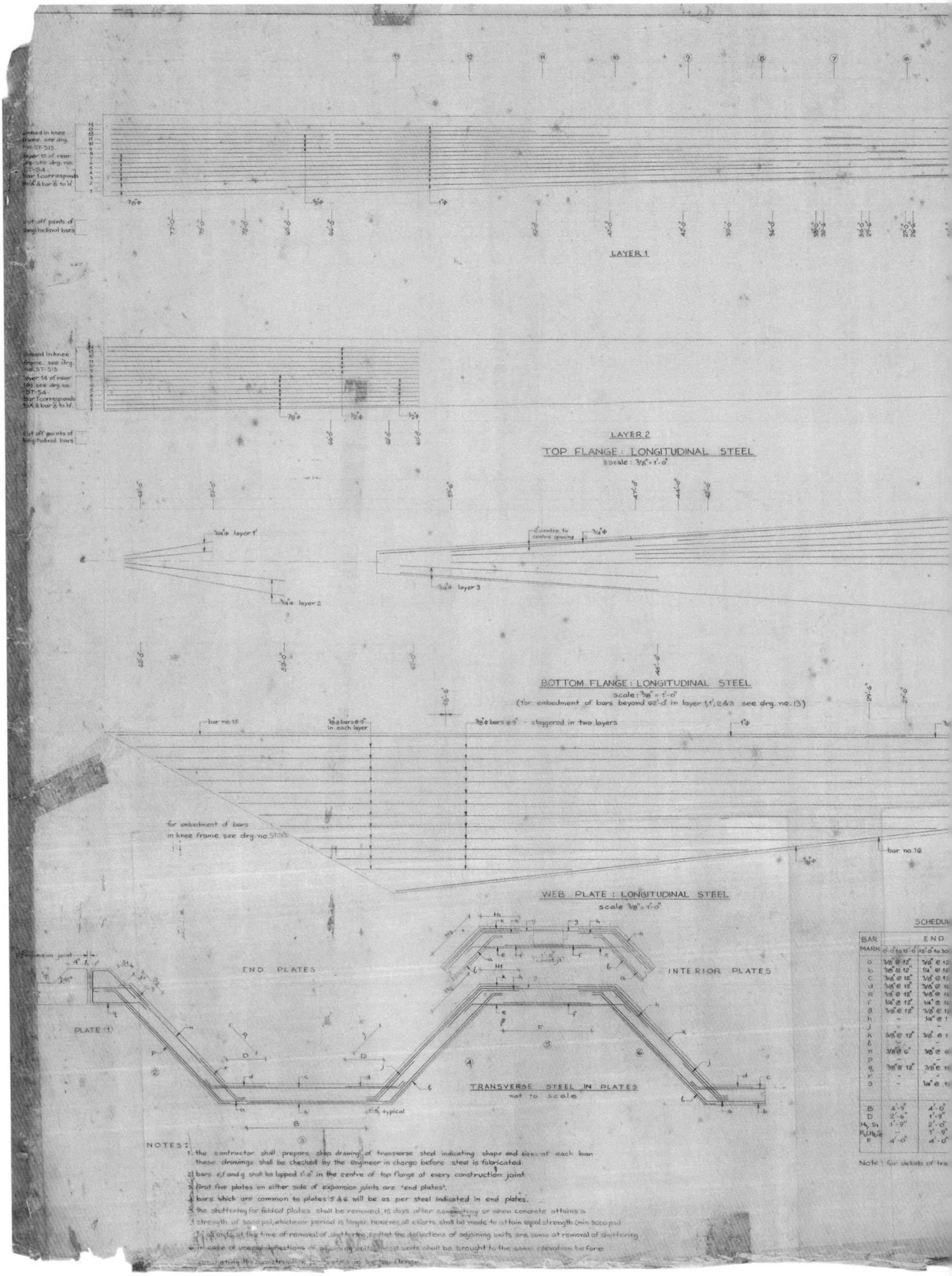
LAYER 1
LAYER 2
TOP FLANGE : LONGITUDINAL STEEL
scale : 3/8" = 1'-0"
BOTTOM FLANGE : LONGITUDINAL STEEL
scale : 3/8" = 1'-0"
(for embedment of bars beyond 92'-0" in layer 1,1',2&3 see drg. no. 13)
3/8"ϕ bars @ 9" - staggered in two layers
for embedment of bars in knee frame see drg. no. ST-513
WEB PLATE : LONGITUDINAL STEEL
scale 3/8" = 1'-0"
END PLATES
INTERIOR PLATES
PLATE (1)
TRANSVERSE STEEL IN PLATES
not to scale
NOTES :
1. the contractor shall prepare shop drawing of transverse steel indicating shape and size of each bar
these drawings shall be checked by the engineer in charge before steel is fabricated.
2. bars e, f and g shall be lapped 1'-0" in the centre of top flange at every construction joint.
3. first five plates on either side of expansion joints are 'end plates'.
4. bars which are common to plates 5 & 6 will be as per steel indicated in end plates.

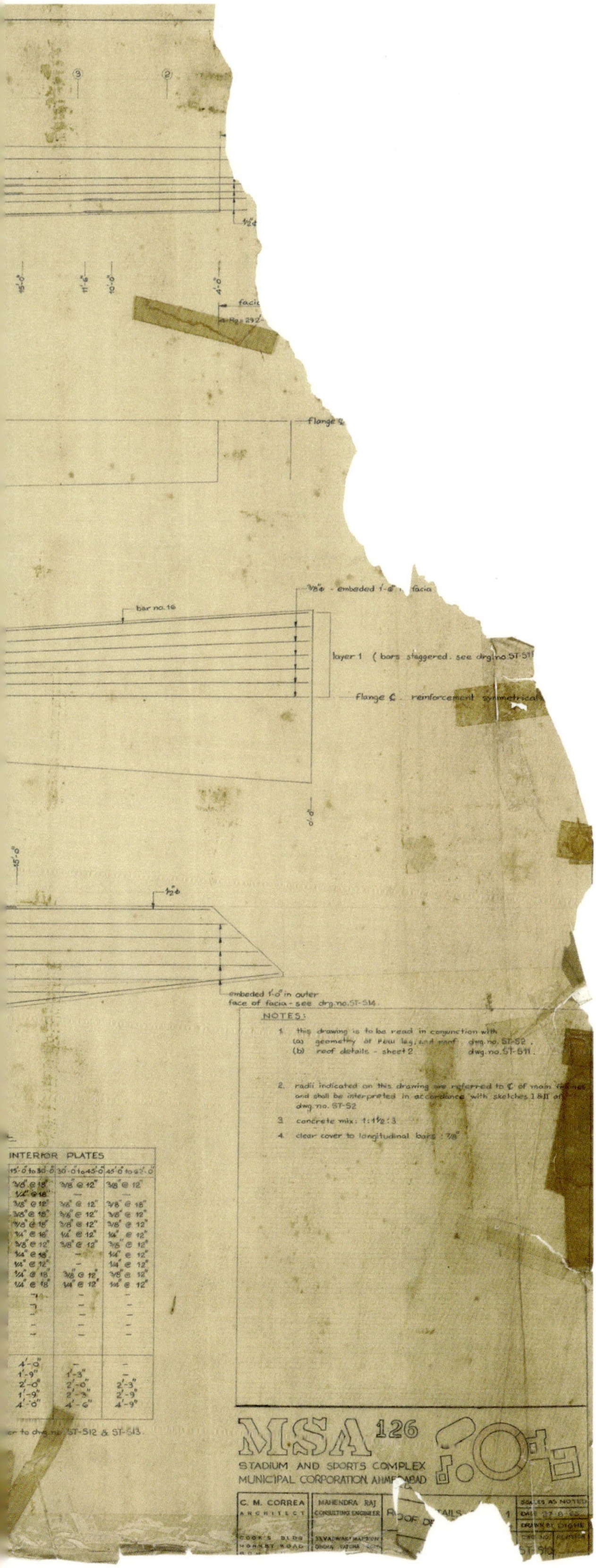

The construction joints in the horizontal folds.

View of the seating with the cantilevered roof folds above.

Drawing no. ST-S 10: Roof Details: 27 August 1965

Reinforcement drawings of the 20m cantilever roof of 0.9–2.7m tapering fold depth with 8–15cm thick plates. During construction the cantilever folds required a bottom form and a manually movable top form to pour and stabilise concrete at the inclined parts of the folds.

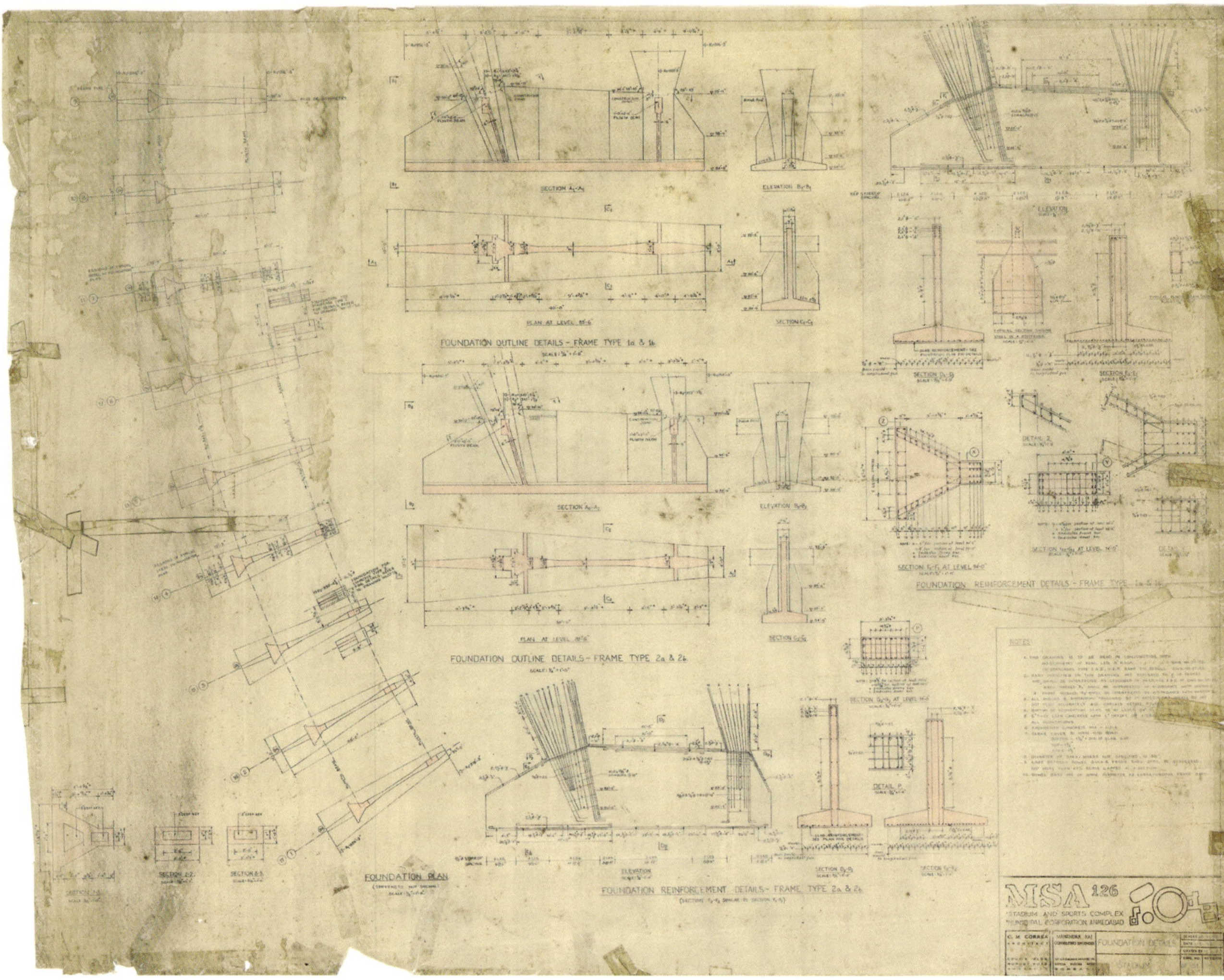

Drawing no. ST-S 1: Foundation Details: 1 August 1963
Details of the combined strip-footing of the inclined legs of the main frame and the seating frame. The tendon-like forms seen connecting the two legs in the strip-footing are an example of the precise shaping of all elements in Raj's work.

View of the inclined leg frame and the seating frame.

The knee joint between the cantilevered roof and the inclined leg.

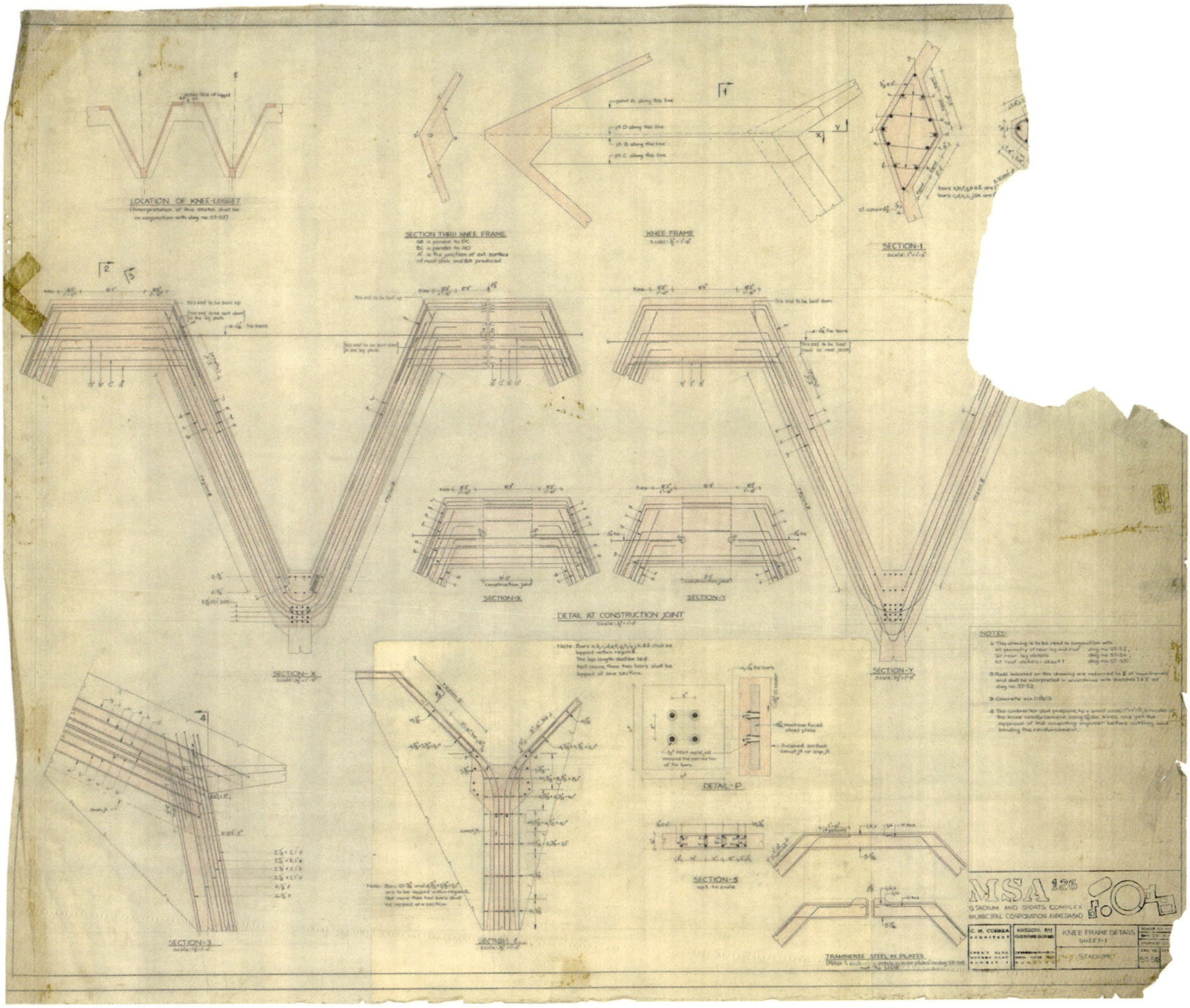

Drawing no. ST-S 12: Knee Frame Details- Sheet1: 27 August 1963

Details of the critical knee joint between the cantilevered roof and the inclined leg. This joint stabilises and transfers complex moments and forces generated by the junction of the 20m cantilever roof and the inclined legs. The forces are resolved elegantly by the dramatic deepening of the folds at this joint making it possible to reduce the required depth as it reaches the combined strip foundation.

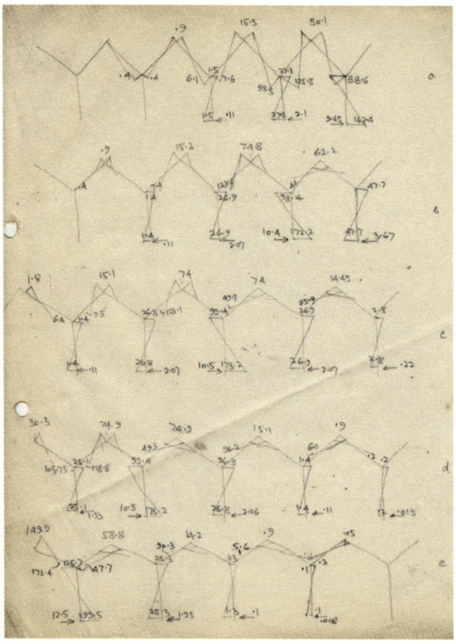

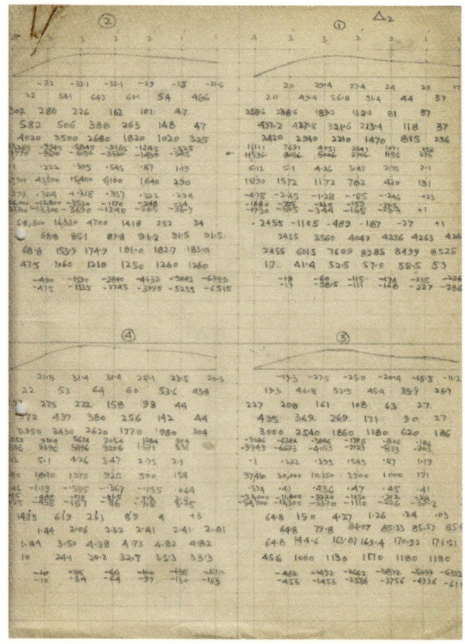

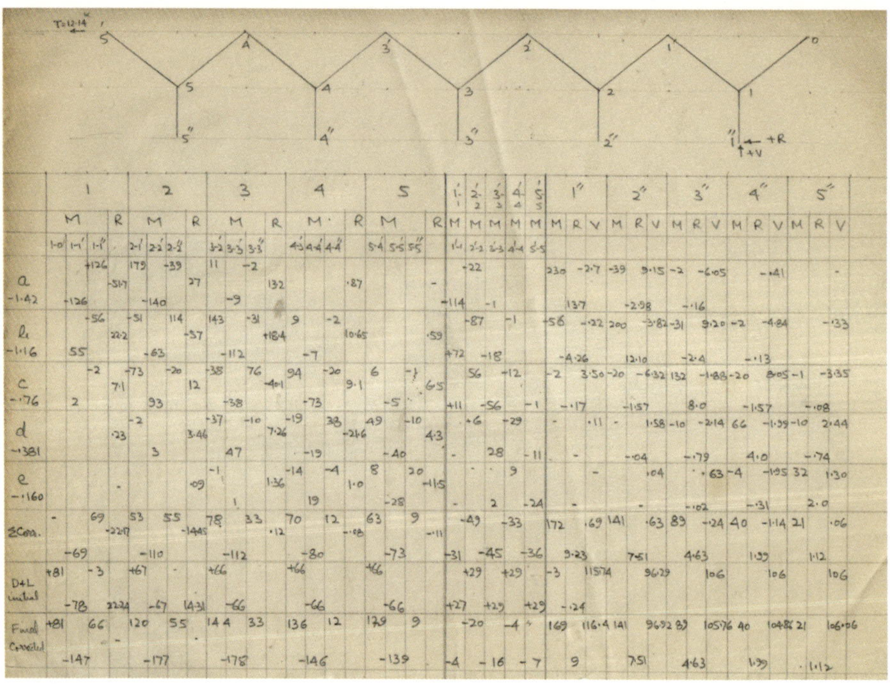

Sample calculation sheets of the project.

Exterior view of the folded inclined leg of the main frame.

ENTRY
EAST END
Rs. 50/-
GATE
11
G
B

TAGORE MEMORIAL THEATRE
AHMEDABAD | 1965

Located on the banks of the Sabarmati River, the Tagore auditorium is only a few hundred metres from Sanskar Kendra, the museum that Le Corbusier designed in Ahmedabad. The sandy soil of the riverbed and the presence of the 'master's' architectural work as the context of this building, for the architect and the engineer were the two main challenges in the initial conceptualisation of the building, recalls Mahendra Raj.

He adds, 'There were different concepts, expressing possible columns, plates, etc. The final concept evolved after a number of studies. When we tried to evaluate them, we used manual calculations for deriving the dimensions. Of course without dimensions, there can be no studies.'

Previous pages and above: The southern facade expresses the rigid frame structural system of 10 trapezoidal folded plates, tapering in as they rise 17m in height, with a flat shear wall section at the stage section of the building.

Finally, a series of rigid frames in folded reinforced concrete plates was used as the outer shell of this hall. The structural elegance and efficiency is the result of carefully proportioned trapezoidal folds whose angles and plate thickness vary according to the stresses, moments and forces. The depth of the vertical folds is equal to the horizontal folds where they meet at the top, decreasing towards the ground. To achieve the vertical taper,

the width of the inner plate was kept constant in the vertical fold while that of the outer plate was increased.

Photo of the eastern (rear) facade of the theatre that faces the Sabarmati River, with Sanskar Kendra, designed by Le Corbusier, in the background.

Vertical folds are 17 metres high, with a constant depth of 1.15 metres up to the lintel level, increasing to 2.4 metres at the top. The horizontal roof folds are 2.4 metres deep, and as they span 33.5 metres, the thickness of the folded plates varies from 10cm to 15cm, increasing near the supports. The lateral forces generated by the frame action of the vertical and horizontal folds is absorbed by ties at the foundation level.

Inside the hall the 700-person seating bowl is supported on an independent structure at the back and is expressed in the foyer area. The sculptural columns and cantilevers of this bowl echo the folded plates of the enclosure while also expressing the lines of forces and moments of its own load.

The hall is supported on individual and strip-footings resting on a sandy river-bed stratum.

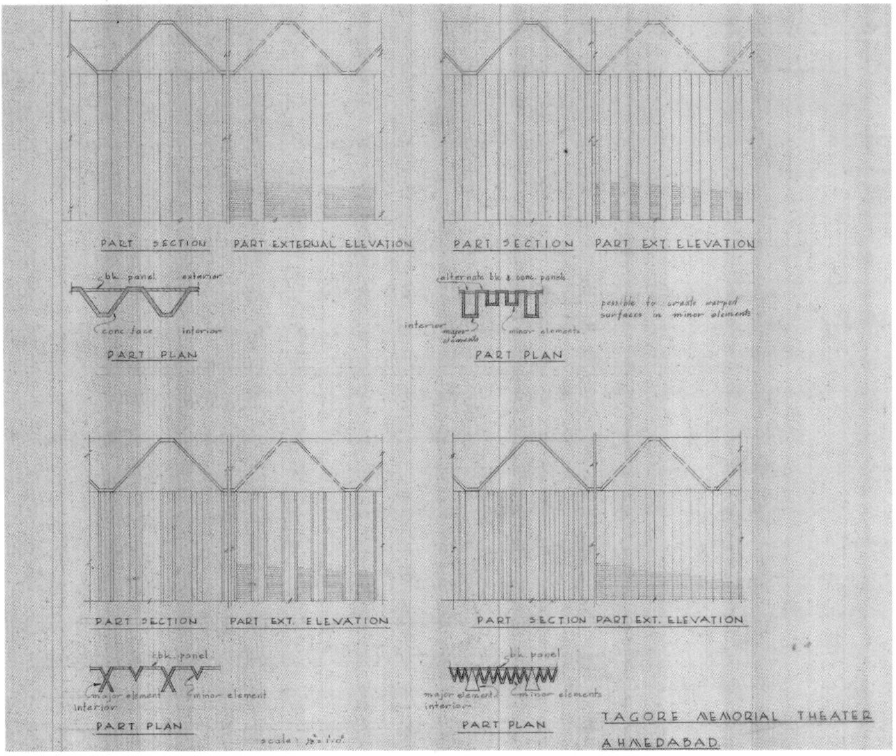

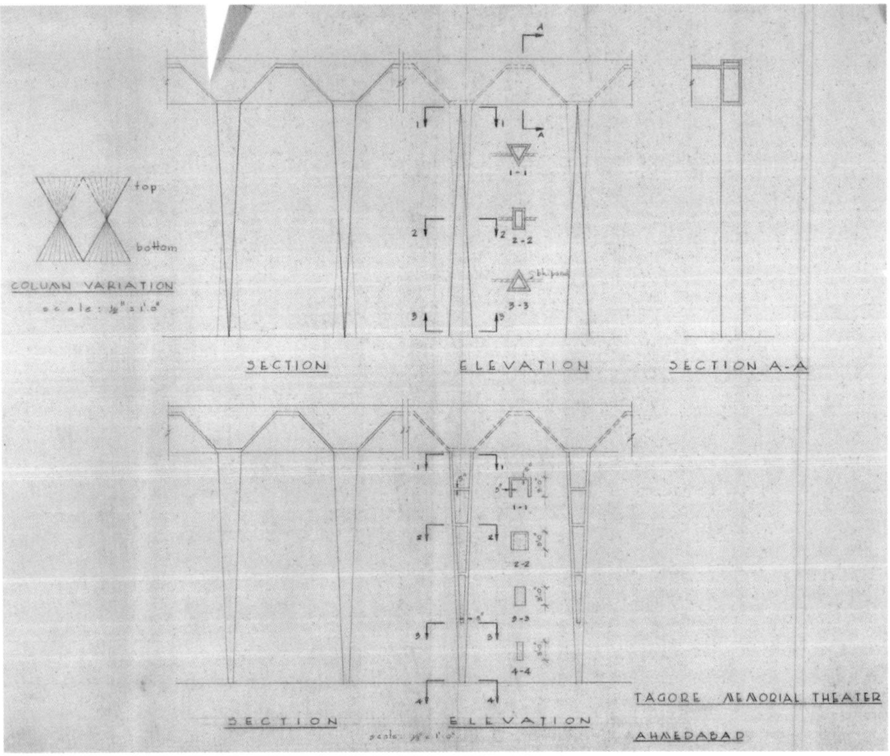

Drawings (on this page and opposite top): More than 12 folded-plate systems were explored for the theatre to arrive at the most optimal geometry. The options included triangular, channel and angular plate cross-sections that progressively became more complex in the way they folded in and out of the surface.

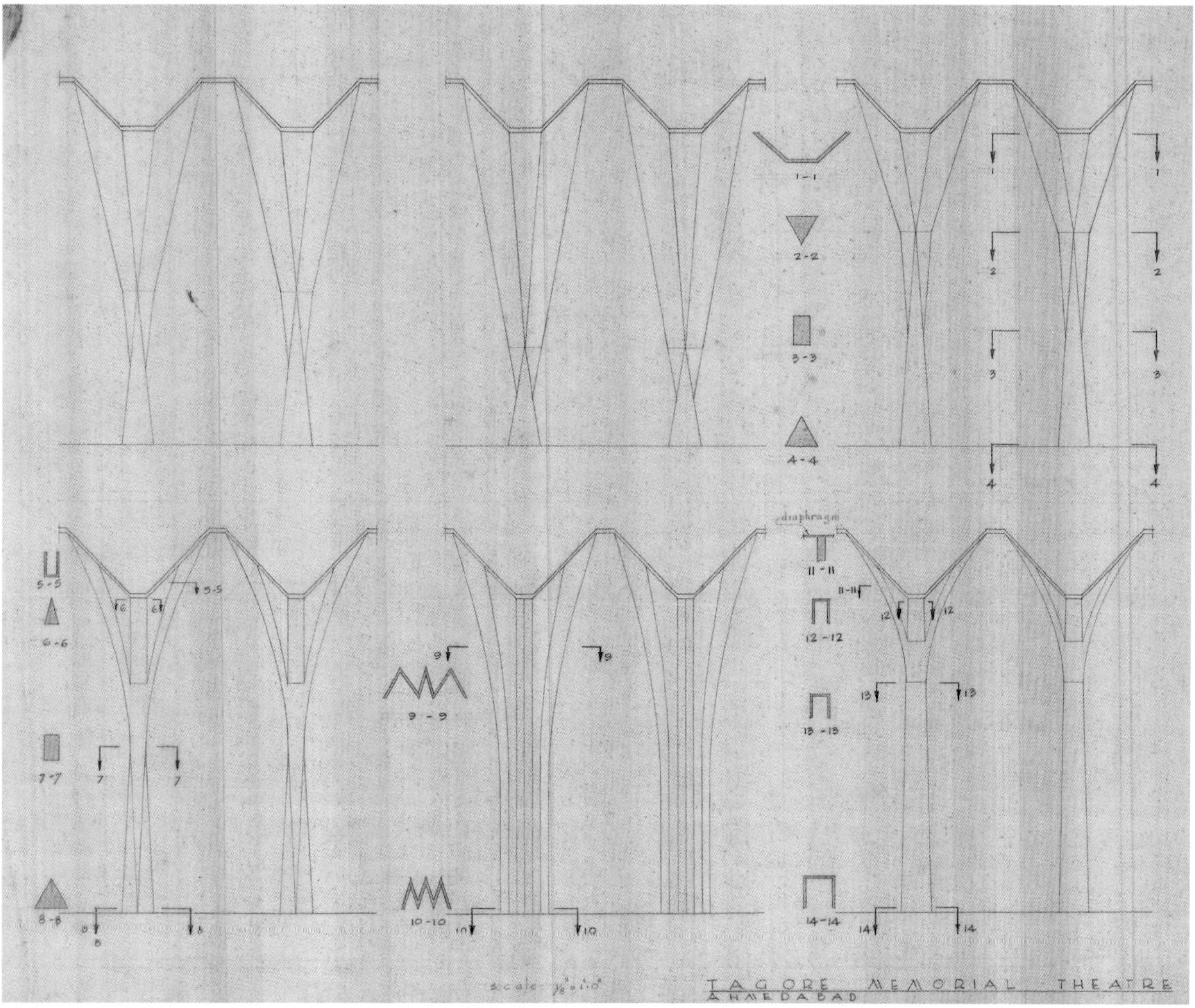

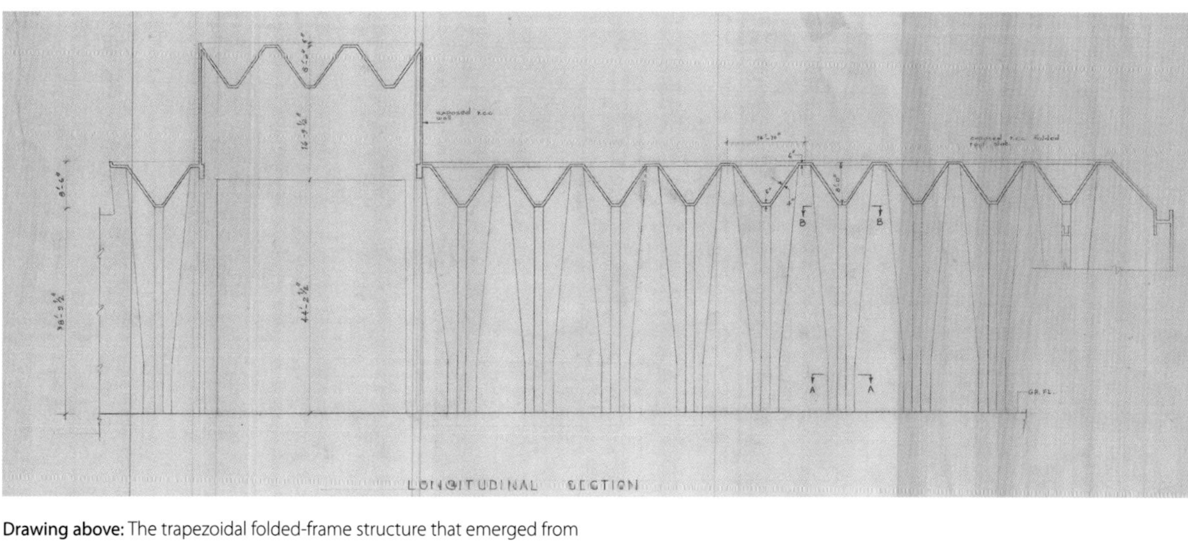

Drawing above: The trapezoidal folded-frame structure that emerged from the various alternatives considered.

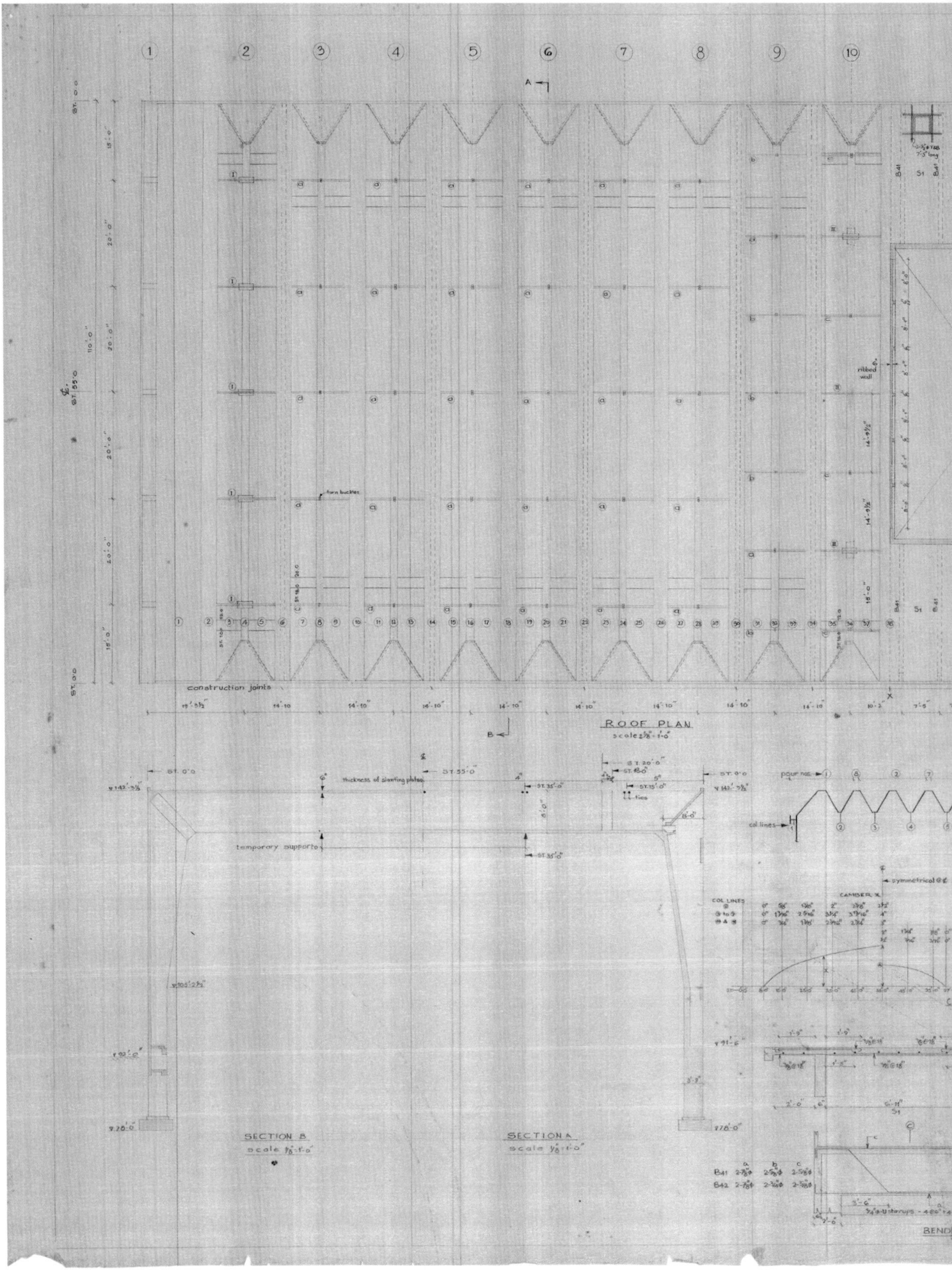
ROOF PLAN
construction joints
turn buckles
ribbed wall
SECTION B
SECTION A
thickness of slanting plates
temporary supports
ties
pour nos.
col lines
symmetrical
CAMBER
COL LINES
BENDI

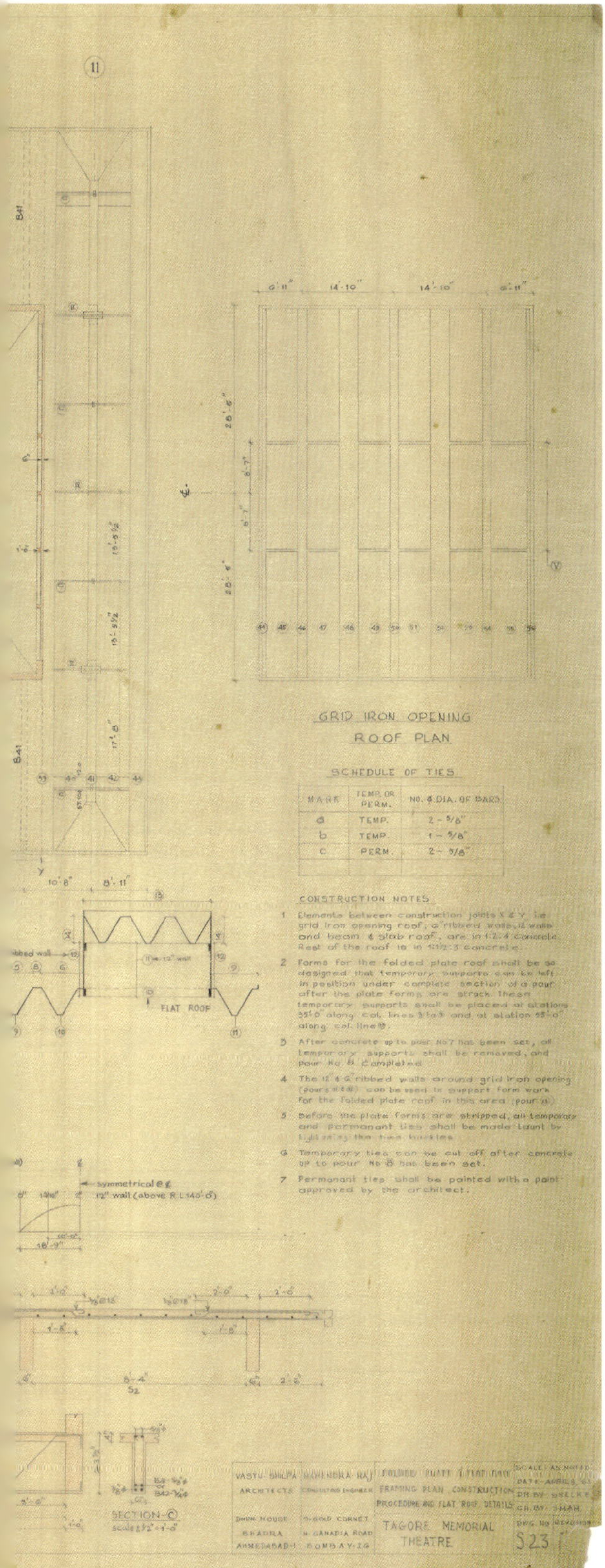

Overall view of the theatre from the south-west corner.

View from the north-east side facing the Sabarmati River.

Drawing no. S23: Folded Plate & Flat Roof Construction Framing Plan, Construction Procedure & Flat Roof Details: 8 April 1963
Construction notes specifying the pouring sequence and the supports and ties that need to be provided for the stability of the individual folded plates at different stages of concreting, until the entire rigid frame structure comes together.

Following pages:
Drawing no. S25: Folded Plate Roof Transverse Steel in Plates: April 1963
Drawing detailing the transverse steel at the roof fold of typical as well as various end conditions. The 'Schedules' specify transverse steel details and plate thicknesses, varying from 4" to 6". The latter demonstrates Raj's constant attention to efficiency and economy in material use. It also details out the exact shapes, lengths and angles of each bar.

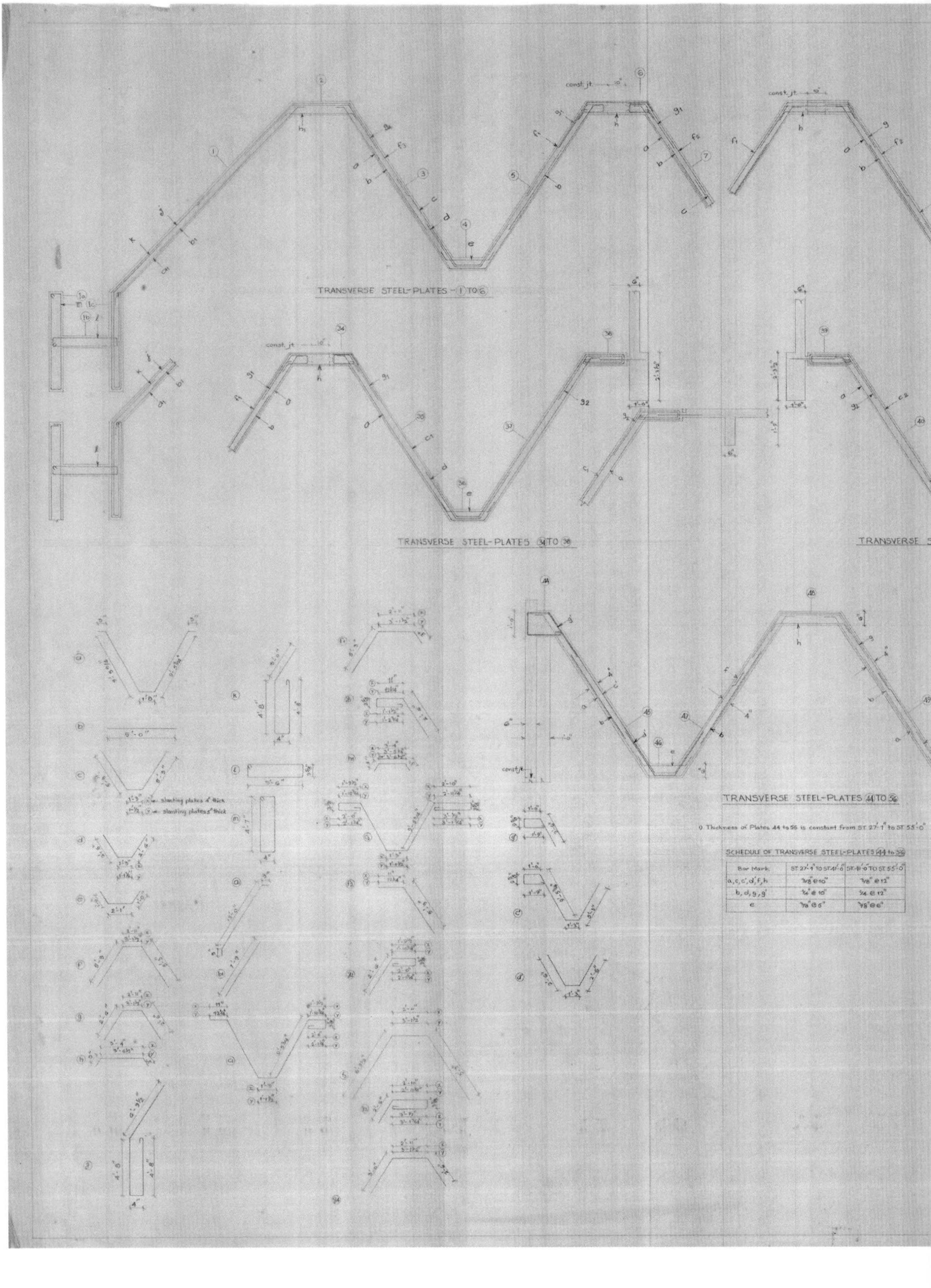
TRANSVERSE STEEL-PLATES 1 TO 6
TRANSVERSE STEEL-PLATES 34 TO 38
TRANSVERSE STEEL-PLATES 44 TO 56
0 Thickness of Plates 44 to 56 is constant from ST 27'-1" to ST 55'-0"
SCHEDULE OF TRANSVERSE STEEL-PLATES 44 to 56
Bar Mark | ST 27'-1" TO ST 41'-0" | ST 41'-0" TO ST 55'-0"
a, c, c', d', f, h | 3/8" @ 10" | 3/8" @ 12"
b, d, g, g' | 1/4" @ 10" | 1/4 @ 12"
e | 7/8" @ 5" | 7/8" @ 6"

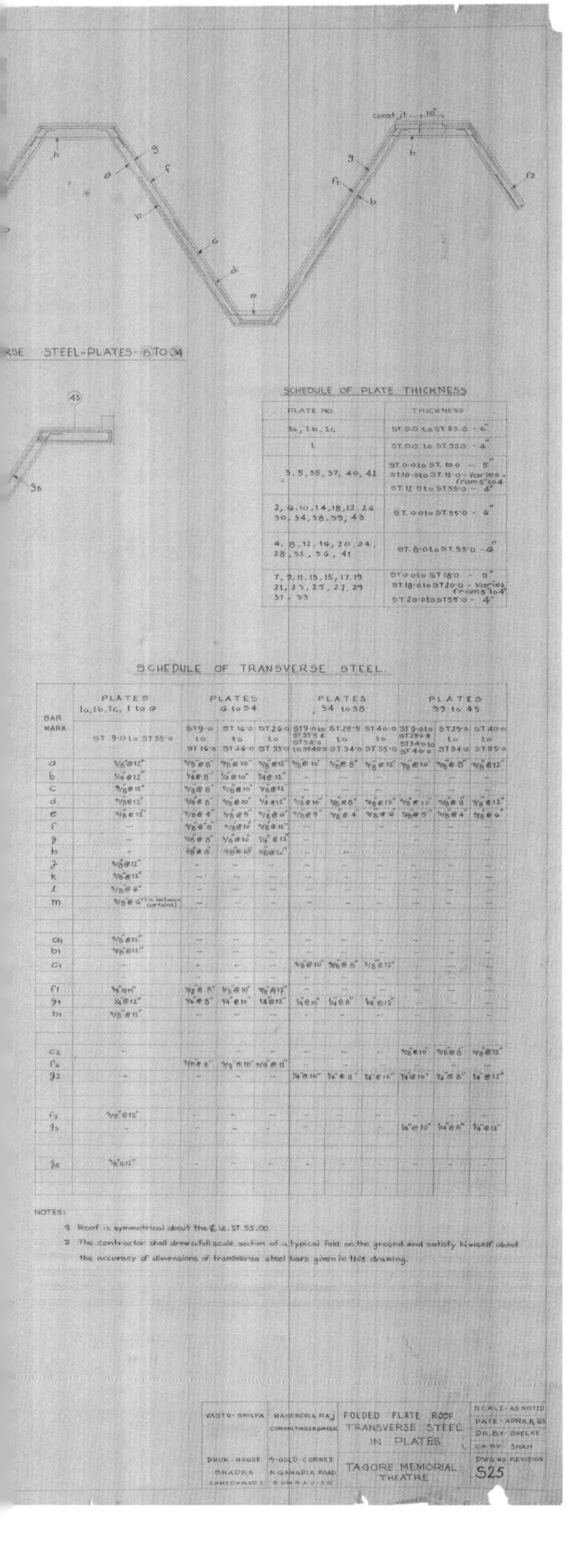

Construction of flat wall sections, staircase and base of folded-plate sections in progress.

Roof folds with temporary and permanent ties during different stages of construction of the folds.

Shuttering and steel layout at the roof level of the western edge of the roof.

Scaffolding for manual movement of materials via two pulley shafts seen as the roof folds are being cast.

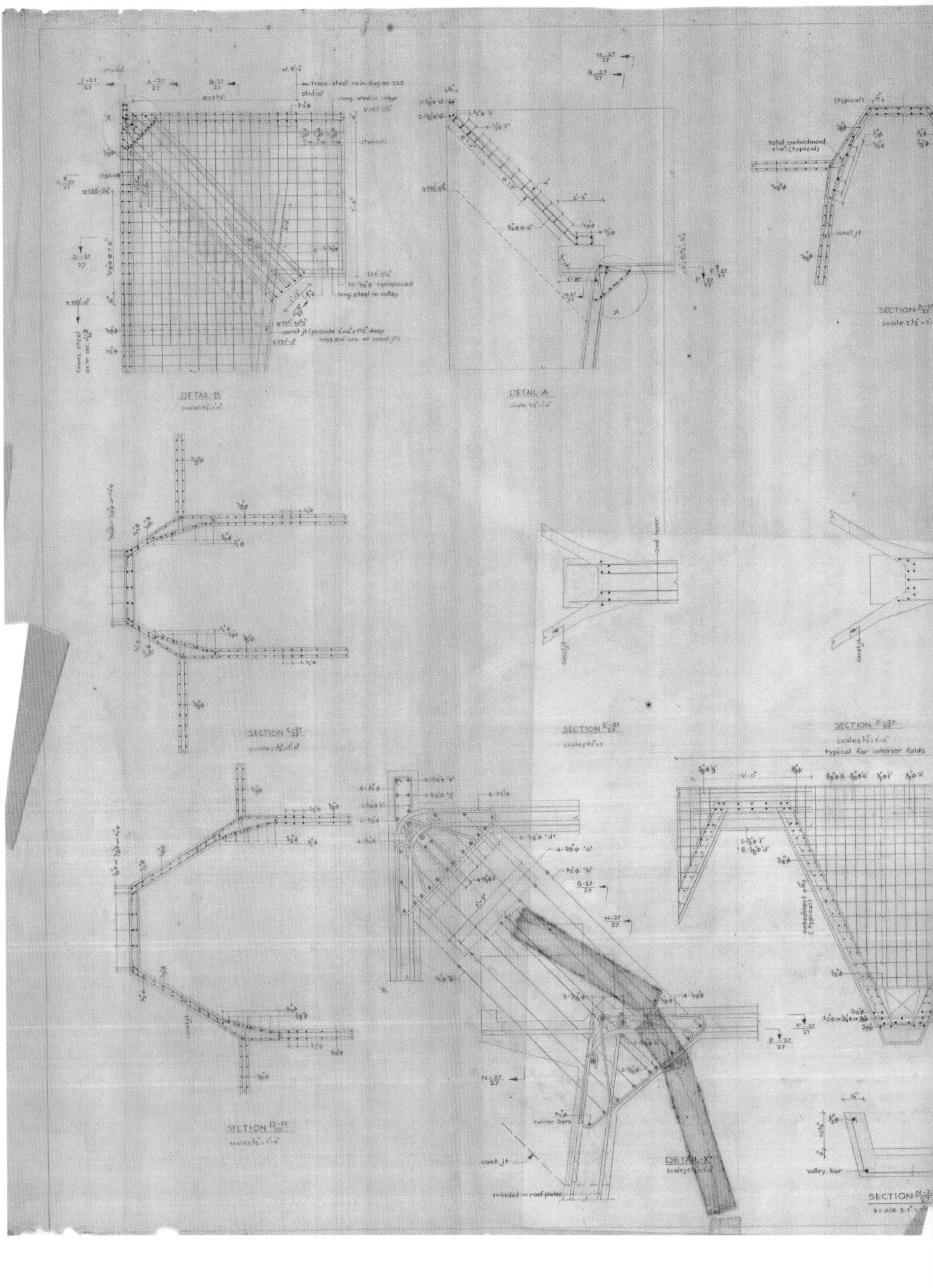
DETAIL-B
DETAIL-A
SECTION A-27/27
SECTION C-27/27
SECTION E-27/27
SECTION F-27/27
typical for interior folds
SECTION D-27/27
DETAIL-X
SECTION N-27/27
long. steel in ridge
long. steel in valley
total embedment 1'-0" (typical)
const. jt.
carrier bars
embeded in roof plates
valley bar
2nd layer

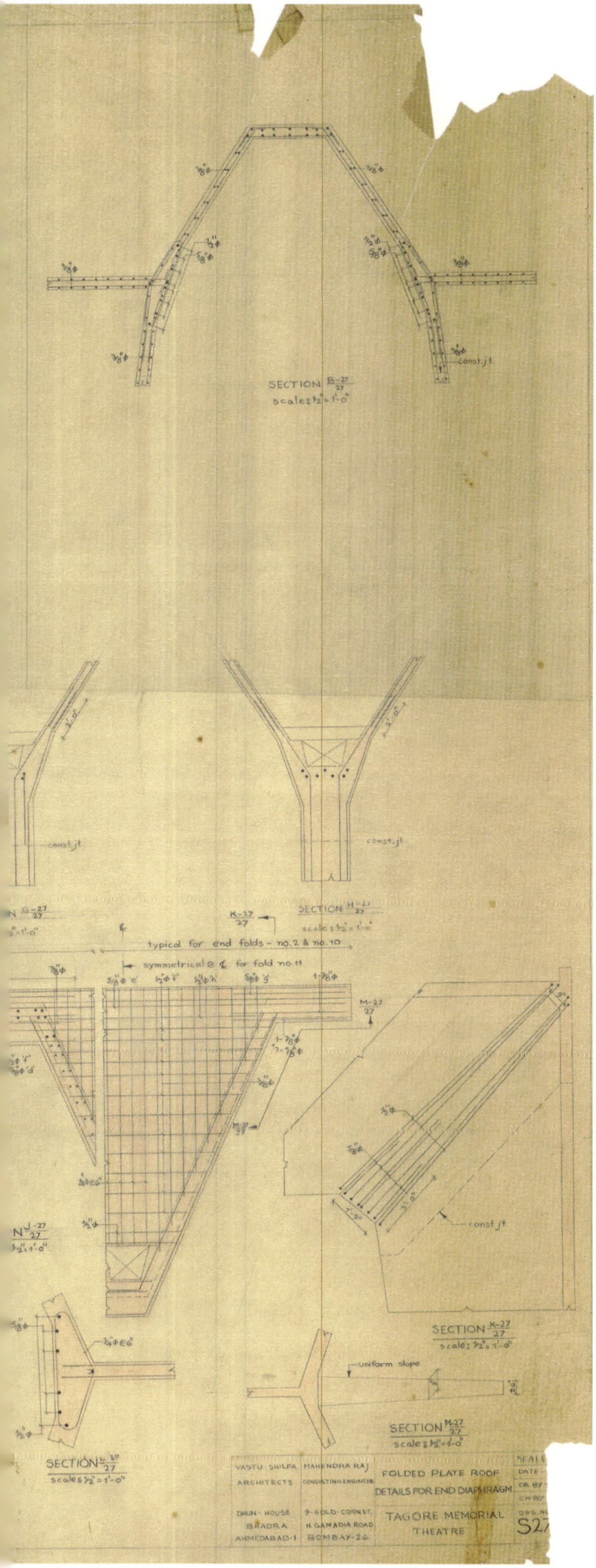

Seen in this photo, diaphragm plates between the folds at the roof level with gutter spouts.

Drawing no. S27: Folded-Plate Roof Details for End Diaphragm

Details of the end diaphragm where the folded-plate roof intersects with the vertical folded plates. This edge also caters to the additional strength required at the junction of the folded frame and functions as a gutter with spouts at every fold.

Foyer interior with the independent structure of the seating bowl which echoes the sculptural quality of the folded-plate building envelope. Inspired by Nervi, the twists of the column express the load transfer and lines of forces from the beam through the column down to the ground plane.

Top and opposite: Exterior views of the theatre building.

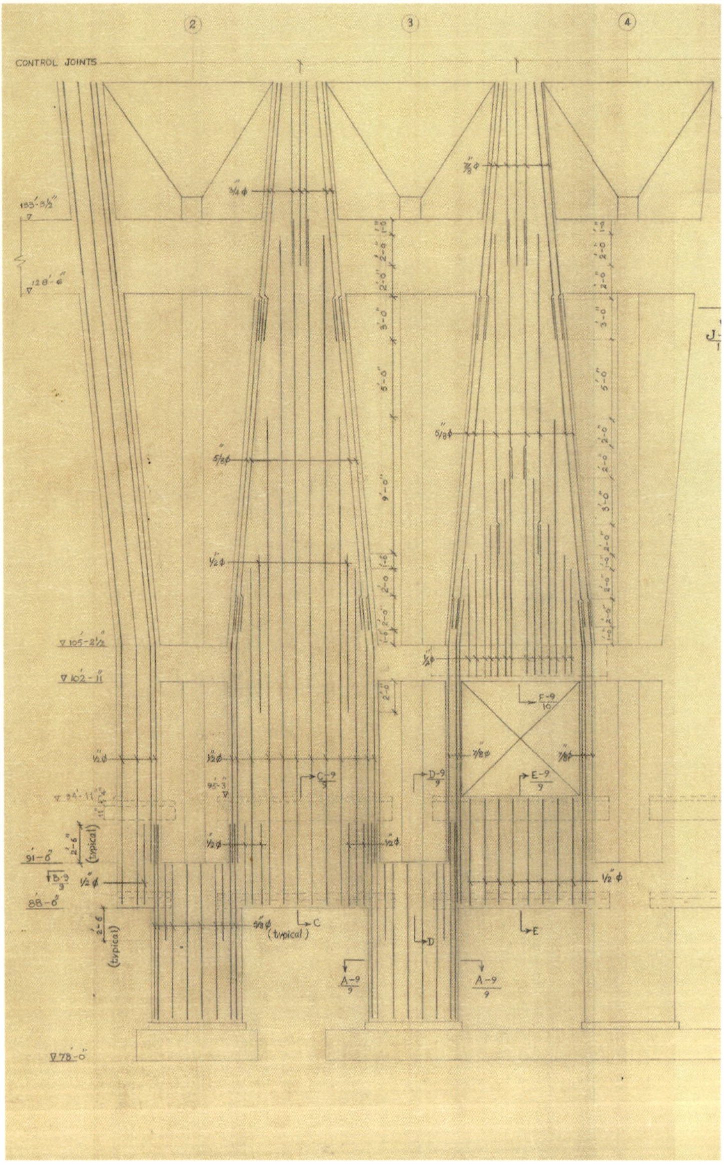

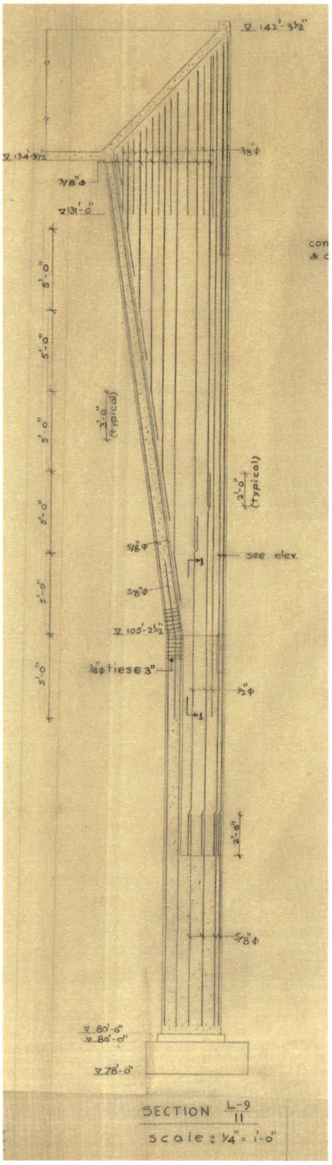

(Partial) Drawing no.S9: Elevations and Details of Frame 'A': 23 May 1962,
(Partial) Drawing no.S11: Frame 'A' and Frame 'B' Details: 8 June 1962,
showing typical reinforcement layouts for vertical folded plates in elevation and section.

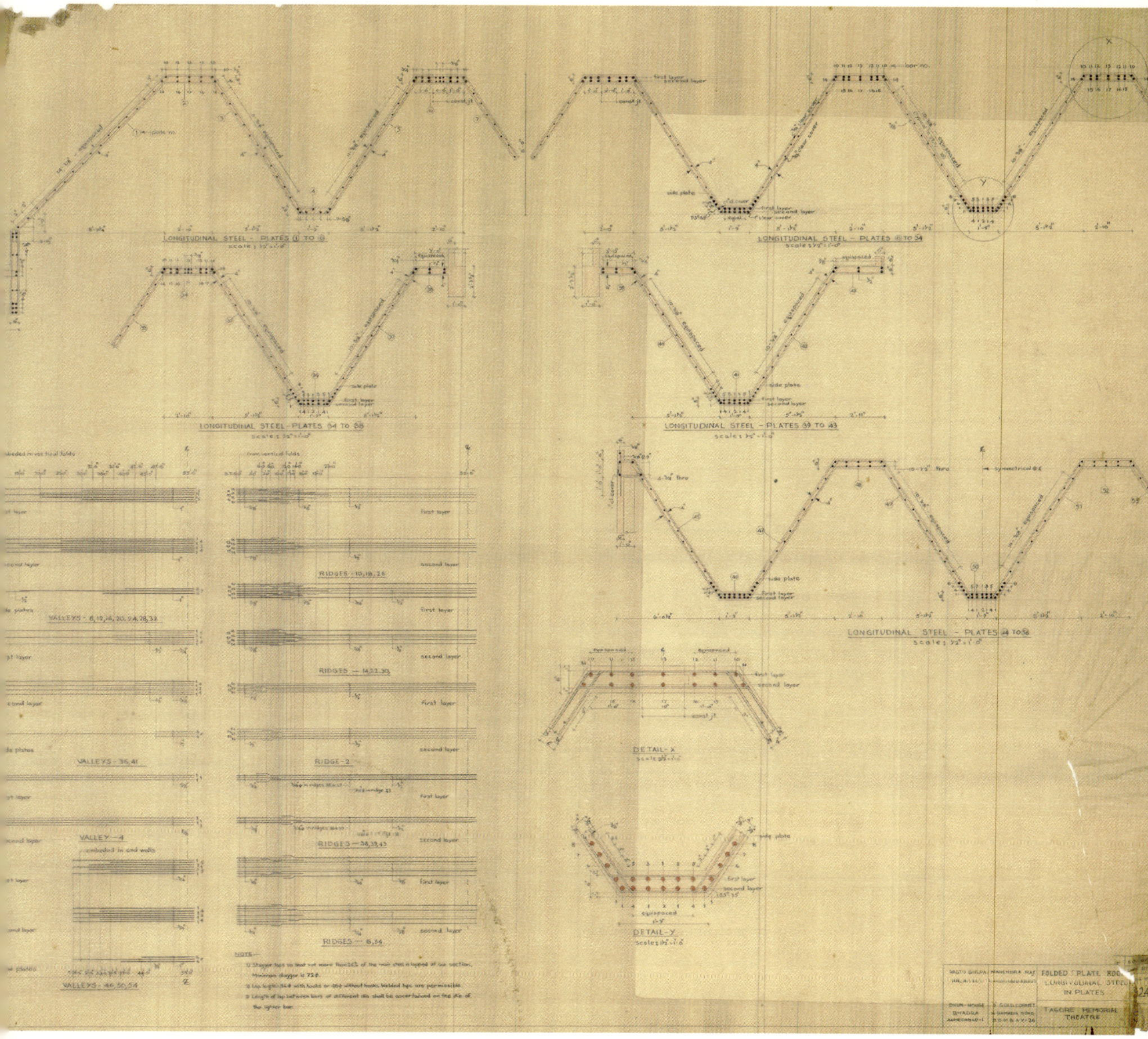

Drawing no. S24: Folded-Plate Roof–Longitudinal Steel in Plates,
shows longitudinal steel layouts of the valley and the ridges for the roof folds.

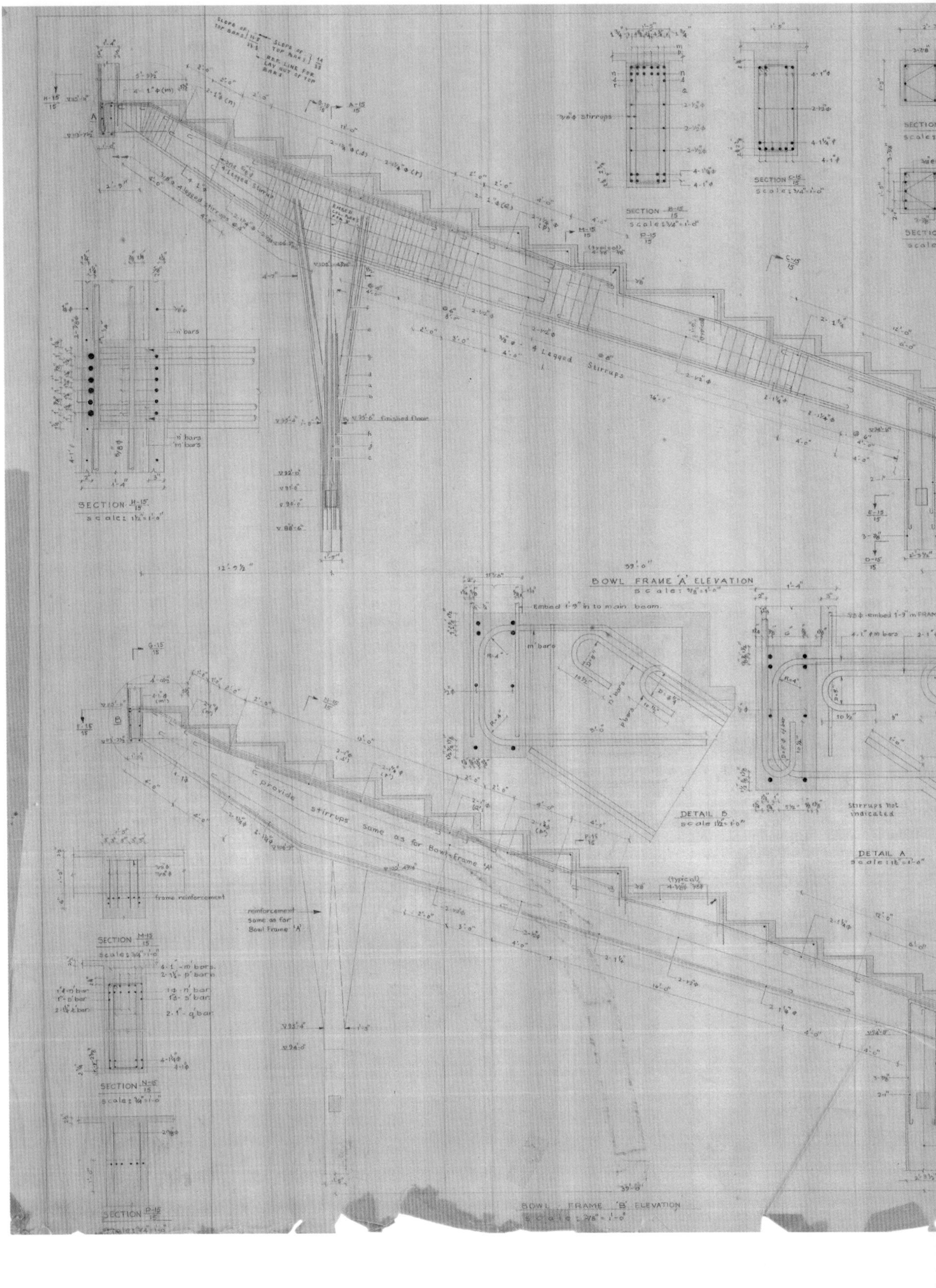
BOWL FRAME 'A' ELEVATION
Embed 1'-9" in to main beam.
m' bars
n' bars
4 Legged Stirrups
finished floor
provide stirrups same as for Bowl-Frame 'A'
reinforcement same as for Bowl Frame 'A'
frame reinforcement
DETAIL B
DETAIL A
Stirrups not indicated
SECTION H-15/15
SECTION M-15/15
SECTION N-15/15
SECTION P-15/15
SECTION B-15/15
SECTION C-15/15
BOWL FRAME 'B' ELEVATION

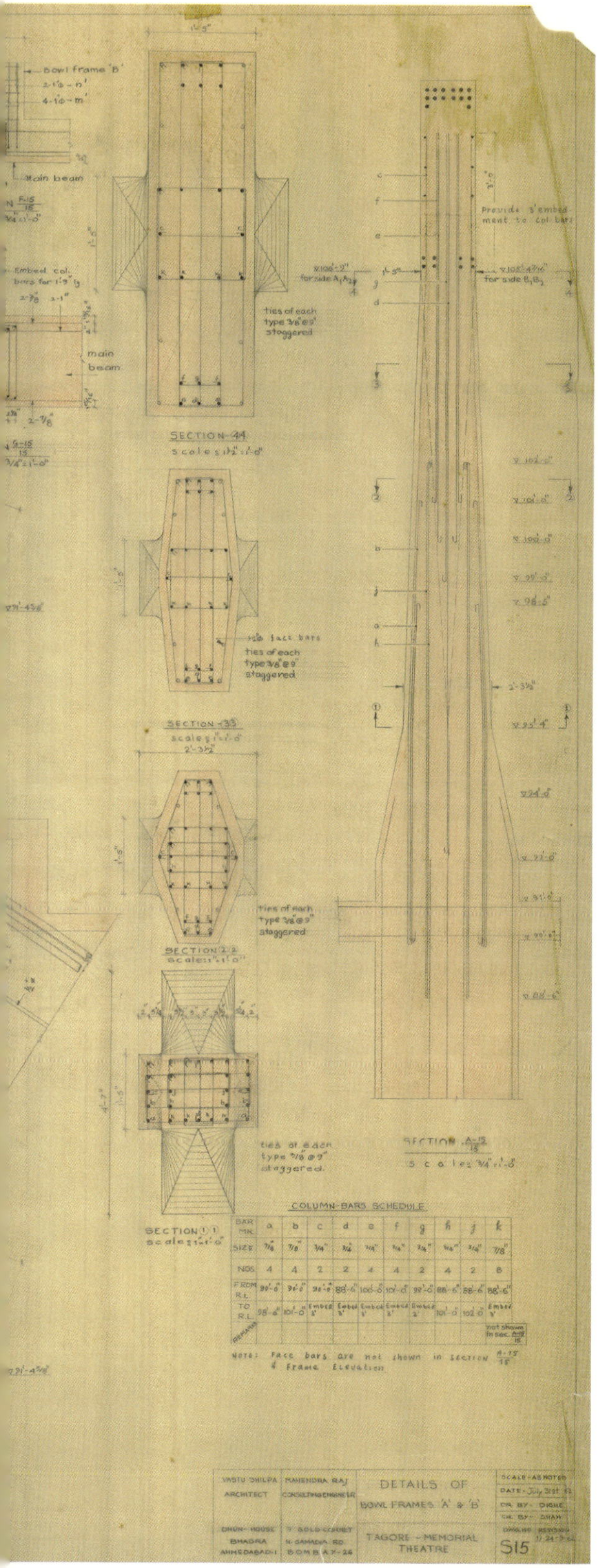

Construction photo of the sculptural twists of one of the columns supporting the seating bowl, expressing the lines of forces it transfers.

Drawing no. S15: Details of Bowl Frames 'A' & 'B': 31 July 1962

The underside of the seating bowl with its structural play exposed in the foyer space.

Interior view of the vertical and roof trapezoidal folded-plate structure in the foyer. The vertical folded plates flare inward from a depth of 1.15m to 2.4m as they go up to meet the roof folds, also 2.4m deep. The centreline distance between the folds is 4.5m. The structural taper of the vertical folds efficiently expresses the forces and moments, lending a distinct dynamic quality to the building.

NO SPITTING
EXIT

SHRI RAM CENTRE

NEW DELHI | 1968

Interior photograph of the radial beams visible in the foyer.

Previous pages: External view of Shri Ram Centre and its distinct structural forms of the circular drum lifted off the ground plane and the floating parasol on top.

The Shri Ram Centre stands for an expression of structural and programmatic integrity. Cast-in-situ pure forms for an auditorium and rehearsal hall are resolved through two distinct structural systems. Although supported on circumferential and radial beams cantilevering out of six columns, the 650-person drum-shaped auditorium appears to be floating above the ground, anchored in space by the parasol roof, which houses the 31m x 31m rehearsal hall.

The hall is supported on four independent cross-shaped columns placed at 19 metres centres with 6 metres cantilevers all around. The auditorium is 21 metres in diameter and is enclosed by a hollow concrete wall that also supports its balcony. The hollow wall is pinned at the bottom on the circumferential beam and at the top to the parasol slab creating a balancing couple that counteracts the cantilever moment generated by the balcony.

The final product that we see is quite different from the designs originally brought by architect Shiv Nath Prasad. For instance the floating effect of the building, now its most notable feature, was introduced by the structural changes suggested by Mahendra Raj.

In Raj's words, 'Shiv Nath Prasad was a very appreciative man. When given the suggestion of the balcony and the separation of the structures on the four columns, he was very happy and integrated these ideas in the design.'

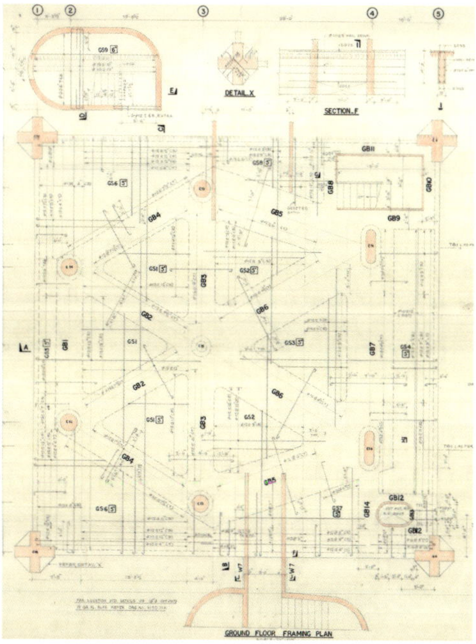

(Partial) Drawing no. 31.TD.10A: Ground Floor Framing Plan: 28 May 1968

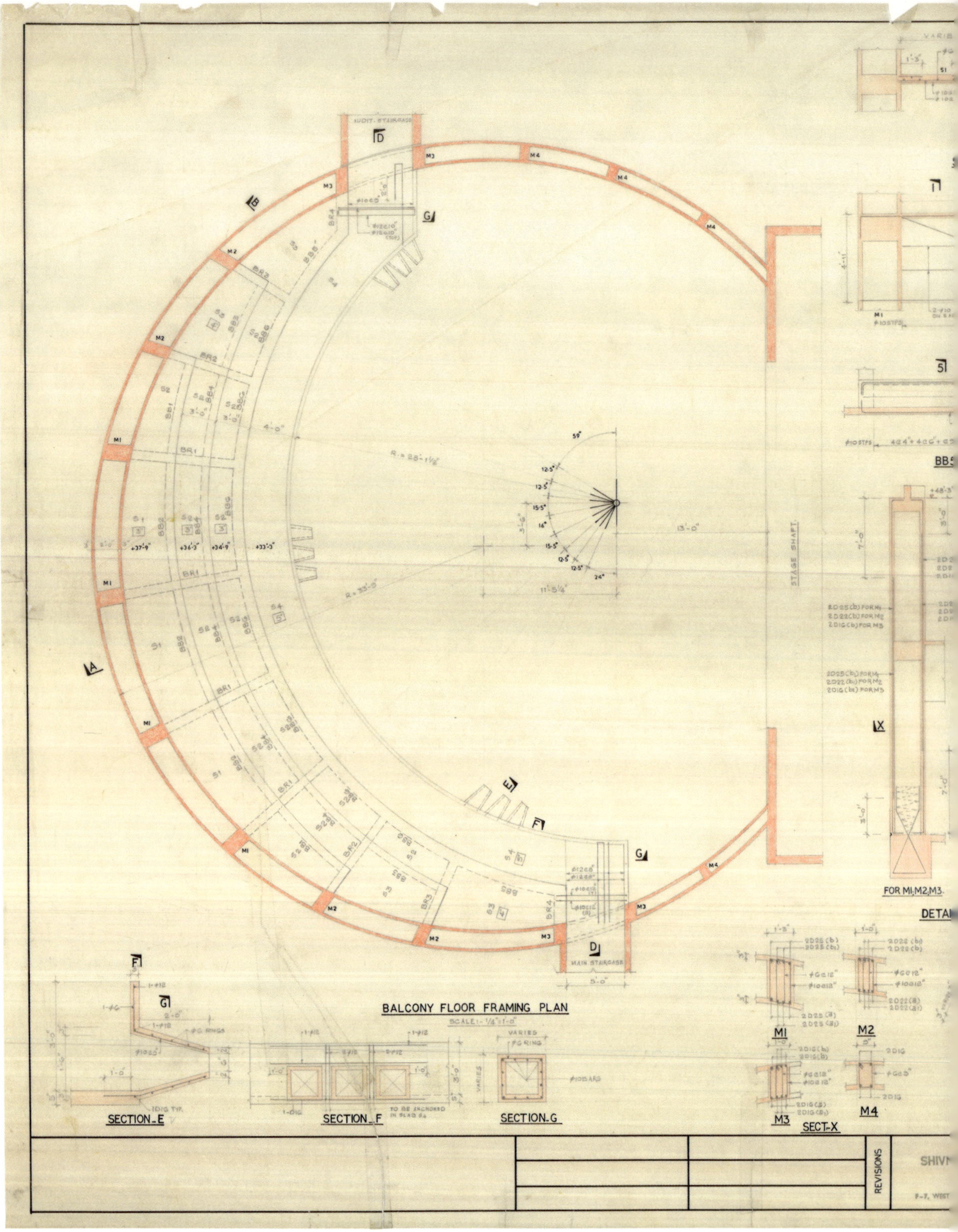
BALCONY FLOOR FRAMING PLAN
STAGE SHAFT
MAIN STAIRCASE
SECTION_E
SECTION_F
SECTION_G
SECT-X
REVISIONS

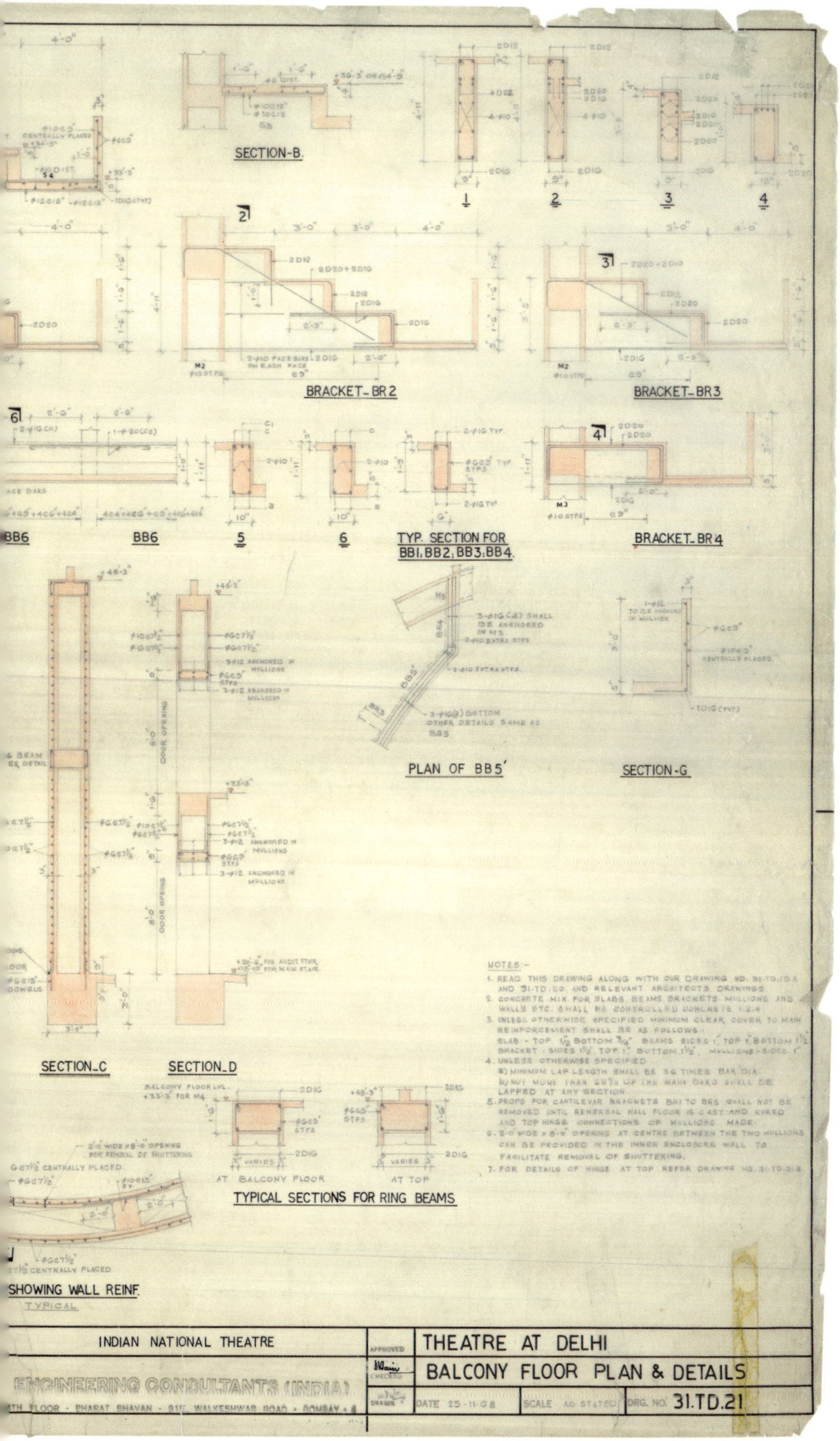

Drawing no. 31.TD.21: Balcony Floor Plan & Details: 25 October 1968, showing how the balcony floor cantilevers from pilasters in the hollow drum wall, which are innovatively pinned to the top and bottom to eliminate the torsional moments on the supporting elements.

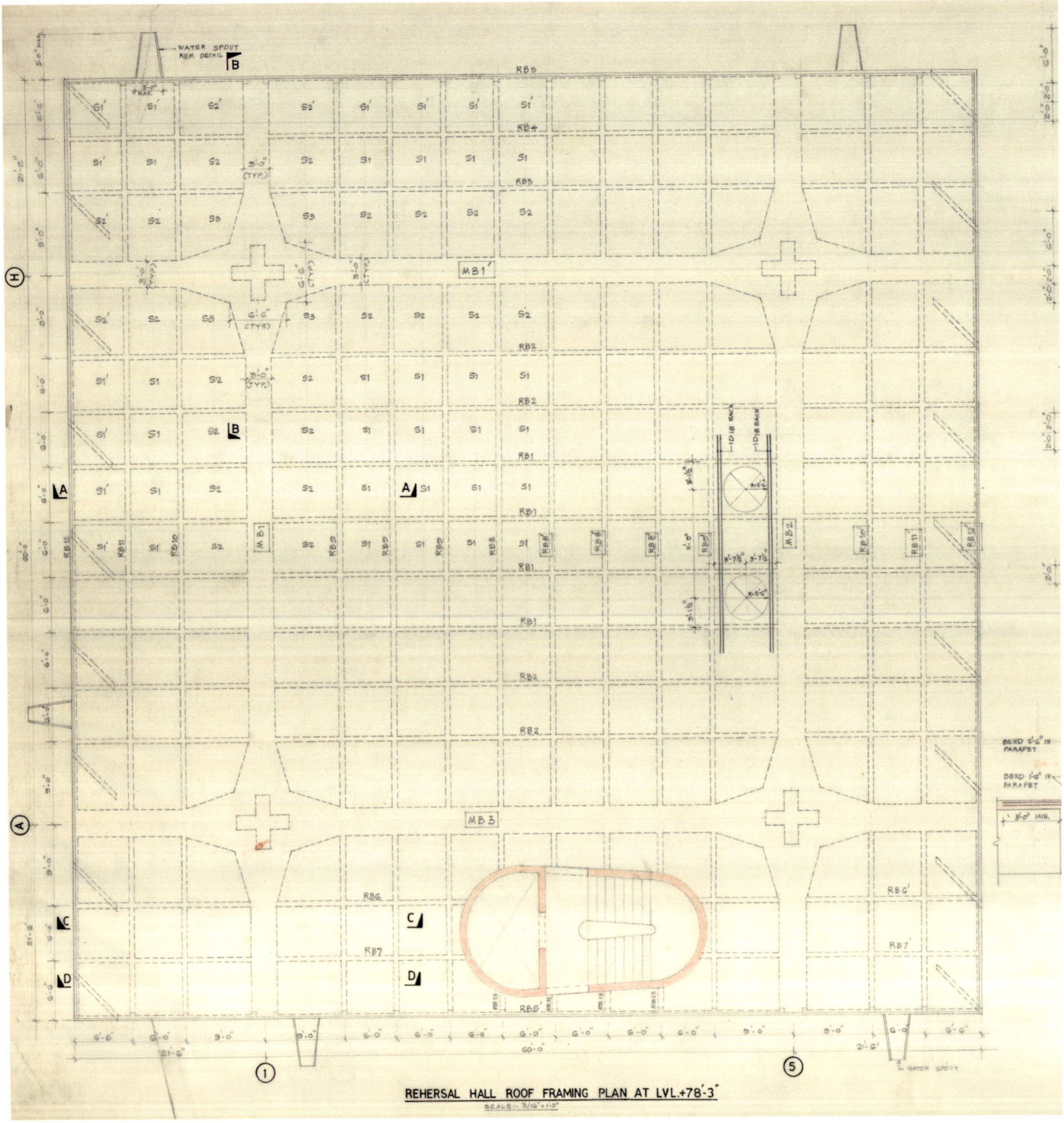

(Partial) Drawing no. 31.TD.27C: Rehearsal Hall Roof Framing Plan: 7 March 1969

The rehearsal hall roof is 31m x 31m with slender tapering two-way ribs at 2m centres, supported on four cross-shaped columns placed 19m apart with 6m cantilevers on all sides.

Exterior photo of the beams flowering out of the cross-shaped column and the delicate ribbed structure of the cantilevered rehersal hall above.

Street view of the theatre.

Seen in this exterior view, the pinned hollow drum wall, the radial beam structure of the auditorium and the cross-shaped columns that lift and separate the parasol above it.

Shankar Lal Murli Dhar Auditorium

CENTRAL BANK

AHMEDABAD | 1969

Central Bank of India
Regional Office
Central Bank of India

A six-storey rectangular office tower emerges from a four-storey podium base with an irregularly shaped plan. The rectangular tower is lifted from the podium, carried on 1.5m-deep, concrete box-shaped transfer girders, to create a large open public plaza and entrance to the building. A mezzanine floor is also suspended from the first floor of the tower, retaining the clear space for the plaza.

From six column units on the lower podium floors, four corner column units rise up to the full height of six floors in the tower block.The box transfer girders carry mullion-like columns on each floor, which in turn carry the ribbed floors above. This system where two transfer girders carry the load of the entire building through columns aligned with the rhythm of the ribs creates a beautiful slender structure with column-free floors of 18.3m x 36.6m. Thinner transfer girders carry a system of sun-breaker fins around the tower block on all six storeys. The total floor area of the building is approximately 11,000 square metres.

View from the mezzanine level of the large uninterrupted interior office spaces created within the podium.

Previous pages: Street view of the solid exposed concrete podium and tower block of the Central Bank with its brise soleil facades.

View of the hung mezzanine floor visible from the plaza.

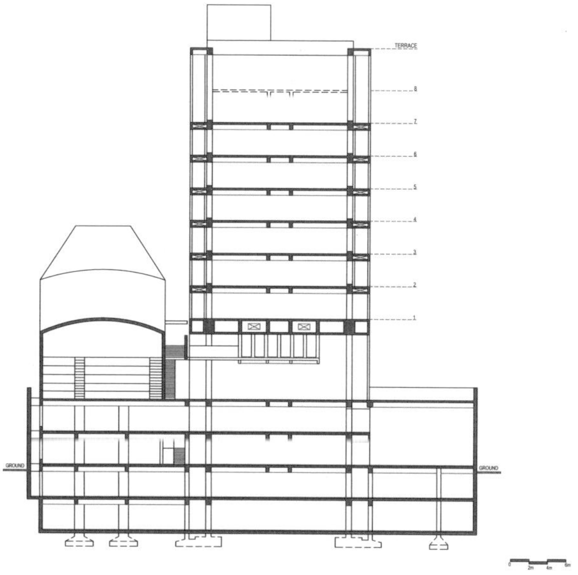

Building section showing the podium, the tower and the hung mezzanine.

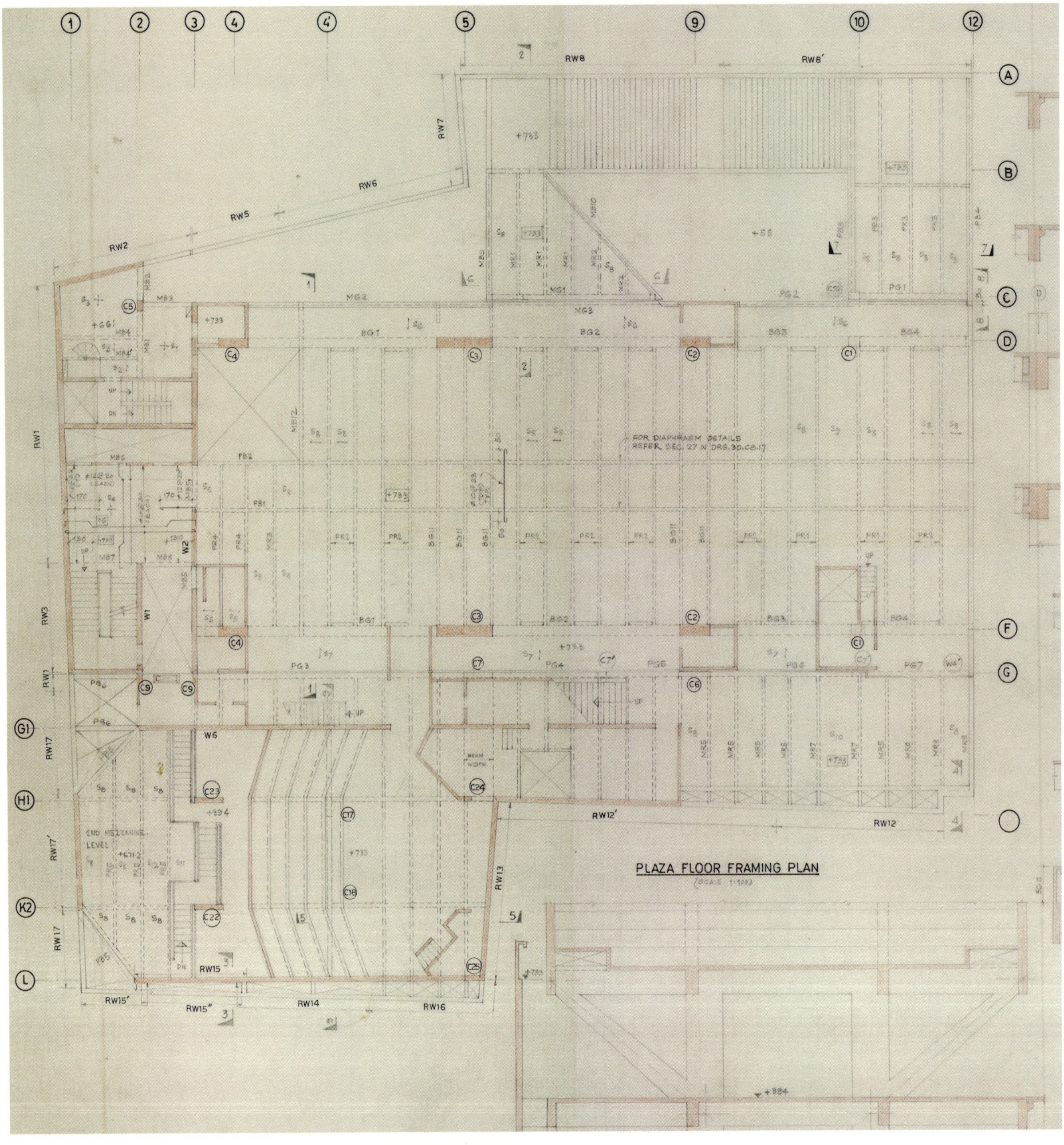

(Partial) Drawing no. 30.CB.33C: Plaza Floor Framing Plan: 19 August 1968

Photo of the unobstructed double-height space created within the podium by a clean and expressive structural system of column bays at the edges and a ribbed spanning system.

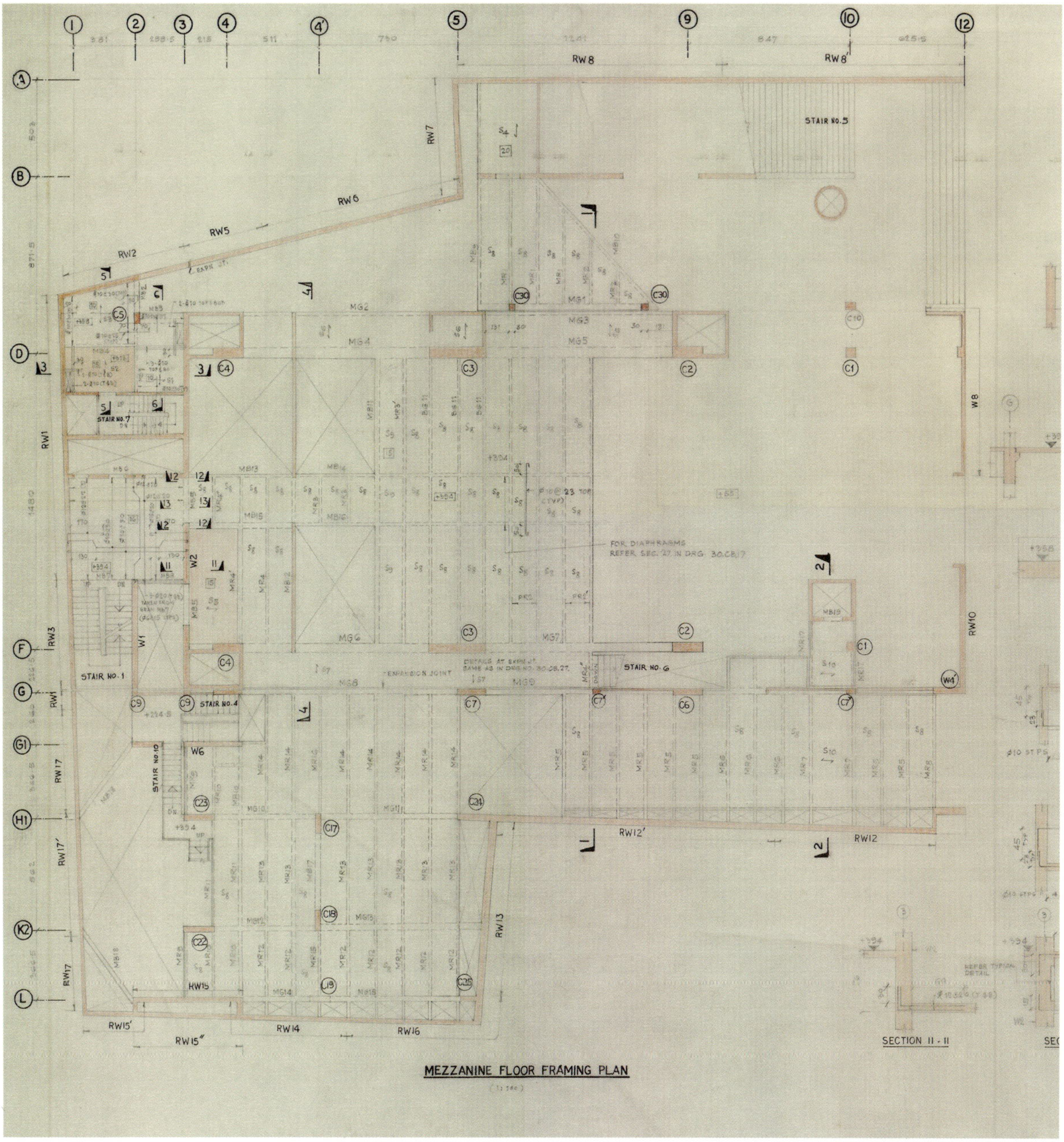

(Partial) Drawing no. 30.CB.30-E: Mezzanine Floor Framing Plan: 13 August 1968

Construction photo of the completed podium and the four corner column bays rising up to carry the parallel box transfer girders. Also seen here, the construction of the mullion-like columns on the first floor.

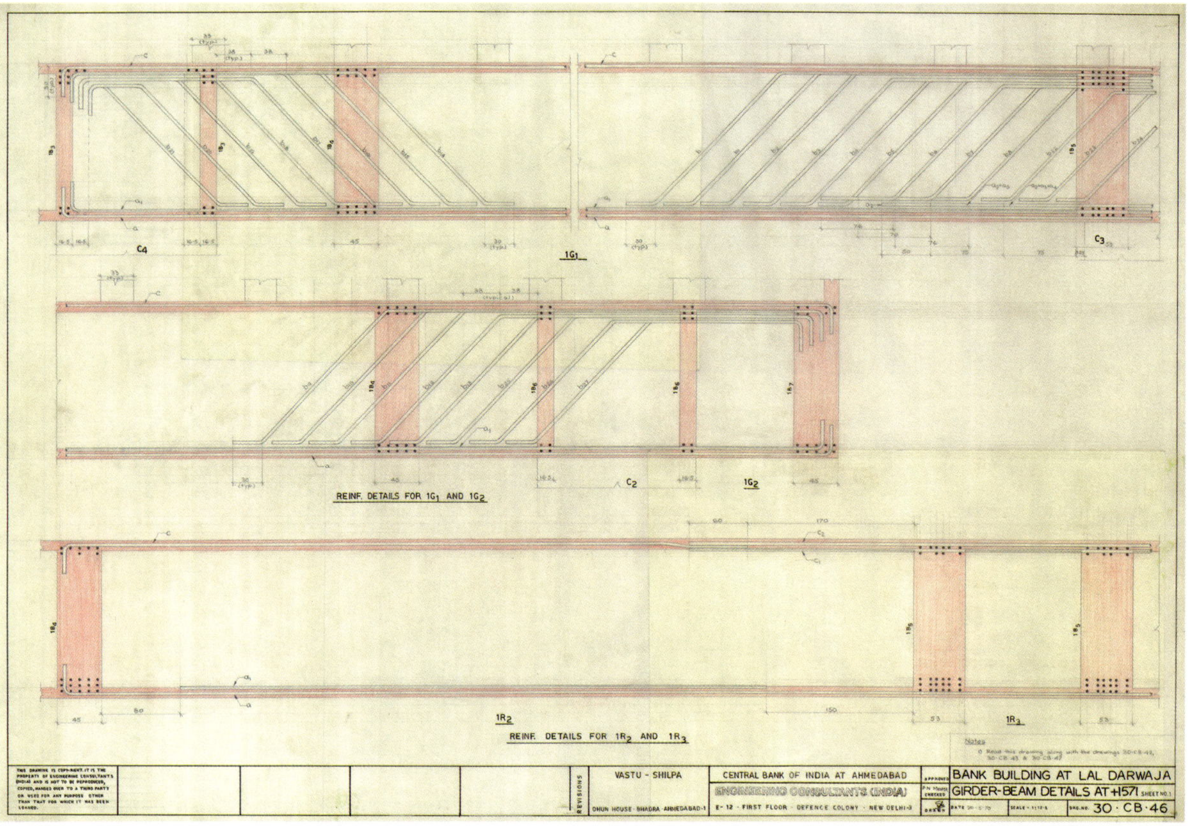
1G1
C4
C3
C2
1G2
REINF. DETAILS FOR 1G1 AND 1G2
1R2
1R3
REINF. DETAILS FOR 1R2 AND 1R3
VASTU - SHILPA
DHUN HOUSE · BHADRA · AHMEDABAD-1
REVISIONS
CENTRAL BANK OF INDIA AT AHMEDABAD
ENGINEERING CONSULTANTS (INDIA)
E - 12 · FIRST FLOOR · DEFENCE COLONY · NEW DELHI-3
BANK BUILDING AT LAL DARWAJA
GIRDER-BEAM DETAILS AT +1571
30 · CB · 46

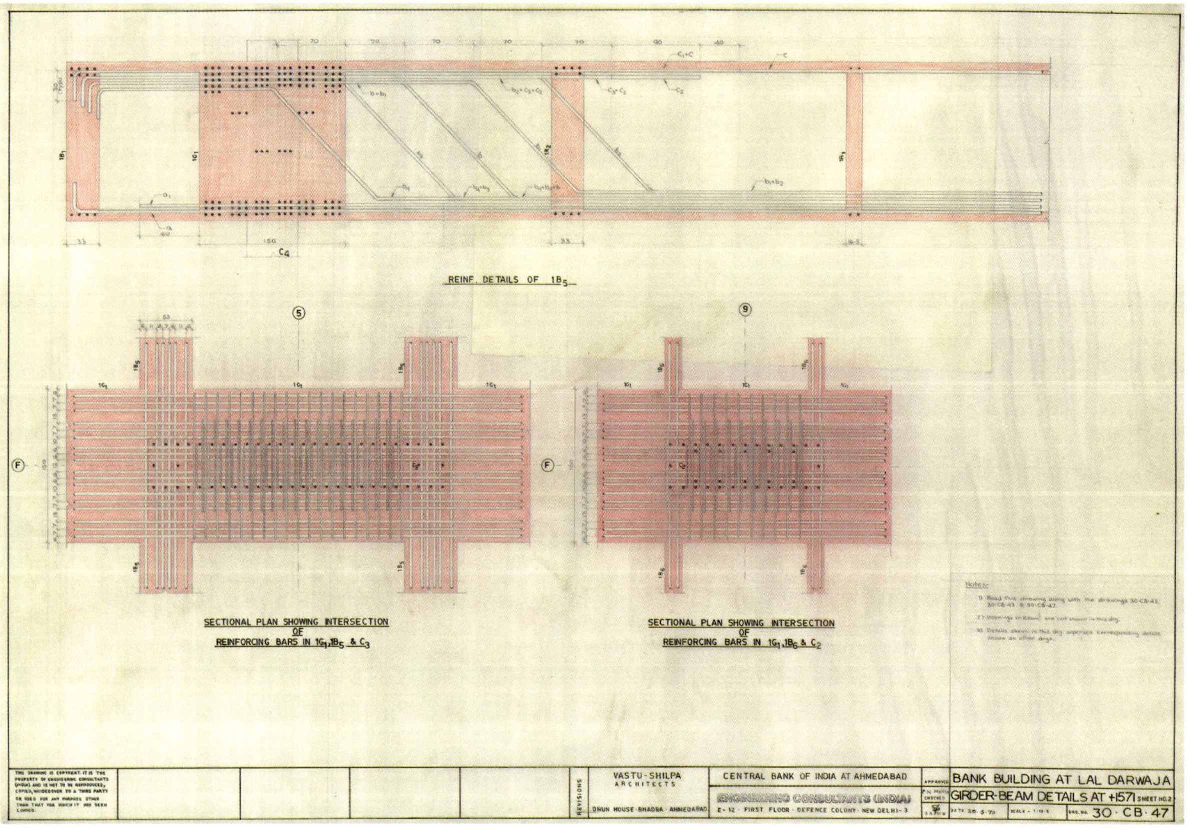
C4
REINF. DETAILS OF 1B5
SECTIONAL PLAN SHOWING INTERSECTION OF REINFORCING BARS IN 1G1, 1B5 & C3
SECTIONAL PLAN SHOWING INTERSECTION OF REINFORCING BARS IN 1G1, 1B6 & C2
VASTU - SHILPA
ARCHITECTS
DHUN HOUSE · BHADRA · AHMEDABAD
REVISIONS
CENTRAL BANK OF INDIA AT AHMEDABAD
ENGINEERING CONSULTANTS (INDIA)
E - 12 · FIRST FLOOR · DEFENCE COLONY · NEW DELHI - 3
BANK BUILDING AT LAL DARWAJA
GIRDER-BEAM DETAILS AT +1571
30 · CB · 47

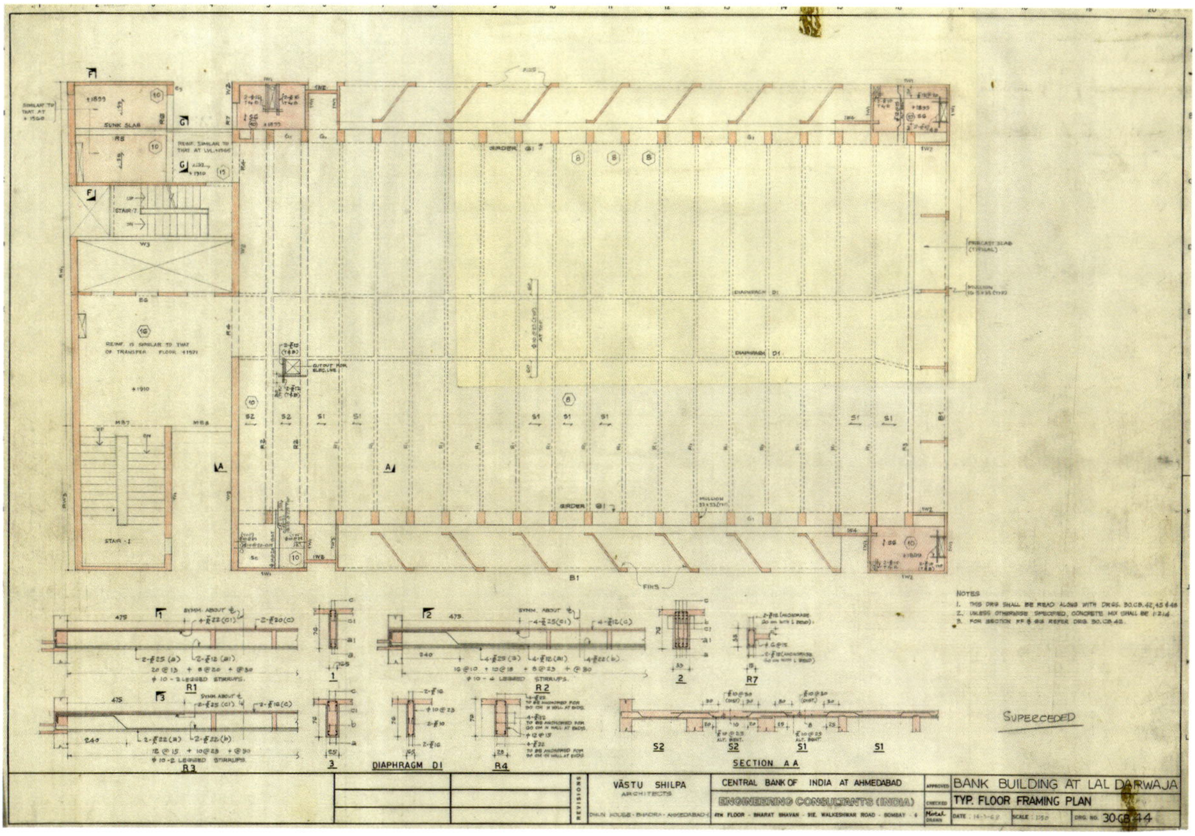

Drawing no. 30.CB.44: Typical Floor Framing Plan–Tower: 14 March 1968

Opposite:

Drawing no. 30.CB.46. Girder-Beam Details at +1571 sheet 1: 20 May 1970

Drawing no. 30.CB.47: Girder-Beam Details at +1571 sheet 2: 20 May 1970

View of a corner column and one of the transfer girders carrying the building above.

Rear end of the building containing the main service cores helps contain the horizontal seismic forces.

View of the mezzanine hung from the first floor.

Looking into the double-height reception area from the mezzanine within the podium.

Opposite: Exterior view of Central Bank.

OF INDIA

AKBAR HOTEL

NEW DELHI | 1969

This dramatic, elegantly proportioned modernist hotel building, 208 x 60 feet in dimension and 14 storeys high, is supported on two rows of columns at 18 x 60 feet centres, two end shear walls and service cores. The top ten floors house 300 hotel rooms, while the lower four floors contain public areas and service facilities.

When Mahendra Raj started working on this project, all the columns from the floors above were being brought down to the lower floors where the foyer and dining room were located. Shiv Nath Prasad, the architect, welcomed his suggestion of achieving a clear floor space for the lower public levels by using a transfer girder to carry the load to the edges. The post-tensioned transfer girders spanning 60 feet carry the columns of the upper ten floors. Two service floors below are hung from the girders, separating the structural and service requirements of the hotel rooms from the lower level. This is perhaps the first hotel tower in India where a transfer girder was used to such effect.

The webbed ceiling of the lobby area.

Previous pages: The Akbar Hotel building in exposed concrete, made possible with an astute set of structural solutions.

The large span achieved in the foyer, however, required deep beams
at the first floor slabs that clashed with the architect's proportions of the facade.

Prasad wanted to restrict the thickness of the periphery beam to no more than 12 inches. To achieve this result Raj introduced tapered ribbed beams converging at the columns in a webbed pattern. These ribs are shallow and flat near the columns to transfer shear, and narrow and deep at the centre to carry the moment, thereby minimising the thickness of the peripheral beam connecting the columns to the architect's desired slenderness.

The structural system of the hotel rooms is a flat plate that is solid at the columns and made hollow in the central sections using pre-cast shallow waffles. All horizontal forces are contained by the end shear walls and interior service cores.

Another special feature of the hotel is a free-standing stairway, cantilevered from a central pin wall and connected to the main tower block at two floors only.

Photo of the architectural building model.

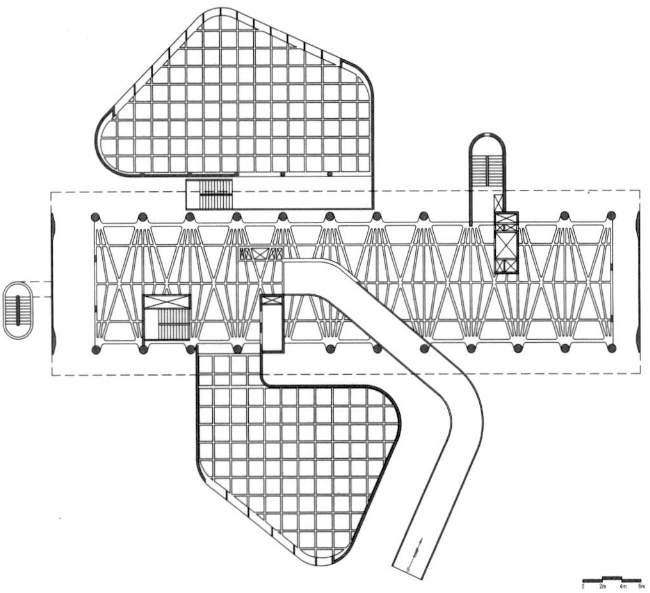

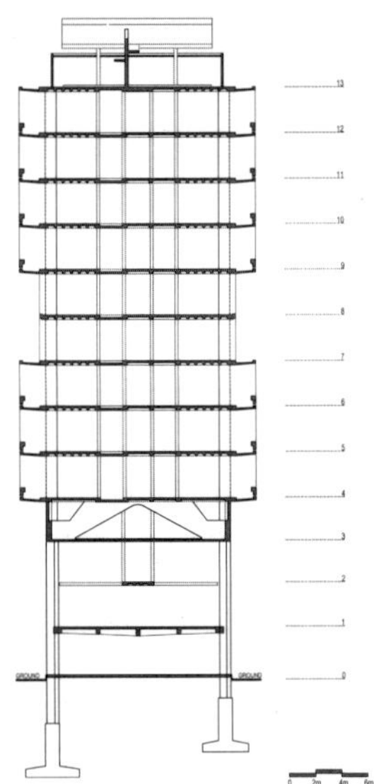

Plan showing flared web-like rib beams and columns and a diagrammatic section of the building with the transfer girder and clear span achieved at the lobby level. The building was founded on disintegrated rock explaining the different foundation levels of the end columns.

Poolside view of the hotel.

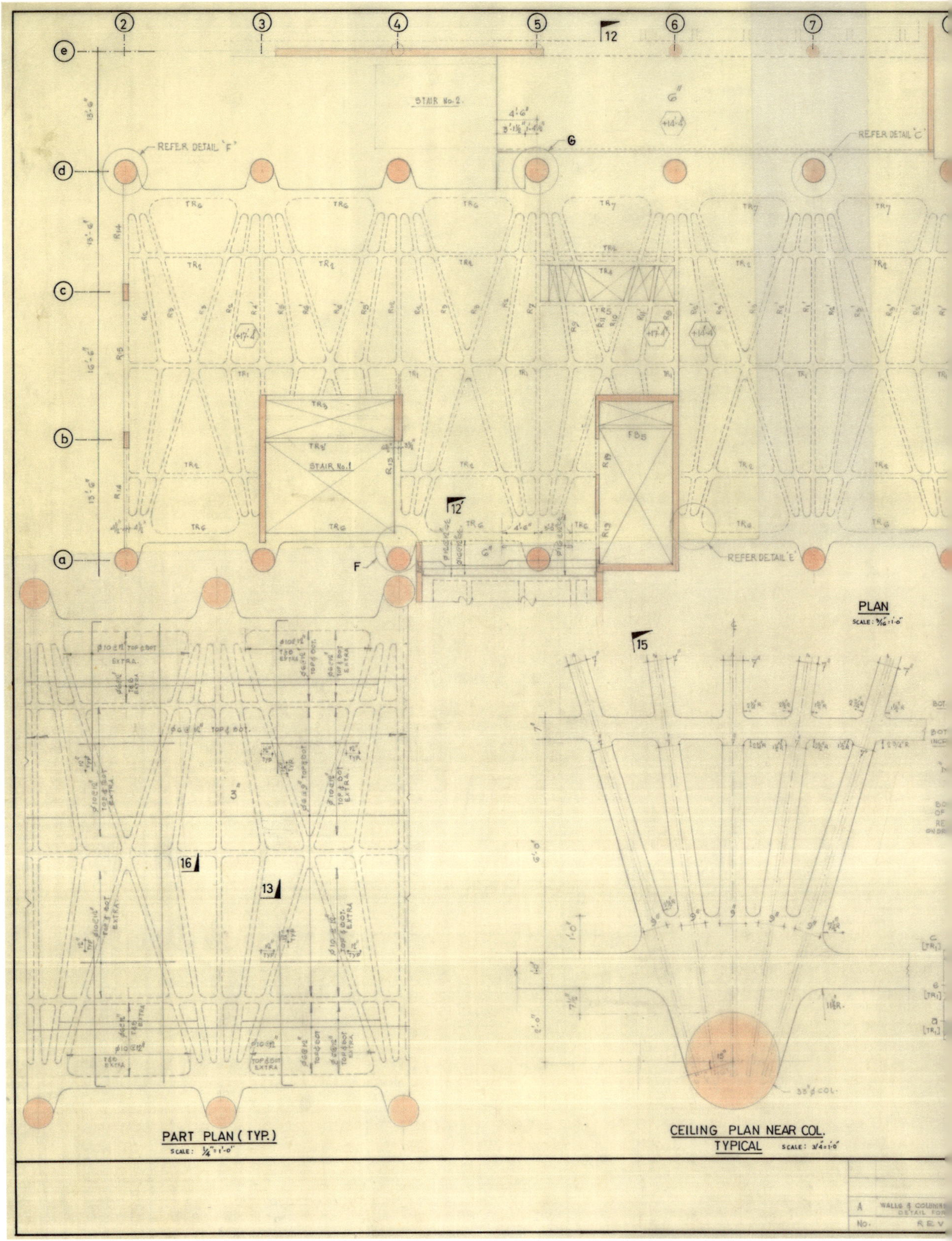
PLAN
STAIR No. 2
STAIR No. 1
REFER DETAIL 'F'
REFER DETAIL 'C'
REFER DETAIL 'E'
PART PLAN (TYP.)
CEILING PLAN NEAR COL.
TYPICAL

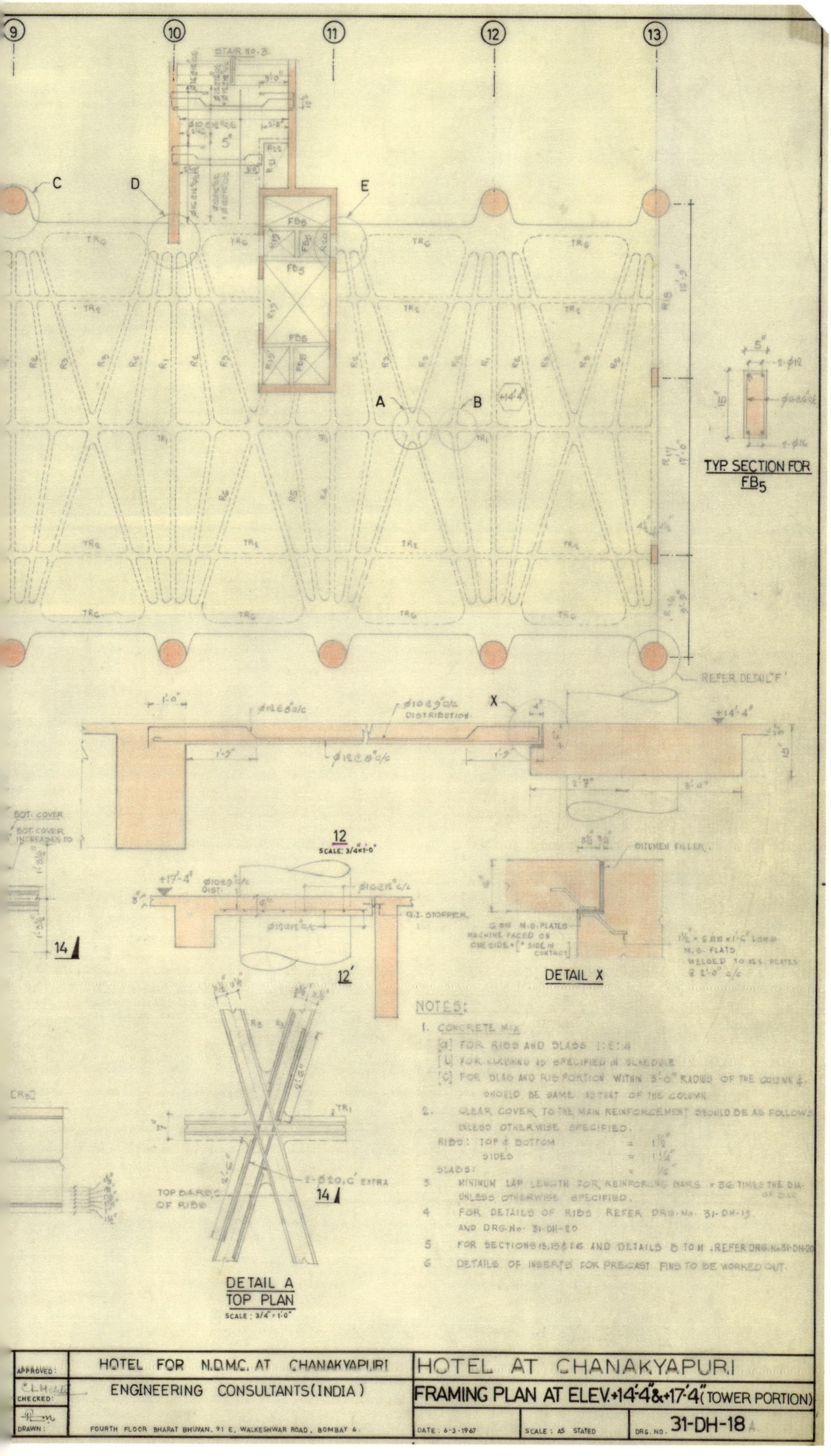

Drawing 31-DH-18 A: Framing Plan at Elev 14'-4" & 17'-4" (Tower portion): 6 March 1967

The solution for the slender, one-foot-deep peripheral beam requirement set by the architect for the lobby was achieved by using flared ribs that transfer the load on the peripheral circular columns on both long sides of the building placed 60 feet apart.

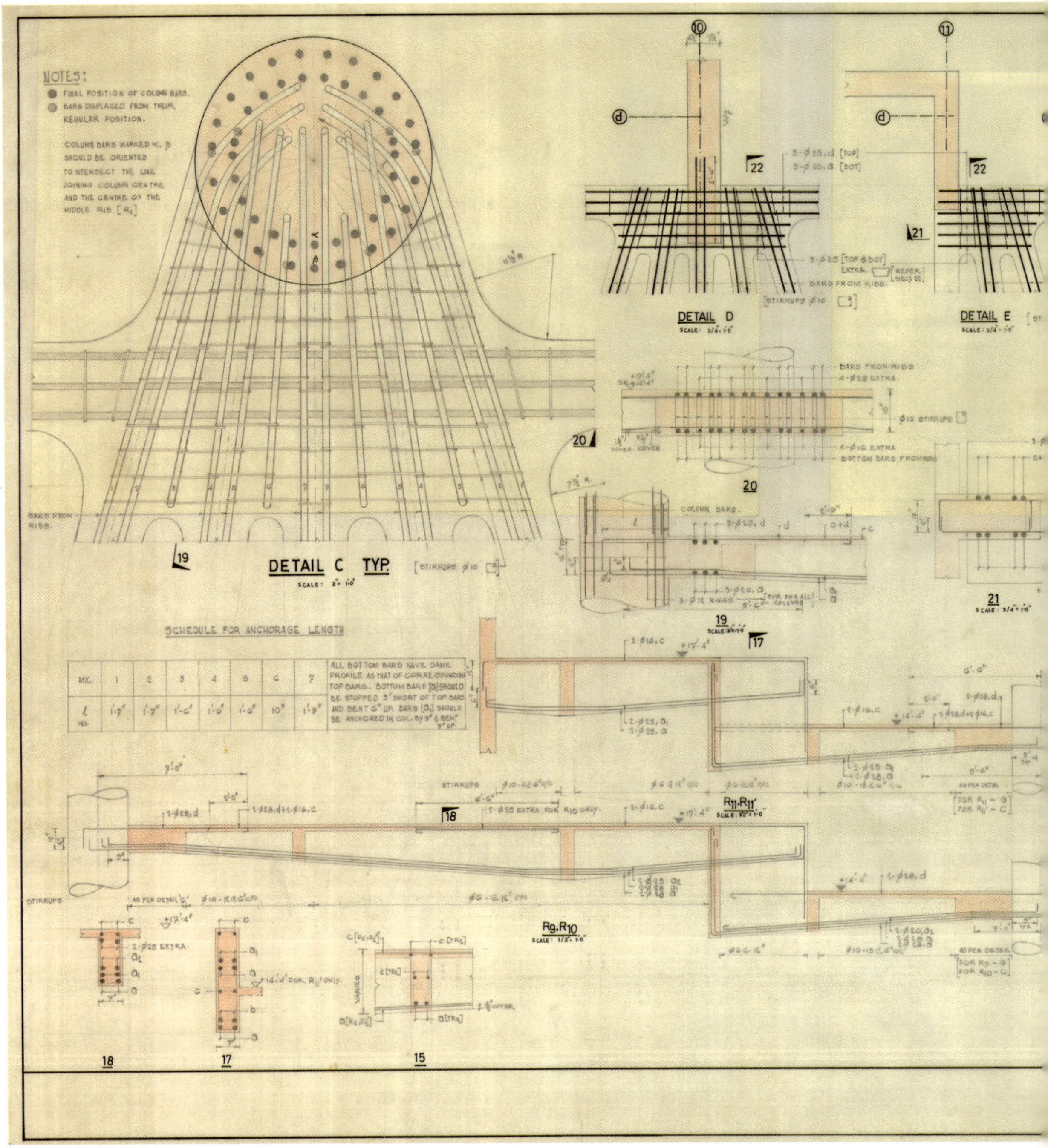
NOTES:
FINAL POSITION OF COLUMN BARS.
BARS DISPLACED FROM THEIR REGULAR POSITION.
COLUMN BARS MARKED α, β SHOULD BE ORIENTED TO INTERSECT THE LINE JOINING COLUMN CENTRE AND THE CENTRE OF THE MIDDLE RIB [R1]
DETAIL C TYP.
[STIRRUPS ϕ10]
DETAIL D
DETAIL E
BARS FROM RIBS
COLUMN BARS
SCHEDULE FOR ANCHORAGE LENGTH
MK. 1 2 3 4 5 6 7
l 1'-7" 1'-7" 1'-6" 1'-6" 1'-6" 10" 1'-9"
ALL BOTTOM BARS HAVE SAME PROFILE AS THAT OF CORRESPONDING TOP BARS. BOTTOM BARS [a] SHOULD BE STOPPED 3" SHORT OF TOP BARS AND BENT 6" UP. BARS [a1] SHOULD BE ANCHORED IN COL. BY 9" & BENT 3" UP.
R11-R11'
R9-R10
STIRRUPS
AS PER DETAIL
18
17
15
19
20
21
22

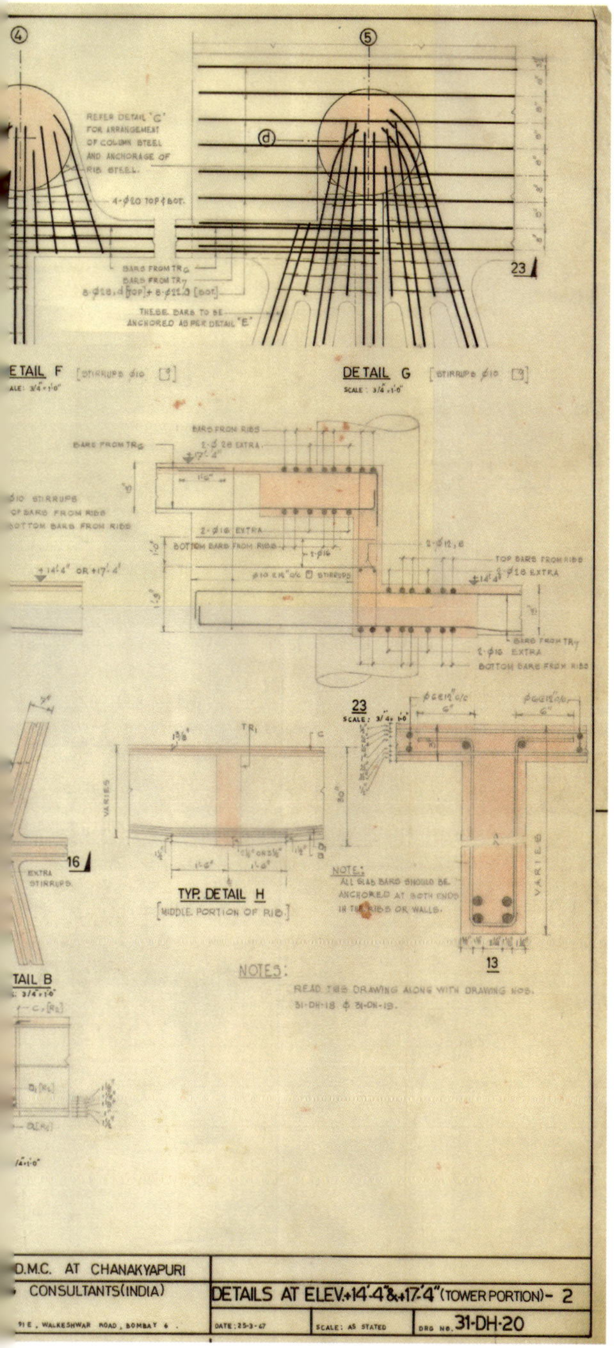

Seen here, the ribbed beams converging at the column.

Drawing no. 31.DH.20: Details at Elev 14'-4" & 17'-4" (Tower Portion)-2: 25 March 1967

Details of the column and section details of the flared rib beams of the lobby ceiling. The ribs are shallow (12") and wider at the columns where they merge, and deeper (30") and slender at the centre.

The lobby ceiling, webbed with exposed flared ribs.

CASHIER

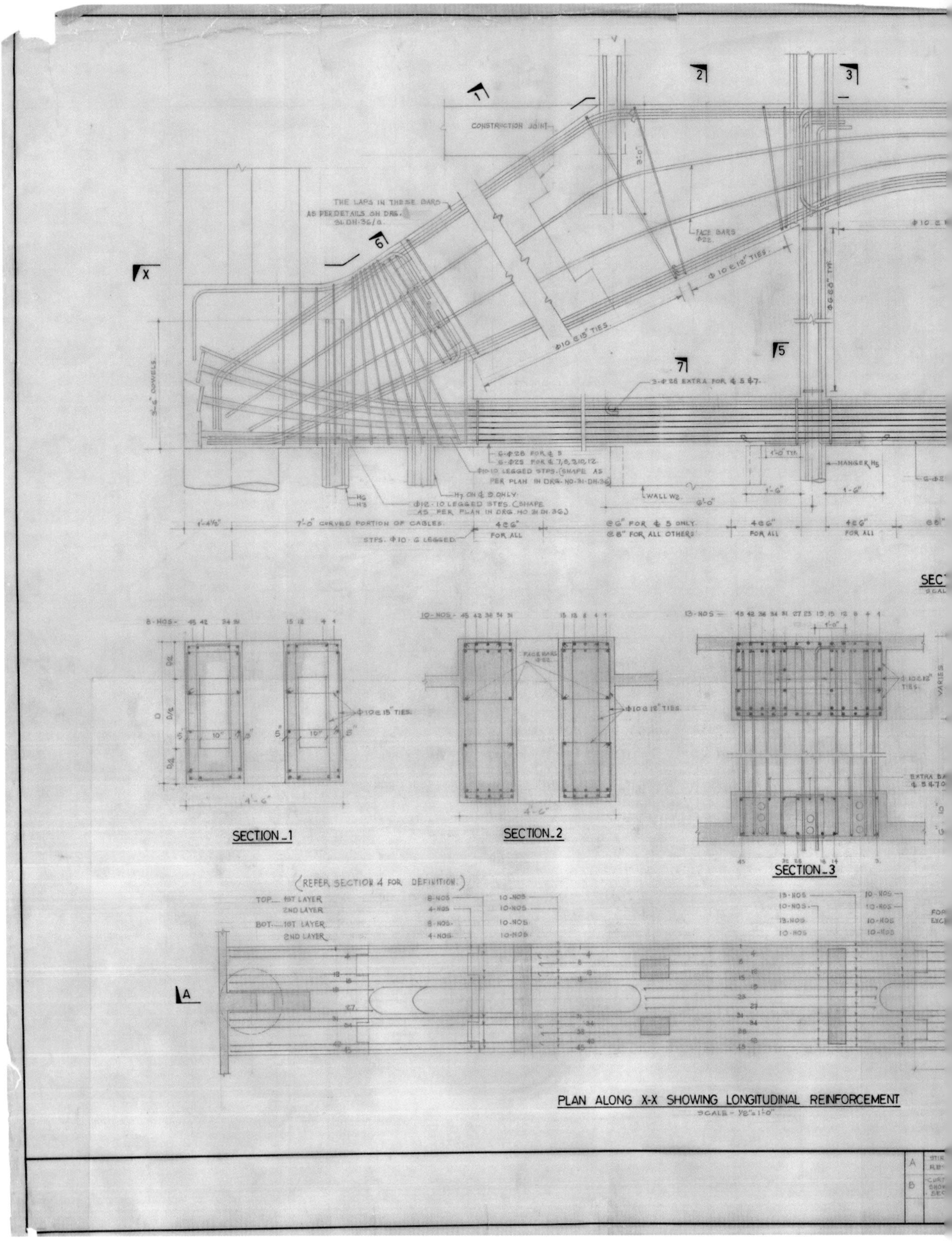
CONSTRUCTION JOINT
THE LAPS IN THESE BARS AS PER DETAILS ON DRG. 31.DH.36/a.
FACE BARS Φ22
Φ10 @ 12" TIES
Φ10 @ 15" TIES
3-Φ28 EXTRA FOR ℄ 5 & 7.
HANGER H5
WALL W2
7'-0" CURVED PORTION OF CABLES.
STPS. Φ10 - 6 LEGGED
FOR ALL
SECTION_1
SECTION_2
SECTION_3
(REFER SECTION 4 FOR DEFINITION.)
TOP... 1ST LAYER
2ND LAYER
BOT... 1ST LAYER
2ND LAYER
PLAN ALONG X-X SHOWING LONGITUDINAL REINFORCEMENT

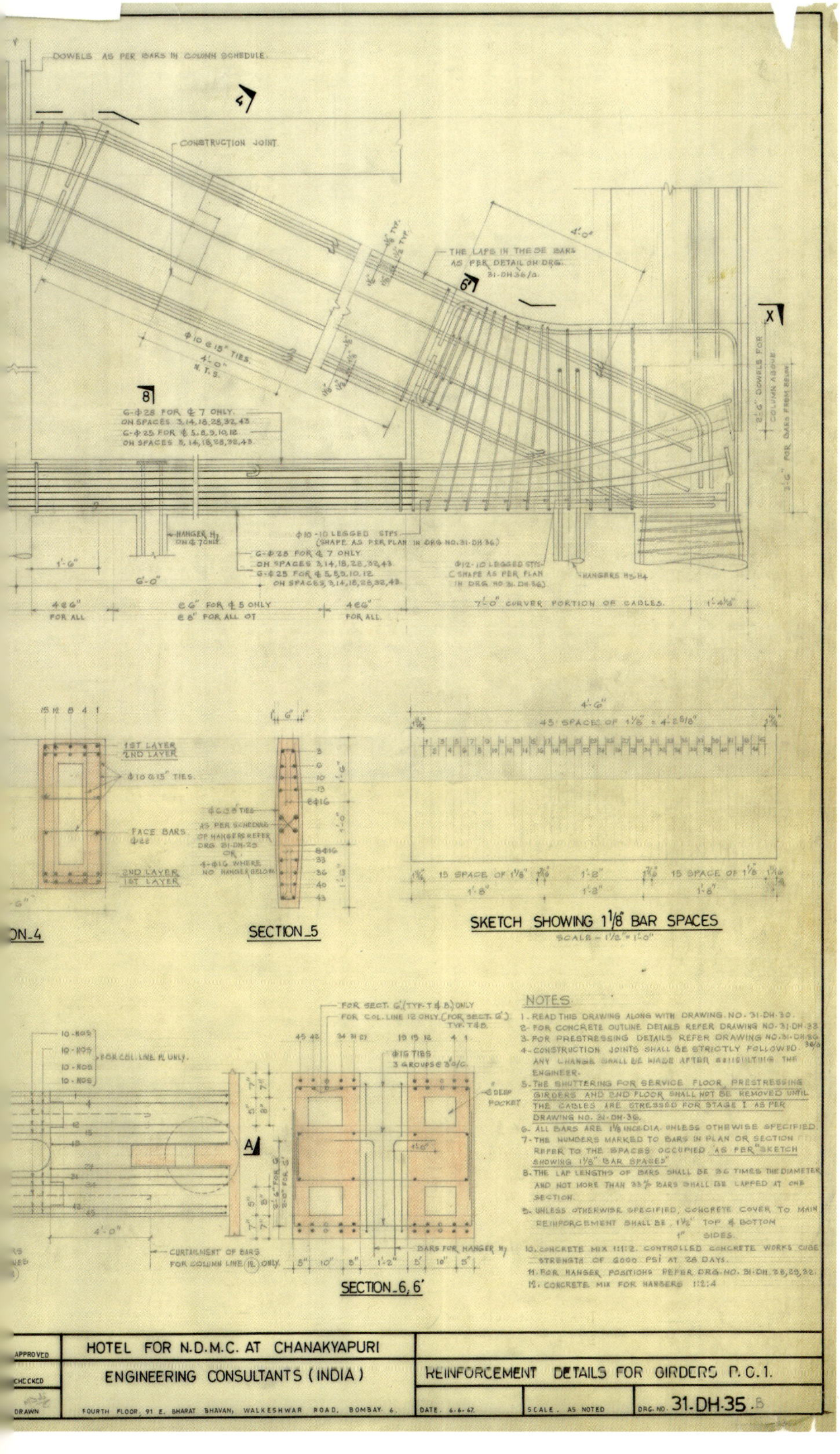

Drawing 31.DH.35.B: Reinforcement Details for Girder PG1: 6 June 1967
Drawing of reinforcement details of a transfer girder and the post-tensioned tie-beam that free the lower public floors from the structural loads and service pipes of the hotel rooms above.

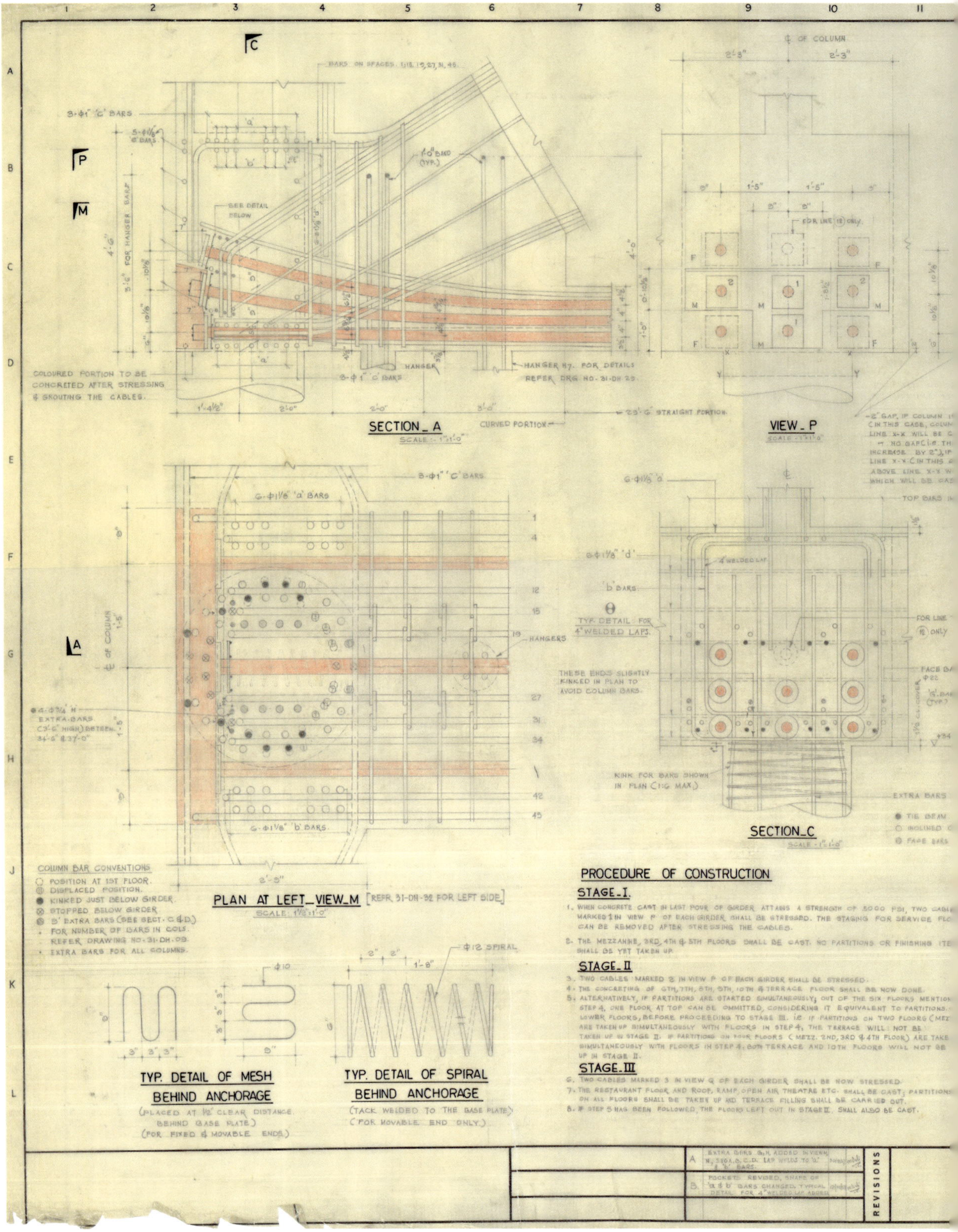
SECTION_A
SCALE :- 1"=1'-0"
COLOURED PORTION TO BE CONCRETED AFTER STRESSING & GROUTING THE CABLES.
VIEW_P
PLAN AT LEFT_VIEW_M
[REFR. 31-DH-32 FOR LEFT SIDE]
SECTION_C
COLUMN BAR CONVENTIONS
POSITION AT 1ST FLOOR.
DISPLACED POSITION.
KINKED JUST BELOW GIRDER.
STOPPED BELOW GIRDER.
TYP. DETAIL OF MESH BEHIND ANCHORAGE
(PLACED AT 1/2" CLEAR DISTANCE BEHIND BASE PLATE)
(FOR FIXED & MOVABLE ENDS)
TYP. DETAIL OF SPIRAL BEHIND ANCHORAGE
(TACK WELDED TO THE BASE PLATE)
(FOR MOVABLE END ONLY)
PROCEDURE OF CONSTRUCTION
STAGE_I.
1. WHEN CONCRETE CAST IN LAST POUR OF GIRDER ATTAINS A STRENGTH OF 5000 PSI, TWO CABL MARKED 1 IN VIEW P OF EACH GIRDER SHALL BE STRESSED. THE STAGING FOR SERVICE FLO CAN BE REMOVED AFTER STRESSING THE CABLES.
2. THE MEZZANINE, 3RD, 4TH & 5TH FLOORS SHALL BE CAST. NO PARTITIONS OR FINISHING ITE SHALL BE YET TAKEN UP.
STAGE_II
3. TWO CABLES MARKED 2 IN VIEW P OF EACH GIRDER SHALL BE STRESSED.
4. THE CONCRETING OF 6TH, 7TH, 8TH, 9TH, 10TH & TERRACE FLOOR SHALL BE NOW DONE.
5. ALTERNATIVELY, IF PARTITIONS ARE STARTED SIMULTANEOUSLY; OUT OF THE SIX FLOORS MENTION STEP 4, ONE FLOOR AT TOP CAN BE OMMITTED, CONSIDERING IT EQUIVALENT TO PARTITIONS. LOWER FLOORS, BEFORE PROCEEDING TO STAGE III. i.e. IF PARTITIONS ON TWO FLOORS (MEZ ARE TAKEN UP SIMULTANEOUSLY WITH FLOORS IN STEP 4, THE TERRACE WILL NOT BE TAKEN UP IN STAGE II. IF PARTITIONS ON FOUR FLOORS (MEZZ. 2ND, 3RD & 4TH FLOOR) ARE TAKE SIMULTANEOUSLY WITH FLOORS IN STEP 4, BOTH TERRACE AND 10TH FLOORS WILL NOT BE UP IN STAGE II.
STAGE_III
6. TWO CABLES MARKED 3 IN VIEW Q OF EACH GIRDER SHALL BE NOW STRESSED.
7. THE RESTAURANT FLOOR AND ROOF, RAMP, OPEN AIR THEATRE ETC. SHALL BE CAST; PARTITIONS ON ALL FLOORS SHALL BE TAKEN UP AND TERRACE FILLING SHALL BE CARRIED OUT.
8. IF STEP 5 HAS BEEN FOLLOWED, THE FLOORS LEFT OUT IN STAGE II, SHALL ALSO BE CAST.
REVISIONS

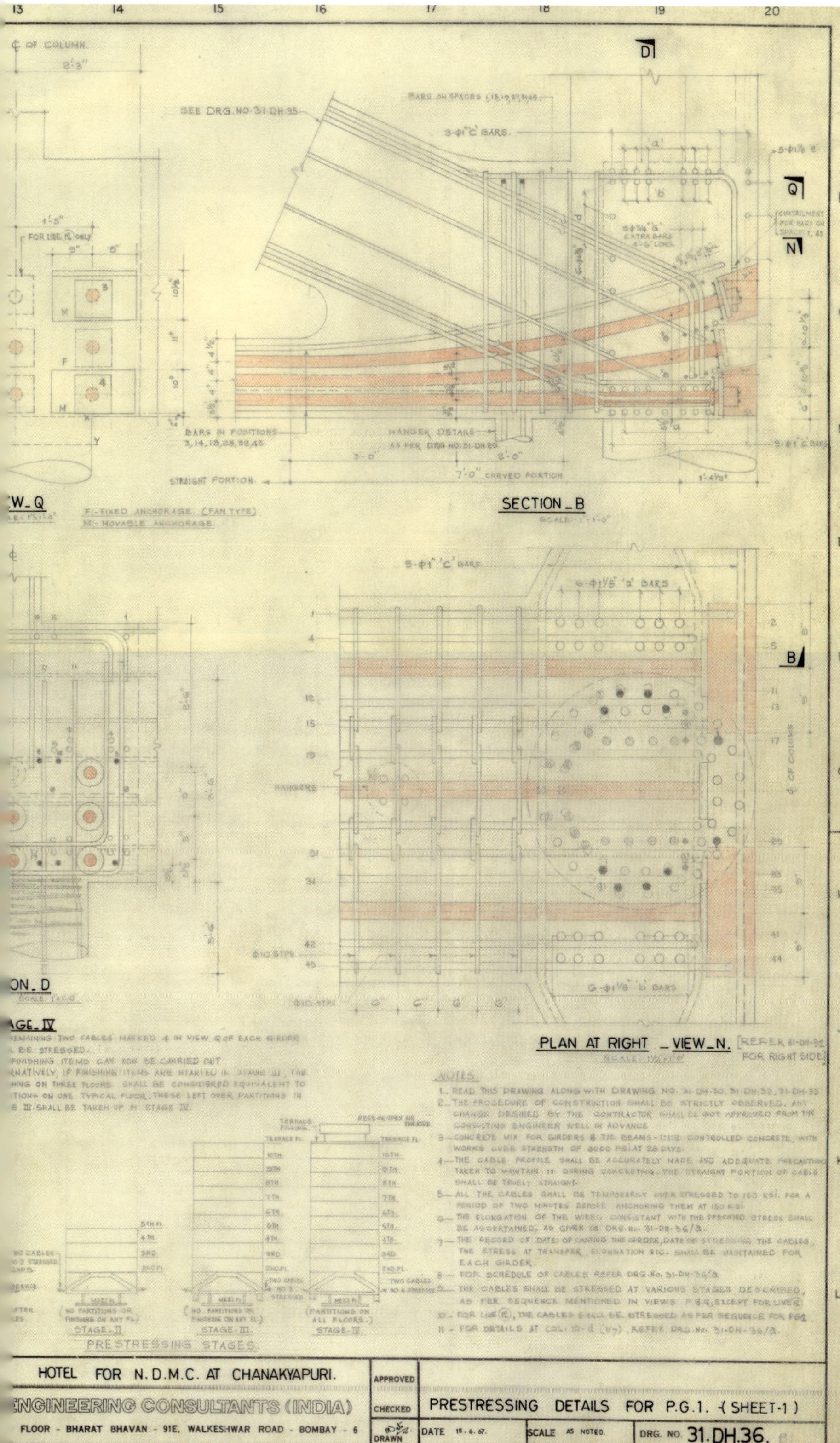

Drawing: 31.DH.36.B: Prestressing details for PG1 (Sheet 1): 15 June 1967

Trapezoidal girders transfer the load of the corridor and peripheral columns of the floor above. The girder itself is held by a post-tensioned tie-beam. The post-tensioning of the tie rods is done sequentially, balancing the load of the tower floors added. These girders also carry two service floors below that are hung from it and allow for all service pipes from the hotel bathrooms above to be collected and carried horizontally to different service cores.

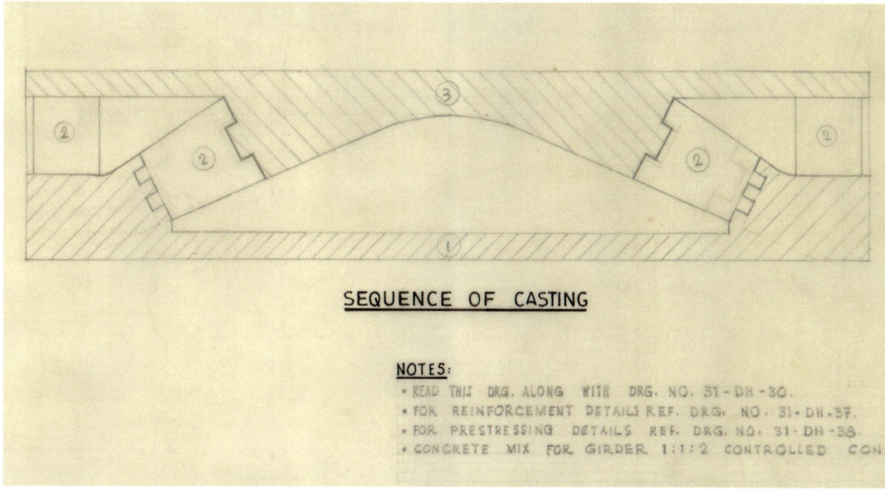

(Partial) Drawing no.31.DH.33 A: Outlines of Prestressed Concrete Girder PG.2: 8 September 1967

View from below looking up at the columns carrying the service floor, expressed in the deep concrete band and the hotel block rising above.

Opposite: View of the end shear wall. The depth of the service floor with the transfer girder is expressed on the exterior facade and the ends of the post-tensioned cables expressed on the column head.

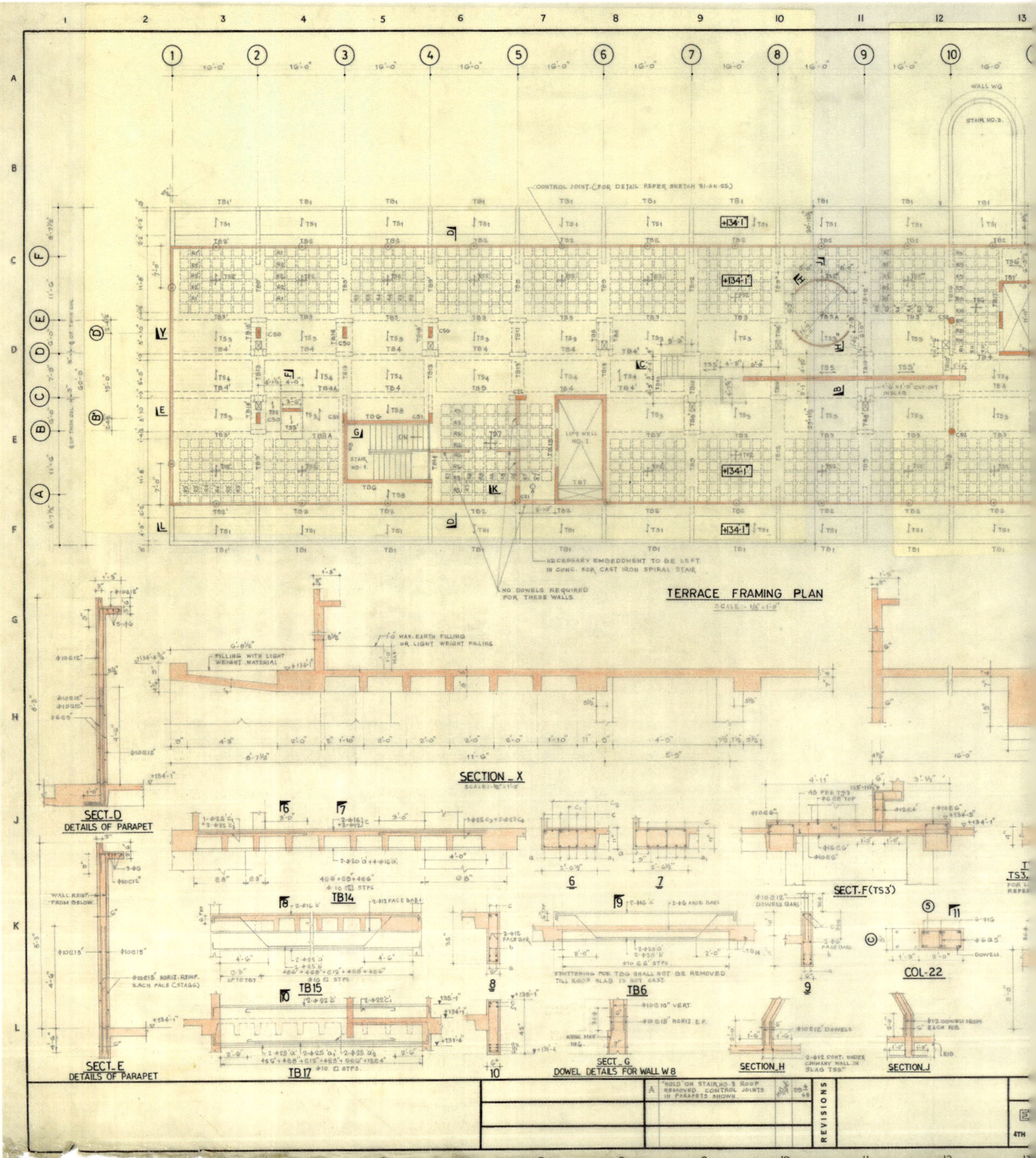
TERRACE FRAMING PLAN
SECTION_X
SECT_D
DETAILS OF PARAPET
SECT_E
DETAILS OF PARAPET
SECT_F(TS3')
SECT_G
DOWEL DETAILS FOR WALL W8
SECTION_H
SECTION_J
COL-22
TB14
TB15
TB17
TB6
REVISIONS

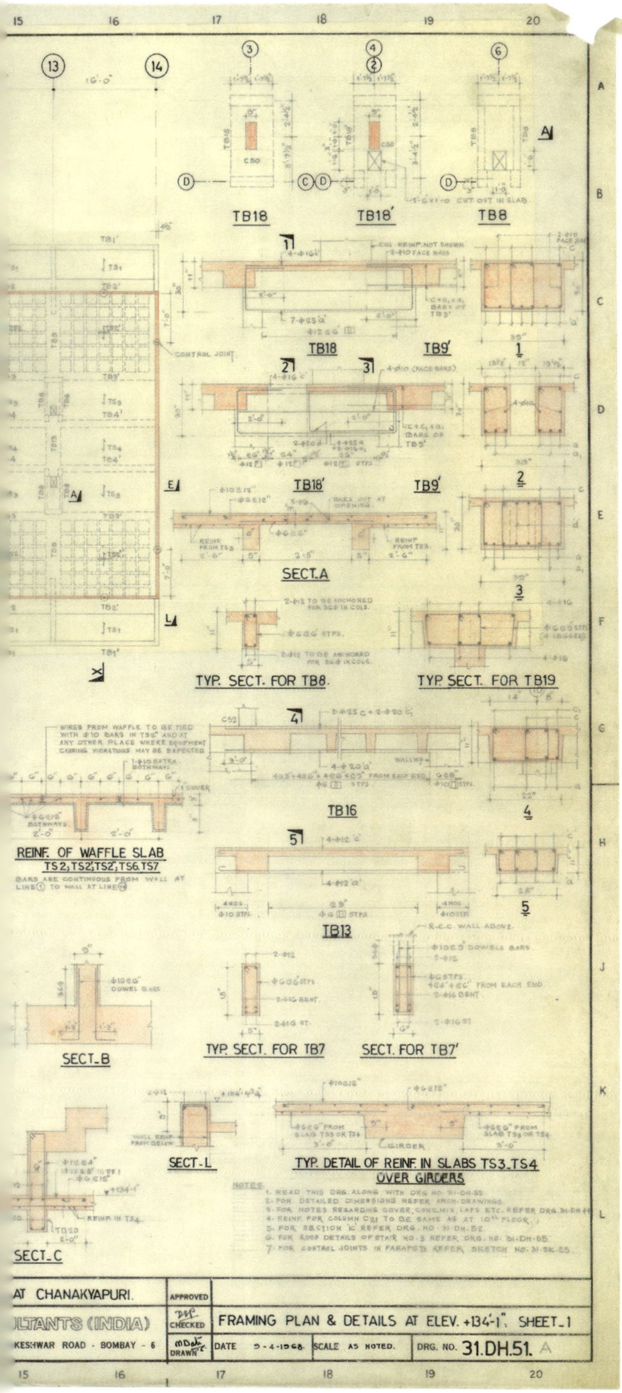

Interior view of a room.

Drawing no. 31.DH.51.A: Framing Plan & Details at Elev. +134'-1" (Sheet 1): 9 April 1968

Floors with hotel rooms have a flat plate slab along the central corridor and peripheral balconies, while the rest of the floor is cast in exposed pre-cast shallow waffles.

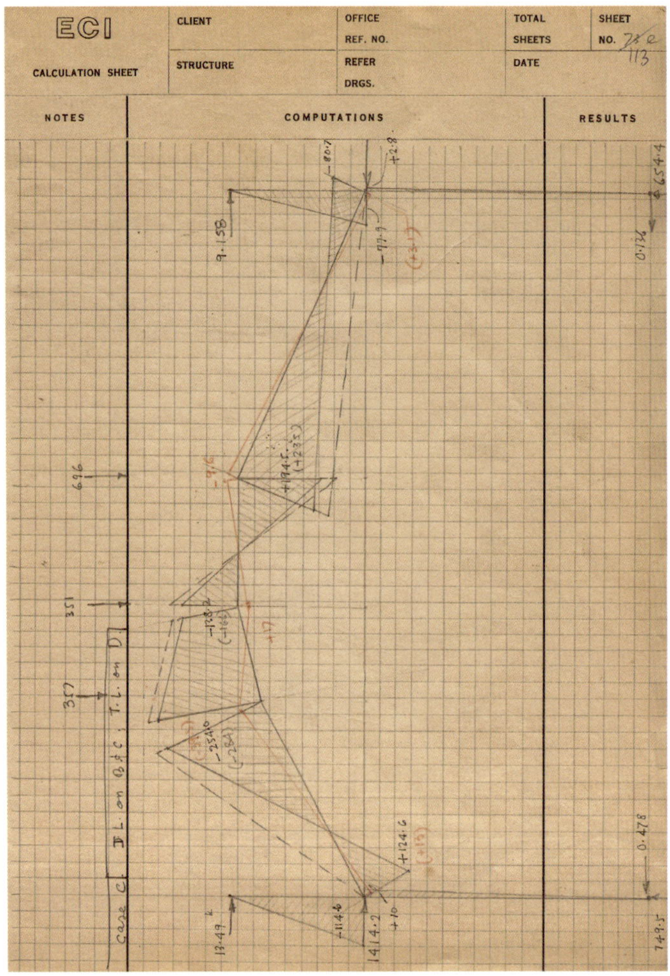

Sample calculation sheet analysing the transfer girder.

View of the end shear wall that contains the horizontal seismic forces with the free-standing stair cantilevered from a central pin wall and connected to the main tower only at two locations.

1970s

Hall of Nations & Halls of Industries

LARGE EXHIBITION HALL COMPLEX | A CASE STUDY

Mahendra Raj

The Trade Fair Authority of India (TFAI) had to create five large permanent exhibition halls to hold an International Trade Fair in Delhi in 1972.

Requirement: Large unobstructed enclosures

The main hall was required to have free, un-obstructed space of 6700 sq.m with an approximate height of 30m and was named 'Hall of Nations'.

Four smaller halls, together, were required to provide free exhibition area of another 7500 sq.m with an approximate height of 16m and were named as 'Halls of Industries'.

On the basis of a competition held in 1970 the partnership firm Raj Rewal & Kuldip Singh was commissioned as architects for these halls and the partner Raj Rewal looked after the project. The author was the structural consultant.

The total period available for planning, designing and construction of all the halls was a little less than two years.

Considering the nature of the job and a very tight time schedule, the architects and the engineers worked in very close co-ordination right from the beginning. It was recognised that while evolving a solution, apart from considering planning and aesthetic requirements, honouring the structural requirements was of prime importance.

Identifying the shape and the form: Truncated Pyramid

For the most economic and appropriate structural solution it was an intuitive direction to find a shape and form that would enable use of the same system for making the enclosure as well as the roof and could also be used for the bigger as well as the smaller halls.

Shells of double curvature, single curvature, hyperbolic paraboloids, folded plates etc. were considered and rejected. Eventually the solution narrowed down to an assembly of truncated pyramids.

The Hall of Nations had a base dimension of 73m x 73m which reduced to almost half at the roof. Each of the four Halls of Industries had a base dimension of 40m x 40m with a similar reduction at the roof level.

Investigating two alternative systems: Lamellas and Space Frames

Two alternative systems were considered for these truncated pyramids.

The first system considered was lamellas made with solid concrete elements or trussed up concrete or steel elements. Investigations indicated that whereas this system was quite appropriate for the smaller halls, it was very unwieldy and cumbersome for the bigger hall. Each side of the bigger hall was a trapezoid with the largest dimension as 78m and an area of approximately 2000 sq.m placed at an inclination of 55°44'8" to the horizontal. Whereas the smaller hall was possible with a system of intersecting ribs, the bigger hall required intersecting heavy trusses.

The other system considered was a space frame. This proved to be more appropriate. It had the flexibility that with a different cross section and different member length the same system could be used for the bigger as well as the smaller halls. Moreover, the enclosure as well as the roof of both types of halls could be created with the same system.

Adopted structural system: Space Frame

While investigating various configurations of the space frame it was identified that the most appropriate system to create truncated pyramids was the one which used the pyramid as the basic element. With the use of this basic element, the truncated pyramid geometry got related to that of a basic pyramid with all equal sides. The openings in the sides could also be formed quite naturally by honoring the geometry. The pyramid geometry adopted for the two types of 'Halls' is shown in Figures 3c and 3d (shown here).

Selection of construction material

The next decision to be taken was the selection of material and the mode of construction. Since we had no access at that time to any of the patented, commercial space frame systems available overseas, whatever the selected material, the responsibility of evolving an appropriate joint and mode of construction was on our shoulders. Materials of construction considered were tubular steel pipes, structural steel members and concrete. On the basis of an approximate analysis the cost of construction with steel pipes, structural steel and reinforced concrete was evaluated.

Adopted material: Reinforced Concrete

This evaluation indicated that the alternative with reinforced concrete was the most economical, with structural steel about 30% more expensive, and with steel pipes even more. Still considering that a space frame in reinforced concrete, at such a scale, had not been done anywhere earlier, for some time we waivered in favour of structural steel or steel pipes which had been used in other parts of the world for fairly large space frames. However, after investigating the availability of both these materials we found it would be difficult to procure any of these materials in the range and quantity needed in the available time frame. It was, therefore, finally decided to use reinforced concrete as the material of construction.

Proposed mode of construction: Pre-cast

While selecting a mode of construction the obvious choice was of a pre-cast construction technique. It was visualised that the members and the joints would be pre-cast separately and put together with the help of bolts and on-site welding of end plates.

Member cross-sectional shape and joint profile

The geometry of the structure demanded considerable effort in the selection and design of the cross-sectional shape of the members and the resultant joint.

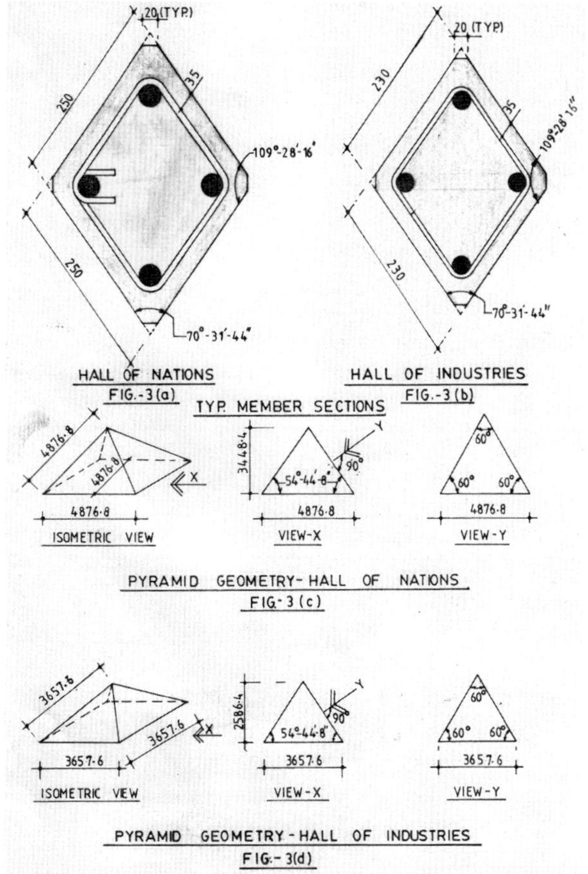

The shape of members, we felt, should be such that all the generated surfaces are flush without protruding edges. The joint also, we felt, apart from containing and transferring the maximum member forces should be neat, proportionate, contained within the flush surfaces, and should be relatively easy to construct.

Selected cross section: Rhombic

Various cross sections were considered and the ensuing joints studied. The standard cross sections like the rectangular, square, circular were found unsuitable. A rhombic section appeared to be most appropriate. The interface angles of the rhombus were the angles obtained from the slopes of the member meeting at the joint. A neat configuration of the joint also emerged quite naturally with this selection of the cross section.

Dimensions of rhombic section

Once the cross-sectional shape was decided the next step was to establish the cross-sectional dimensions.

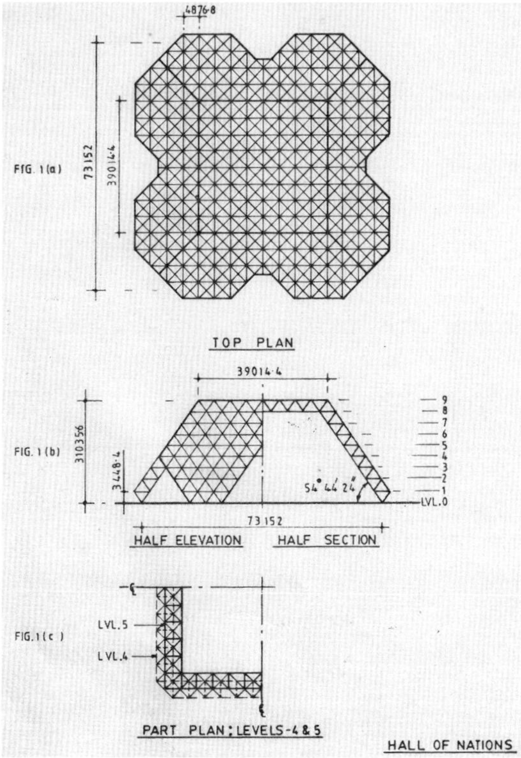

These dimensions had to be such that the member could be given sufficient strength to withstand the maximum forces at the springing point, was not overly redundant for members with smaller forces, could be used in the roof, was neither too slender nor too heavy but just appropriate, with dimensions which were in scale to the entire structure.

Cladding and roofing materials

In order to establish member dimensions it was necessary to have a preliminary analysis. But before such an analysis could be done decisions on the cladding material, the extent of cladding, the roofing material and the sequence of erection of the structure and of the cladding etc. had to be taken.

After considering various materials it was decided to use gunited and glazed triangular cladding plates and light-weight concrete roof.

Cladding system

The architect had to use all his ingenuity to evolve a cladding system which protected the inner space, provided sufficient light, and expressed the space frame from inside as well as outside.

The final form: Hall of Nations

Figure 1a (shown here) indicates the final form of the main pyramid. It has large openings on each face and at each corner. The structure springs out of 24 nodes at ground level–six nodes at each face of the square plan.

Planning the construction sequence

A study of the final size and configuration of the openings indicated that the structure would be stable only after construction had reached at least one level beyond the openings where it gets the first all round continuity–that is at level 5 (indicated in Figures 1b and 1c, shown here). Till it reached that level it would have to be kept supported by a system of scaffoldings. After the construction reached this level the structure could support additional construction loads, thereby completely eliminating the use of scaffolding erected from the ground.

This mode of construction released the ground level for other trades of work at an early stage of construction of the pyramid. However it was evident that if we followed this mode of construction our analytical work would increase tremendously.

Because, first, after completion of each additional ring it was a different structure; second, it had to be decided as to which of these different structures would carry what extent of the additional load of cladding plates.

We realised that we would have to analyse four different configurations of the structure, each with a number of loading conditions, merely to establish forces during the construction stage. This of course, had to be followed by a detailed analysis of the fifth configuration i.e. the completed structure.

The only way to reduce this extent of analytical work was to keep the entire structure supported till it was completed, including the roof. Adopting this technique only to save analytical work was senseless, because the cost and completion time of the entire structure would be much more. Moreover it was planned to construct about 1200 sq.m of mezzanine floor on independent supports within the hall. Construction of the mezzanine could be taken in hand only after the scaffolding etc. in that area had been cleared.

Analysis of five different structural configurations

It was finally decided to increase our own analytical work–anticipate as many loading conditions as we could with the five different configurations of the structure and provide sufficient strength so that a certain amount of flexibility was available in the mode and sequence of construction.

However, considering that such large volumes of analytical work could not be related to a member section which did not work and which had later to be increased or decreased and the entire analysis redone, it was necessary to arrive at a cross-section with a fairly high level of accuracy prior to all this work.

This was successfully achieved with the help of a few preliminary analyses.

While making a selection of the cross section, apart from considerations discussed earlier in this paper, it was also kept in mind that not more than four reinforcing bars could be placed in one member. This was because on an average nine members met at one joint and even with four bars per member, connections for as many as thirty-six bars had to be planned at the joint.

Adopted member size and dimensions

After intensive study of all these variables a rhombic section with chamfered edges, a face dimension of 25cm, and cross-sectional area of 585 sq.cm was selected for the Hall of Nations. The node-to-node dimension was 4.877m.

Analysis of different loading conditions assuming pinned joints

The detailed analysis required investigating fourteen different loading conditions on five different structural configurations. In addition, analysis was also done for wind and earthquake loads as well as for creep conditions. All these analyses were done considering pinned joints. After all these analyses it was found that the section adopted was adequate and was used for the Hall of Nations.

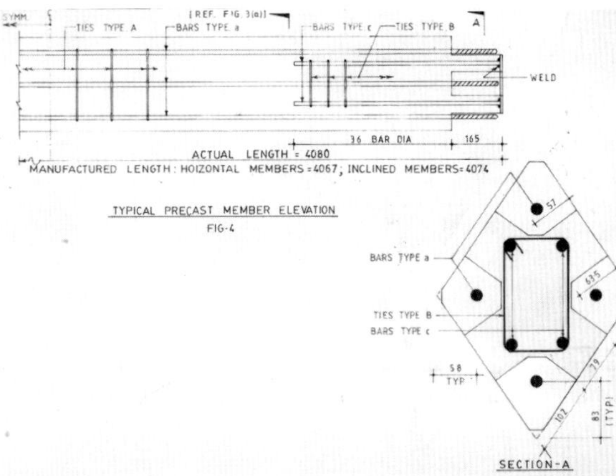

TYPICAL PRECAST MEMBER ELEVATION
FIG-4

FIG-5(c)

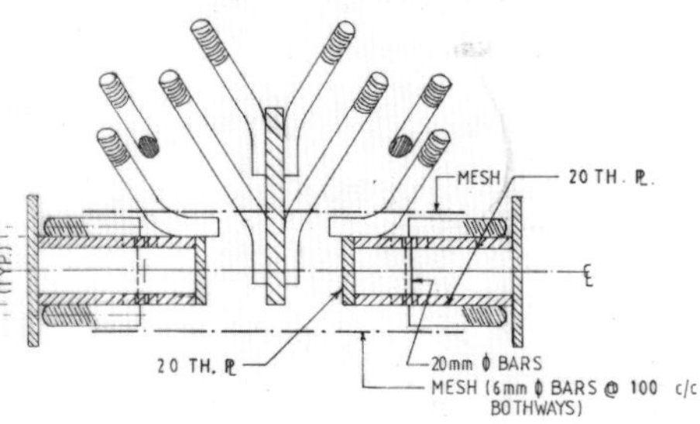

PLATES OF MEMBERS PROJECTING ABOVE NOT SHOWN

FIG.-5(b)

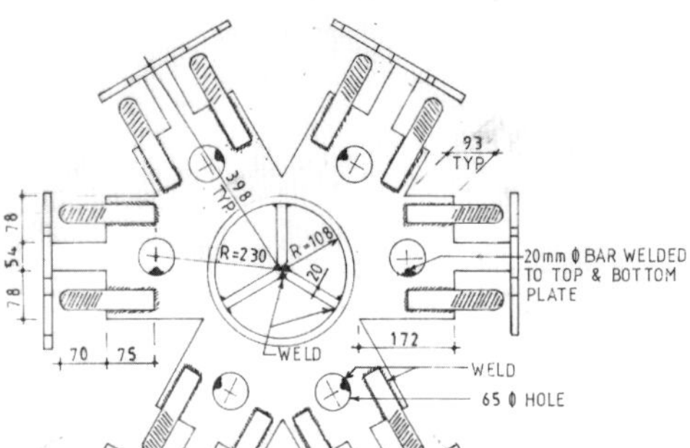

PLAN SHOWING ARRANGEMENT OF ANCHOR PLATES AND BOLTS AT PRECAST JOINTS

Verification of effect of rigid joints

Even though construction was started on the basis of analysis with pinned joints we still wanted to confirm our design using an analysis with rigid joints. Unfortunately at that time we did not have access to a computer program which could take in such a large problem. IIT Kanpur was commissioned to develop such a program and reanalyse the structure. By the time this analysis arrived, a considerable part of the structure had already been built. However, as anticipated, it was found that member moments were very small and whatever strength had been given in the members was adequate.

Typical pre-cast joint

Figure 4 (opposite, top) shows a typical detail of a pre-cast member. The configuration of the joint element that emerged is shown in Drawing (page 158, left, middle) and Figures 5b, 5c (opposite, bottom). These joint elements were planned to absorb and transfer the forces applied at the joints by the various members having different magnitudes and different nature of forces–either tension or compression.

Preferred mode of construction: In-situ

When the job was put to tender, except for one contractor who gave a bid for the pre-cast construction all others preferred to use the in-situ construction technique, and one of them was the lowest.

Development of a joint for in-situ construction

Before this technique could be accepted it was necessary for us to work out a feasible joint for transference of forces. Considering that thirty-six bars, some carrying tension and some carrying compression were meeting at the joint it was not an easy task. In the pre-cast joint element assistance had been taken from steel plates embedded in concrete. But it was evident that in the in-situ construction that would not be possible. After intensive study, a system was evolved whereby only twenty bars passed through the joint element. Bars coming from four lower members were lapped with four upper members, which accounted for sixteen bars catering for eight members. Four bars of the ninth member were passed through the joint and anchored in one of these eight members. While evolving a configuration for the placement of bars it was kept in mind that containment and development of requisite forces had to be achieved. On a study of the forces in various members at different joints, a few alternative configurations of lapping and anchorage of bars were evolved.

Typical in-situ joint

Drawing (page 172, bottom) shows a typical joint profile for in-situ construction and drawings (pages 174, 175) shows the arrangement of reinforcements at several such joints.

Sequence of construction of members and joints

Because of the complexity of threading reinforcing bars through the joint it was planned that members would have straight length bars which would be lap-welded to short-length curved and bent bars placed in the joint. The sequence of construction adopted was to cast the straight length of members from joint to joint, place in position partially pre-assembled formwork of the joint along with placed in position short length curved and bent bars to be lap welded to members below and above, and after welding to the bars of members below, concrete the joint. Straight length bars of members above were then welded, formwork of members upto the joint above erected, and the members concreted. This sequence was followed to concrete the hall right upto the roof level.

Adopted construction sequence of structure

Construction of all the four sides of the pyramid was taken in hand simultaneously. As explained earlier, the structure had to be kept supported till the completion of level 5, i.e. completion of one ring above the openings at level 4, when it could be released to support itself. However the rate of progress of work on each side was not equal. By the time all the four sides were brought to the same level, construction had reached almost upto level 6 on some of the sides. It was, therefore, at this level that the structure was completely released. The remaining two levels were constructed from a platform which was erected on trusses placed at level 6 and supported on a few stanchions resting on ground.

Verification of different than assumed loading conditions

During the course of construction it became necessary to reanalyse a few more configurations of the unfinished structure–for some loading conditions imposed by construction sequences which were different from those assumed in the original analysis and design–and clear them.

Roof construction

At one stage it was thought that the space frame members of the roof will be made in structural steel and then encased with concrete to conform to the same section as of other members. But this turned out to be a very cumbersome exercise–particularly the joint. Eventually it was decided to cast the roofing members also with in-situ concrete. This was done from a platform erected at the roof level supported on the completed structure along with four additional stanchions in the middle. The roofing surface was constructed with pre-cast light-weight reinforced concrete plank units specially manufactured by Hindustan Housing Factory in Delhi.

Foundations

The structure was supported on cast-in-situ driven piles. Horizontal components of forces in one direction were absorbed with reinforced concrete grade beams interconnecting six springing points on each face. In the other direction these were absorbed with post-tensioned grade beams connecting opposite faces 73m apart and were stressed in stages as the structure went up.

Mezzanine floors

Within the hall, a mezzanine floor covering a 1200 sq.m area along with access ramps and stairs etc. was also constructed. The mezzanine floor was a 17m wide waffle slab, 69cm deep, supported on a single row of hollow circular drum-shaped columns 4.87m diametre each, placed at 17m centres. This mezzanine provided additional exhibition space and was also supported on piles.

Member forces and material consumption

Figures 7a and 7b (shown here) show the forces in the inclined members on the outer and inner faces under full load conditions for Hall of Nations. Material consumption and other pertinent data for Hall of Nations is given in Figures 9a and 9b (right, top and bottom).

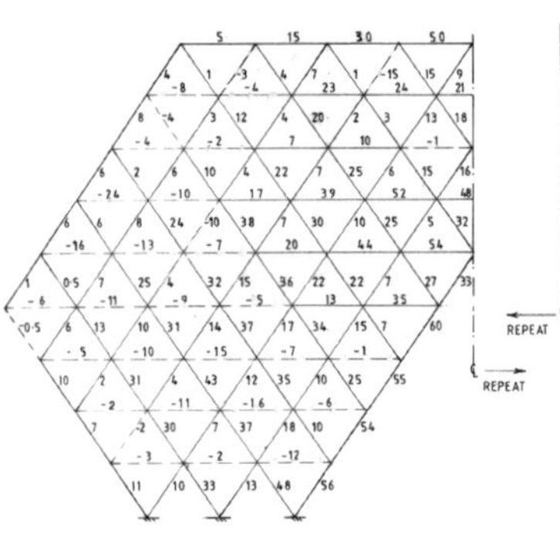

OUTER FACE FORCES (T)

HALF ELEVATION

FIG-7a

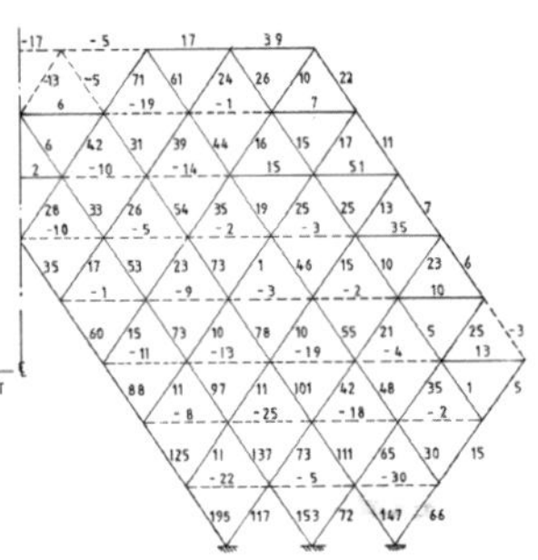

INNER FACE FORCES (T)

HALF SECTION

FIG-7b

HALL OF NATIONS

MATERIAL CONSUMPTION STATEMENT (FIG-9a)

SL NO.	ITEM	SPACEFRAME	MEZZANINE FLOOR	TOTAL
1.	CONCRETE	1485 Cum.	1100 Cum.	2585 Cum.
2.	FORM WORK	17150 Sqm.	6450 Sqm.	23600 Sqm.
3.	REINFORCING STEEL	320 MT.	170 MT.	490 MT.
4.	PRESTRESSING STEEL	10 MT.	—	10 MT.
5.	SHEATHING	2050 M.	—	2050 M.
6.	GROUTING	2100 M.	—	2100 M.
7.	ANCHORAGES	56 NOS.	—	56 NOS.

FLOOR AREA UNDER SPACE FRAME = 5350 Sqm.
MEZZANINE FLOORS = 1350 "
TOTAL 6700 Sqm.

SURFACE AREA OF SPACE FRAME = 9520 Sqm.
ACTUAL BUILT UP AREA = SURFACE AREA OF SPACE FRAME + MEZZANINE FLOOR AREA = 9520+1350 = 10870 Sqm.

PERTINENT DATA

1. PILES —— = 72 NOS., 480 mm dia, 90 TONNE PAY LOAD CAPACITY.
2. NUMBER OF JOINTS = 812
3. NUMBER OF MEMBERS = 3260
4. MAXIMUM NUMBER OF MEMBERS MEETING AT A JOINT = 11
5. CONCRETE GRADES USED ARE
   M-350: IN PRESTRESSED TIE BEAMS
   M-300: IN SPACE FRAME MEMBERS, JOINTS AND PLATES
   M-250: IN SPACE FRAME MEMBERS, JOINTS AND MEZZANINE FLOOR STRUCTURE

HALL OF NATIONS

MATERIAL CONSUMPTION STATEMENT (FIG-9b)

ANALYSIS OF MATERIAL CONSUMPTION :—

1. AVERAGE CONCRETE CONSUMPTION ON BUILT UP AREA ON PLAN
   = 2585/6700 = 0·385 Cum/Sqm.
2. AVERAGE CONCRETE CONSUMPTION ON ACTUAL BUILT UP AREA
   = 2585/10870 = 0·24 Cum/Sqm.
3. AVERAGE STEEL CONSUMPTION ON PLAN AREA
   = 490000/6700 = 73 Kg / Sqm.
4. AVERAGE STEEL CONSUMPTION ON ACTUAL BUILT UP AREA
   = 490000/10870 = 45 Kg / Sqm.
5. AVERAGE STEEL TO CONCRETE RATIO = 490000/2500 = 196 Kg / m^3

CONSUMPTION RATIOS FOR SPACE FRAME ONLY :—

1. AVERAGE CONCRETE CONSUMPTION ON PLAN AREA
   = 1485/5350 = 0·28 Cum/Sqm.
2. AVERAGE CONCRETE CONSUMPTION ON BUILT UP AREA
   = 1485/9520 = 0·156 Cum/Sqm.
3. AVERAGE STEEL CONSUMPTION ON PLAN AREA
   = 320000/5350 = 60 Kg / Sqm.
4. AVERAGE STEEL CONSUMPTION ON BUILT UP AREA
   = 320000/9520 = 34 Kg / Sqm.
5. AVERAGE STEEL TO CONCRETE RATIO = 320000/1485 = 215 Kg / m^3

Halls of Industries

The structural concept and design principles adopted for the smaller 'Halls', namely the Halls of Industries were exactly similar to those of the Hall of Nations. Four such halls were constructed.

The final shape of the halls is shown in Figures 2a, 2b and 2c (shown here). Base dimensions of each hall was approximately 40m x 40m. The structure springs out of the ground from eight nodes–two on each side of a square plan. The space-frame members were 3.658 meters long between two consecutive nodes. The member cross-section adopted was rhombic, 23cm on each side, with similar chamfered edges, of a net cross-sectional area of 495 sq.cm. The adopted cross section of members is shown in Figure 3b (page 143), and the basic pyramid geometry in Figure 3d (page 143). Member details and joint details for both, the pre-cast construction and the final in-situ construction were planned as those for the Hall of Nations.

The forces at the springing level being much smaller than the Hall of Nations, the foundation system adopted for these structures was of simple open footings. The horizontal forces were absorbed by a system of reinforced tie beams.

Unlike Hall of Nations, in which all the four sides and cladding arrangements were identical in shape and form, the four Halls of Industries had different cladding, openings and elevational treatment on different faces. For the sake of simplicity in design and construction the four halls had identical design but had provisions for different types of cladding arrangement.

Each of these halls also had mezzanine floors of a character similar to that provided in the Hall of Nations but smaller in size and resting on open footings.

Conclusion

Reinforced concrete is an economical and viable material for large-span space frames. The material consumption to enclosed volume ratio is low and competitive.

FIG. 2 (a)

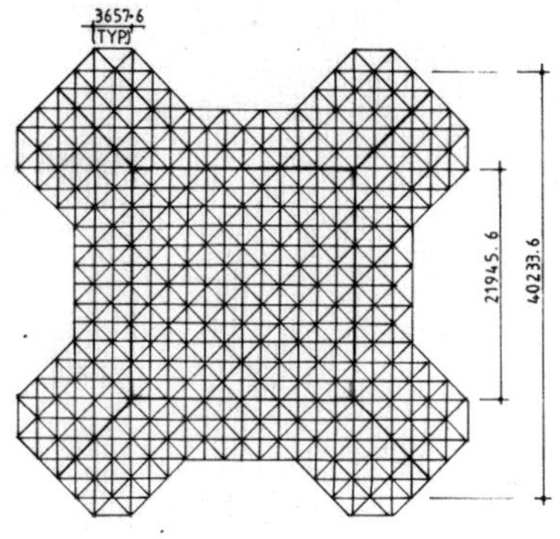

TOP PLAN

FIG. 2(b)

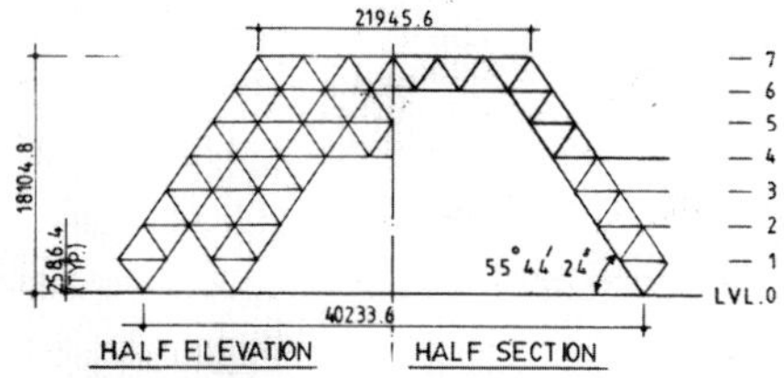

FIG. 2(c)

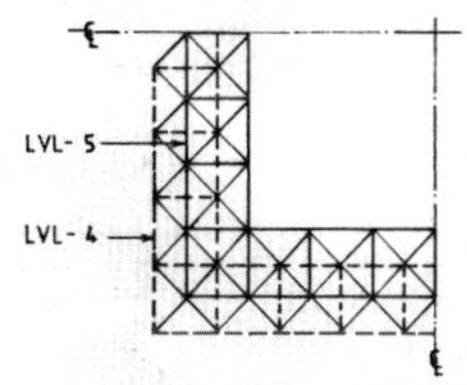

PART PLAN: LEVELS-4 & 5 HALL OF INDUSTRIES

In-situ, pre-cast, or a combination of the two techniques could be used for the construction of such space frames.

The in-situ technique takes less materials, because it eliminates use of steel plates and erection bolts etc. But construction effort and cost of formwork is more. This effort and cost can, however, be reduced by developing and adopting a construction system which eliminates or at least reduces the use of extensive scaffolding, and provides for easy assembly, release and maximum re-use of shuttering.

The pre-cast technique uses more materials in the form of steel plates, bolts and additional welding, but will reduce construction time. The precision requirement with this technique is however very high.

A combination of the two techniques, where the member is pre-cast but the joint is made in-situ might be the most suitable mode of construction. The use of additional joint plates and bolts etc. is eliminated and the precision requirements is not that high. However, in order to minimise the use of scaffolding and maximise the re-use of shuttering, a construction system has to be developed and adopted.

It is also possible to investigate other alternatives of combining these two techniques of construction.

In conclusion, it is important to say that our involvement in planning, designing and the construction of these halls was a unique and exciting experience, particularly because there is no prior example anywhere in the world of reinforced concrete being used in a space frame of equivalent or even smaller size. We had to anticipate and resolve all the analytical, design and even construction problems ourselves.

Originally presented at a seminar organised by the Indian Concrete Institute in Madras in May 1988 by Mr Mahendra Raj.

Exterior view, Halls of Industries.

Page 144: Interior view of the Hall of Nations.

HALL OF NATIONS & HALLS OF INDUSTRIES

NEW DELHI | 1972

This complex of exhibition halls for the 1972 International Trade Fair was planned, designed and constructed in 22 months. Its large-span cast-in-situ concrete space frames were the first of their kind to be built in the world.

The choice of shape and form of the structural system was driven by its potential versatility: it contains an enclosure with a column-free main hall of 6700 sq.m, and four smaller halls adding an additional 7500 sq.m, under its roof structure. After investigating multiple alternatives, such as folded plates and hyperbolic paraboloids, a space frame system with a truncated pyramid as the basic module and overall form was adopted. Steel or pre-cast concrete would have been obvious choices of material for such a space frame, but the constraints of the available construction technology of that time necessitated the use of in-situ concrete.

Aerial view of the exhibition hall complex with the larger Hall of Nations and four smaller Halls of Industries built for the 1972 International Trade Fair.

Previous pages: View of the Hall of Nations from the Halls of Industries, visible in profile on both sides of the image.

After rigorous investigations into pyramid geometry, the rhombic section developed for all the members of this very large building may well be the secret of its ultimate delicate and light appearance. With chamfered faces of just 25cm in the Hall of

Nations and 23cm in the Halls of Industries, and interface angles drawn from the slopes of its members, the robust rhombus allowed for the in-situ construction of the space frame where as many as nine to twelve members come together at one node. The innovative resolution of the nodes in this way provided the requisite strength at the node to absorb the heavy compressive or tensile forces from the members.

The large truncated pyramids of both structures are composites of smaller unit pyramids. The Hall of Nations consists of units of a 4.9m x 4.9m base and 3.5m height which form the 73m x 73m pyramid, truncated at the height of 30m with a roof span of 39m x 39m. The Halls of Industries are four 18m-high pyramids of 40m x 40m base, with truncated roofs of 22m x 22m, made of units that are 2.6m high and 3.6m x 3.6m at the base. The configuration of both these basic pyramids is such that all members of the square base and the four triangular sides have the same length, finally allowing the same building slope and the angle of this basic unit at 54°44'8".

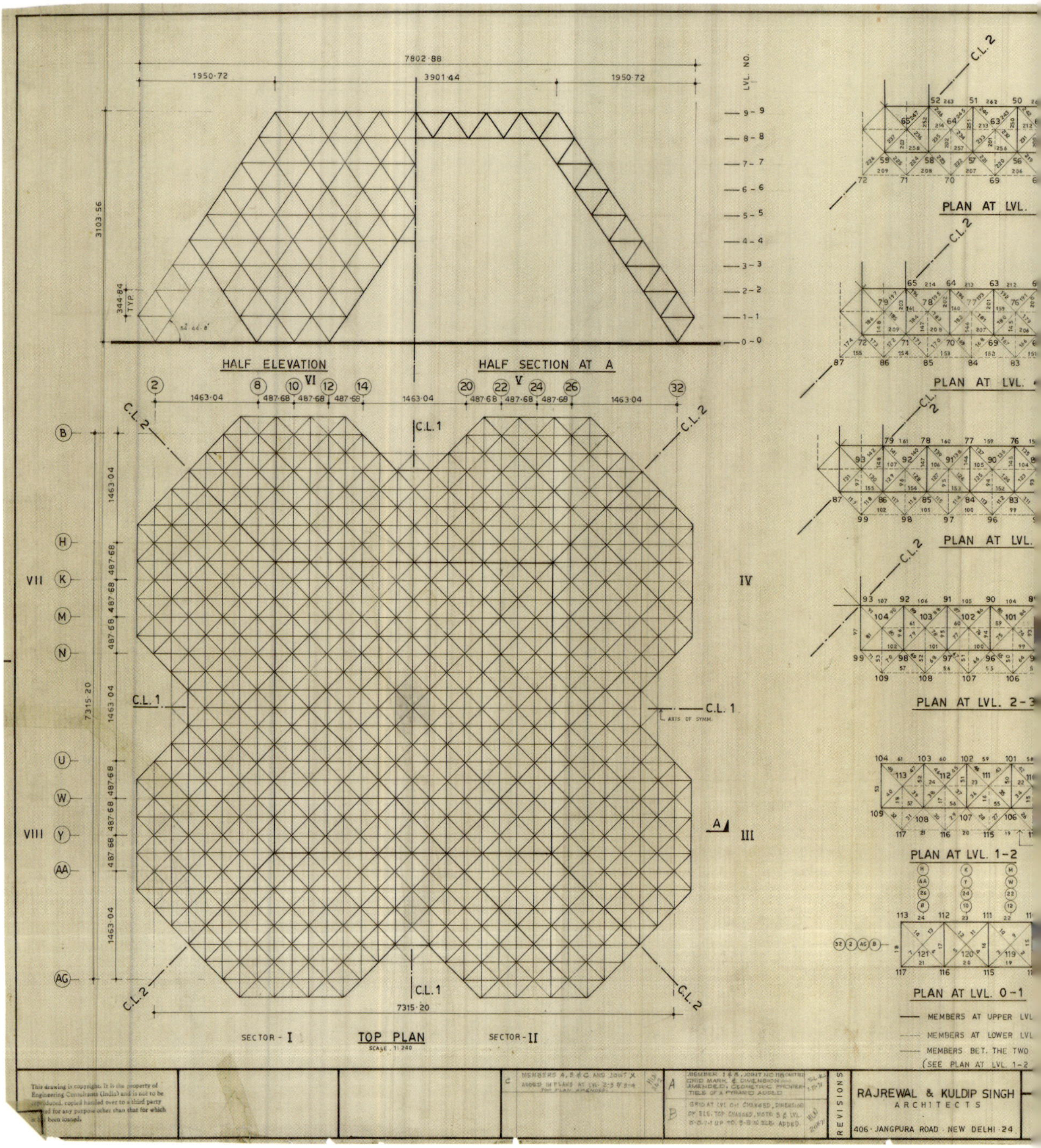

Drawing no. 101.ITF.02D: Plans, Elevation, Section (Hall of Nations): 16 August 1971

Drawing of plans at all nine levels of the Hall of Nations, with an overall section and elevation and also the geometry of the pyramid used as a basic module. Each plan includes two levels to help explain the pyramidical space frame geometry and how it connects to create the truncated pyramid form. The simplicity and clarity of the pyramid geometry is carried through consistently in plans, sections and elevations. The structure blooms out of twenty-four nodes at the ground level, six nodes at each face of the square plan.

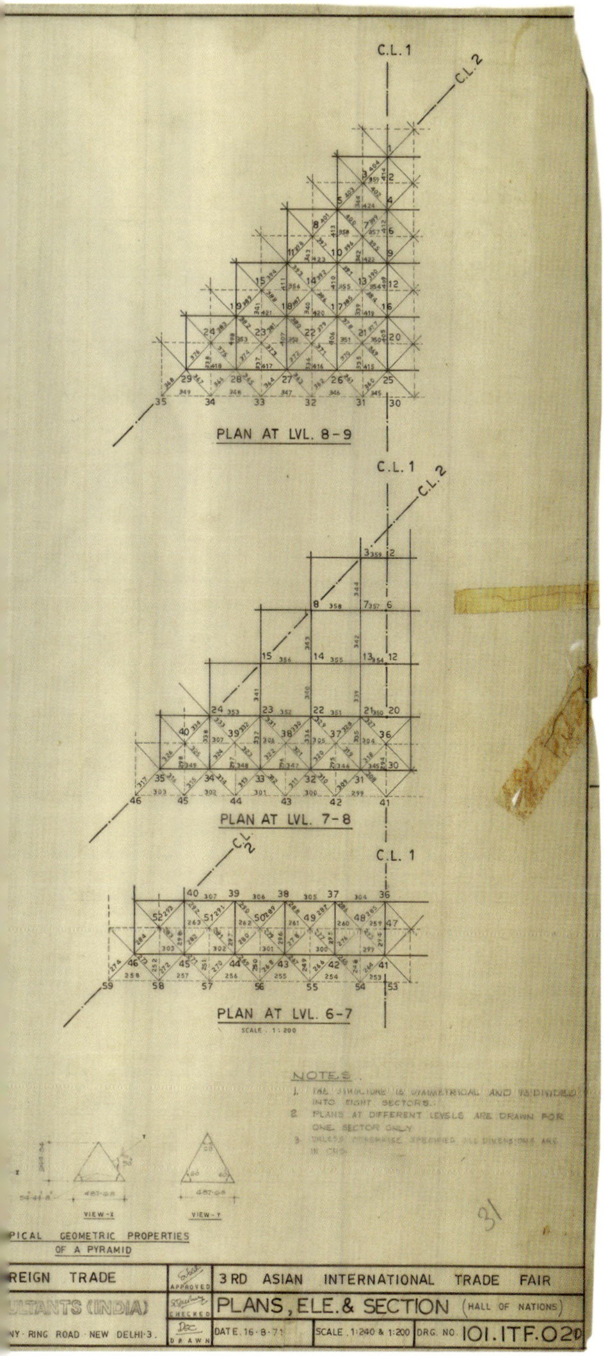

Recent photo of the large interior space of the Hall of Nations diminishing the large trucks parked inside by comparison.

Left, top to bottom: Drawings of a typical joint in steel, precast concrete and in-situ concrete. A complete set of construction documents was developed for each of these material options and finally the cast-in-situ was adopted.

Below and opposite: Construction sequence: drawing and photos.
Raj's careful conceptualisation of the stages of construction helped reduce efforts and costs as the entire structure was not supported with scaffolding till the roof. Based on extensive analysis, a detailed construction sequence was established. A system of scaffolding would support the structure till level 5, one level above all the desired openings in the lower levels. After level 5, the structure itself supports additional construction load, freeing the ground space. With each successive level, four different configurations of structure were envisaged and analysed for all possible loading conditions before the final building was completed.
During construction, however, despite simultaneous construction on all sides, the rate of progress was not the same; one side was up to the sixth level and another to the fourth. This was a new structure not envisaged or analysed before. At this stage, after more analytical work, the scaffolding was removed as the structure was found to be self-supporting.

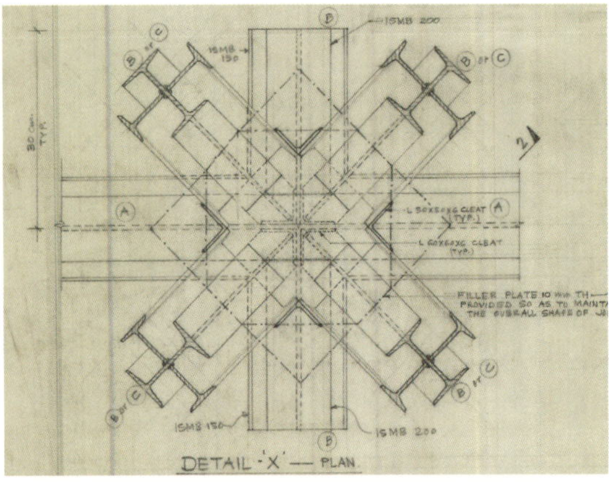

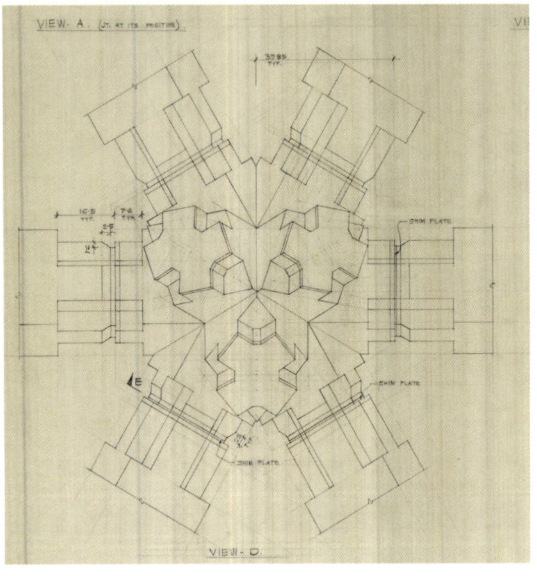

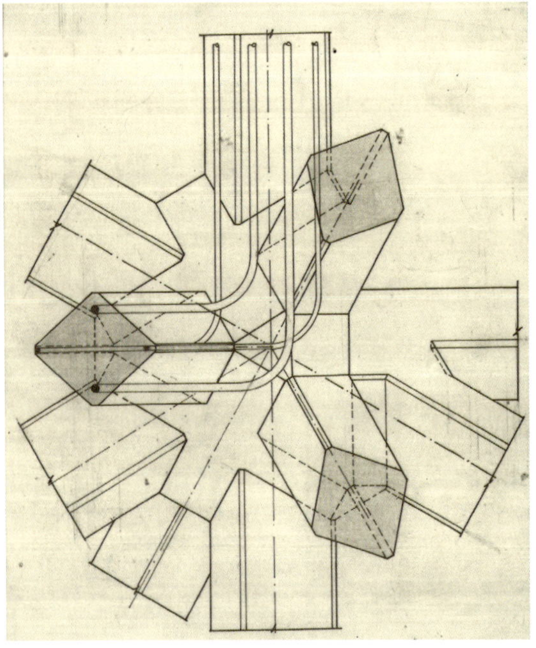

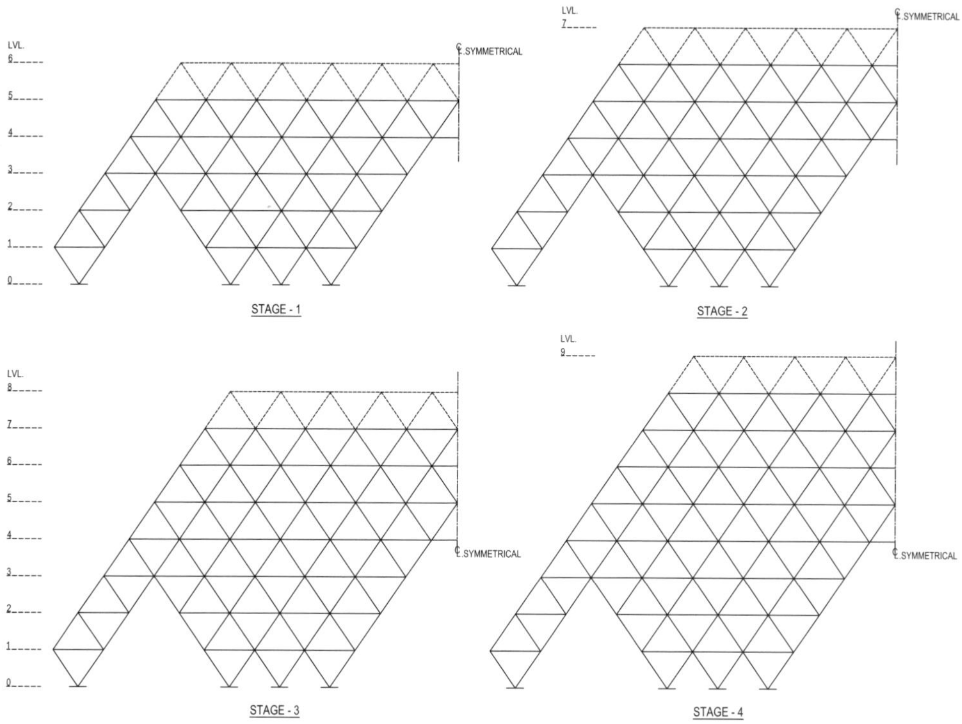

CONSTRCTION SEQUENCE

NOTES :- 1. CONSTRUCTION UP TO LEVEL 5 WITH TEMPORARY SUPPORTS
2. STRUCTURE TOTALLY DEPROPPED ON MATURITY OF CONCRETE AT LEVEL-5

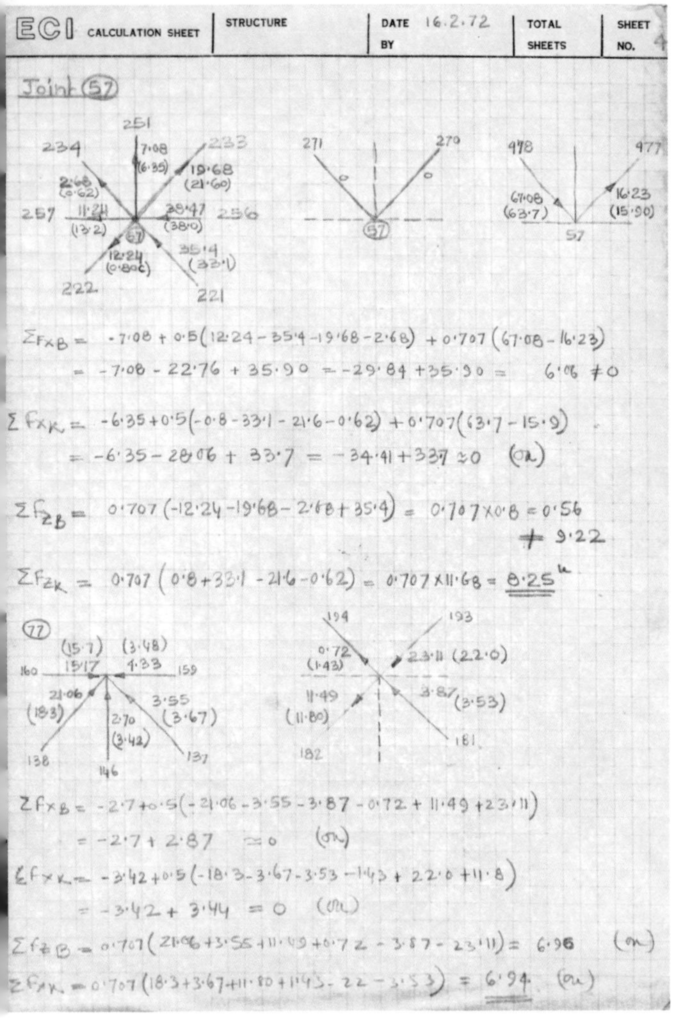
ECI CALCULATION SHEET | STRUCTURE | DATE 16.2.72 | BY | TOTAL SHEETS | SHEET NO. 4

Joint (57)

ΣFxB = -7·08 + 0·5(12·24 - 35·4 - 19·68 - 2·68) + 0·707(67·08 - 16·23)
= -7·08 - 22·76 + 35·90 = -29·84 + 35·90 = 6·06 ≠ 0

ΣFxK = -6·35 + 0·5(-0·8 - 33·1 - 21·6 - 0·62) + 0·707(63·7 - 15·9)
= -6·35 - 28·06 + 33·7 = -34·41 + 33·7 ≈ 0 (ok)

ΣFzB = 0·707(-12·24 - 19·68 - 2·68 + 35·4) = 0·707 × 0·8 = 0·56 + 9·22

ΣFzK = 0·707(0·8 + 33·1 - 21·6 - 0·62) = 0·707 × 11·68 = 8·25

(77)

ΣFxB = -2·7 + 0·5(-21·06 - 3·55 - 3·87 - 0·72 + 11·49 + 23·11)
= -2·7 + 2·87 ≈ 0 (ok)

ΣFxK = -3·42 + 0·5(-18·3 - 3·67 - 3·53 - 1·43 + 22·0 + 11·8)
= -3·42 + 3·44 = 0 (ok)

ΣFzB = 0·707(21·06 + 3·55 + 11·49 + 0·72 - 3·87 - 23·11) = 6·96 (ok)

ΣFzK = 0·707(18·3 + 3·67 + 11·80 + 1·43 - 22 - 3·53) = 6·94 (ok)

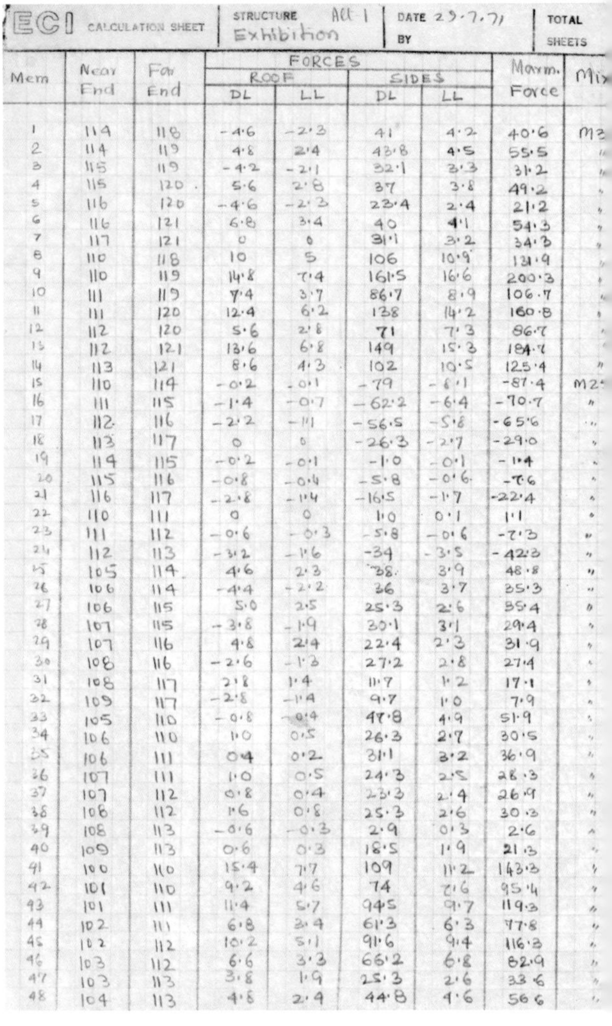
ECI CALCULATION SHEET | STRUCTURE Act 1 Exhibition | DATE 23.7.71 | BY | TOTAL SHEETS

Mem	Near End	Far End	Forces Roof DL	Roof LL	Sides DL	Sides LL	Morm. Force	Mix
1	114	118	-4·6	-2·3	41	4·2	40·6	M3
2	114	119	4·8	2·4	43·8	4·5	55·5	"
3	115	119	-4·2	-2·1	32·1	3·3	31·2	"
4	115	120	5·6	2·8	37	3·8	49·2	"
5	116	120	-4·6	-2·3	23·4	2·4	21·2	"
6	116	121	6·8	3·4	40	4·1	54·3	"
7	117	121	0	0	31·1	3·2	34·3	"
8	110	118	10	5	106	10·9	131·9	"
9	110	119	14·8	7·4	161·5	16·6	200·3	"
10	111	119	7·4	3·7	86·7	8·9	106·7	"
11	111	120	12·4	6·2	138	14·2	160·8	"
12	112	120	5·6	2·8	71	7·3	86·7	"
13	112	121	13·6	6·8	149	15·3	184·7	"
14	113	121	8·6	4·3	102	10·5	125·4	"
15	110	114	-0·2	-0·1	-79	-8·1	-87·4	M2
16	111	115	-1·4	-0·7	-62·2	-6·4	-70·7	"
17	112	116	-2·2	-1·1	-56·5	-5·8	-65·6	"
18	113	117	0	0	-26·3	-2·7	-29·0	"
19	114	115	-0·2	-0·1	-1·0	-0·1	-1·4	"
20	115	116	-0·8	-0·4	-5·8	-0·6	-7·6	"
21	116	117	-2·8	-1·4	-16·5	-1·7	-22·4	"
22	110	111	0	0	1·0	0·1	1·1	"
23	111	112	-0·6	-0·3	-5·8	-0·6	-7·3	"
24	112	113	-3·2	-1·6	-34	-3·5	-42·3	"
25	105	114	4·6	2·3	38	3·9	48·8	"
26	106	114	-4·4	-2·2	36	3·7	35·3	"
27	106	115	5·0	2·5	25·3	2·6	35·4	"
28	107	115	-3·8	-1·9	30·1	3·1	29·4	"
29	107	116	4·8	2·4	22·4	2·3	31·9	"
30	108	116	-2·6	-1·3	27·2	2·8	27·4	"
31	108	117	2·8	1·4	11·7	1·2	17·1	"
32	109	117	-2·8	-1·4	9·7	1·0	7·9	"
33	105	110	-0·8	-0·4	47·8	4·9	51·9	"
34	106	110	1·0	0·5	26·3	2·7	30·5	"
35	106	111	0·4	0·2	31·1	3·2	36·9	"
36	107	111	1·0	0·5	24·3	2·5	28·3	"
37	107	112	0·8	0·4	23·3	2·4	26·9	"
38	108	112	1·6	0·8	25·3	2·6	30·3	"
39	108	113	-0·6	-0·3	2·9	0·3	2·6	"
40	109	113	0·6	0·3	18·5	1·9	21·3	"
41	100	110	15·4	7·7	109	11·2	143·3	"
42	101	110	9·2	4·6	74	7·6	95·4	"
43	101	111	11·4	5·7	94·5	9·7	119·3	"
44	102	111	6·8	3·4	61·3	6·3	77·8	"
45	102	112	10·2	5·1	91·6	9·4	116·3	"
46	103	112	6·6	3·3	66·2	6·8	82·9	"
47	103	113	3·8	1·9	25·3	2·6	33·6	"
48	104	113	4·8	2·4	44·8	4·6	56·6	"

Four sample sheets from the volumes of manual calculations meticulously done for this project.

View from the fourth level of the concrete space frame.

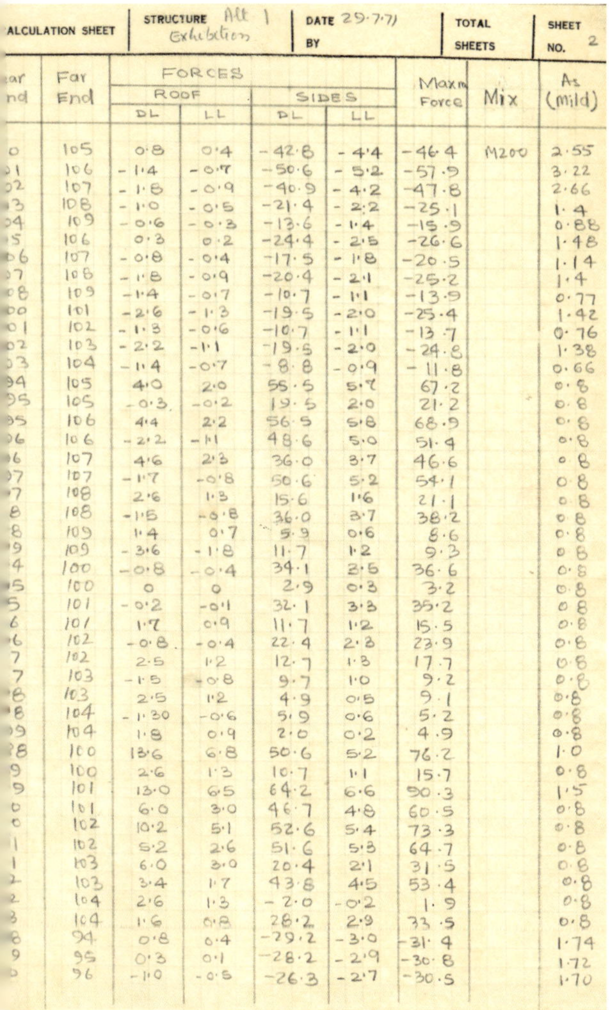

CALCULATION SHEET	STRUCTURE Alt 1 Exhibition	DATE 29·7·71 BY	TOTAL SHEETS	SHEET NO. 2

Near End	Far End	FORCES ROOF DL	ROOF LL	SIDES DL	SIDES LL	Maxm Force	Mix	As (mild)
0	105	0·8	0·4	-42·8	-4·4	-46·4	M200	2·55
1	106	-1·4	-0·7	-50·6	-5·2	-57·9		3·22
2	107	-1·8	-0·9	-40·9	-4·2	-47·8		2·66
3	108	-1·0	-0·5	-21·4	-2·2	-25·1		1·4
4	109	-0·6	-0·3	-13·6	-1·4	-15·9		0·88
5	106	0·3	0·2	-24·4	-2·5	-26·6		1·48
6	107	-0·8	-0·4	-17·5	-1·8	-20·5		1·14
7	108	-1·8	-0·9	-20·4	-2·1	-25·2		1·4
8	109	-1·4	-0·7	-10·7	-1·1	-13·9		0·77
00	101	-2·6	-1·3	-19·5	-2·0	-25·4		1·42
01	102	-1·3	-0·6	-10·7	-1·1	-13·7		0·76
02	103	-2·2	-1·1	-19·5	-2·0	-24·8		1·38
03	104	-1·4	-0·7	-8·8	-0·9	-11·8		0·66
94	105	4·0	2·0	55·5	5·7	67·2		0·8
95	105	-0·3	-0·2	19·5	2·0	21·2		0·8
95	106	4·4	2·2	56·5	5·8	68·9		0·8
96	106	-2·2	-1·1	48·6	5·0	51·4		0·8
6	107	4·6	2·3	36·0	3·7	46·6		0·8
7	107	-1·7	-0·8	50·6	5·2	54·1		0·8
7	108	2·6	1·3	15·6	1·6	21·1		0·8
8	108	-1·5	-0·8	36·0	3·7	38·2		0·8
8	109	1·4	0·7	5·9	0·6	8·6		0·8
9	109	-3·6	-1·8	11·7	1·2	9·3		0·8
4	100	-0·8	-0·4	34·1	3·5	36·6		0·8
5	100	0	0	2·9	0·3	3·2		0·8
5	101	-0·2	-0·1	32·1	3·3	35·2		0·8
6	101	1·7	0·9	11·7	1·2	15·5		0·8
6	102	-0·8	-0·4	22·4	2·3	23·9		0·8
7	102	2·5	1·2	12·7	1·3	17·7		0·8
7	103	-1·5	-0·8	9·7	1·0	9·2		0·8
8	103	2·5	1·2	4·9	0·5	9·1		0·8
8	104	-1·30	-0·6	5·9	0·6	5·2		0·8
9	104	1·8	0·9	2·0	0·2	4·9		0·8
8	100	13·6	6·8	50·6	5·2	76·2		1·0
9	100	2·6	1·3	10·7	1·1	15·7		0·8
9	101	13·0	6·5	64·2	6·6	90·3		1·5
0	101	6·0	3·0	46·7	4·8	60·5		0·8
0	102	10·2	5·1	52·6	5·4	73·3		0·8
1	102	5·2	2·6	51·6	5·3	64·7		0·8
1	103	6·0	3·0	20·4	2·1	31·5		0·8
2	103	3·4	1·7	43·8	4·5	53·4		0·8
2	104	2·6	1·3	-2·0	-0·2	1·9		0·8
3	104	1·6	0·8	28·2	2·9	33·5		0·8
8	94	0·8	0·4	-29·2	-3·0	-31·4		1·74
9	95	0·3	0·1	-28·2	-2·9	-30·8		1·72
0	96	-1·0	-0·5	-26·3	-2·7	-30·5		1·70

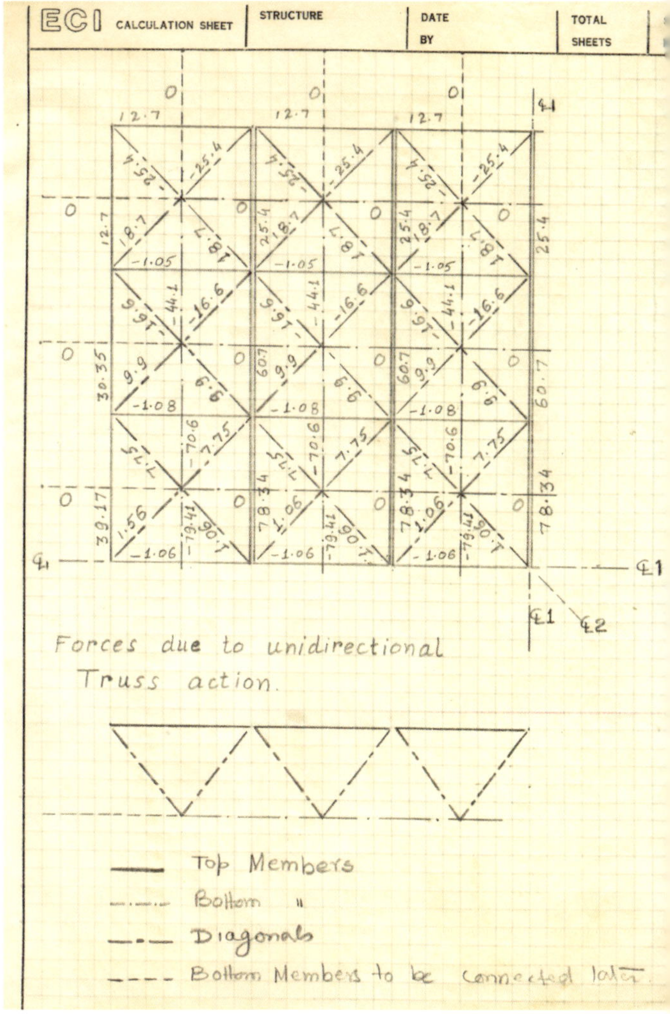

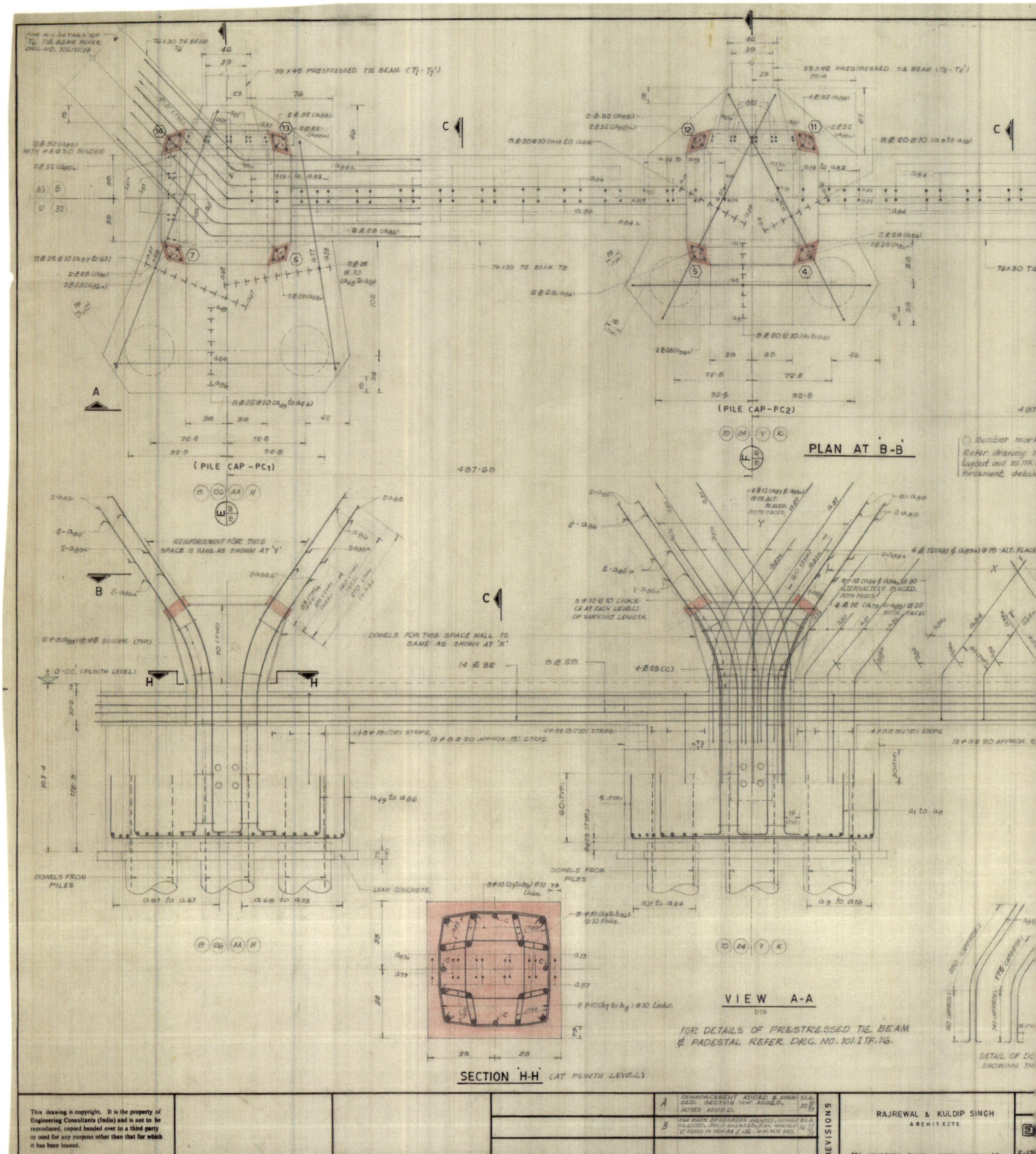

Drawing no. 101.ITF.15B: Details of Pile Cap and Tie Beam (Hall of Nations): 21 October 1971

The Hall of Nations was supported on in-situ-driven piles tied together with grade beams in two directions. First, along the six springing points in one face of the pyramid, and second, along the other direction connecting the opposite faces of the pyramid. The latter were post-tensioned in stages to cater to the large horizontal forces that developed as the structure went up.

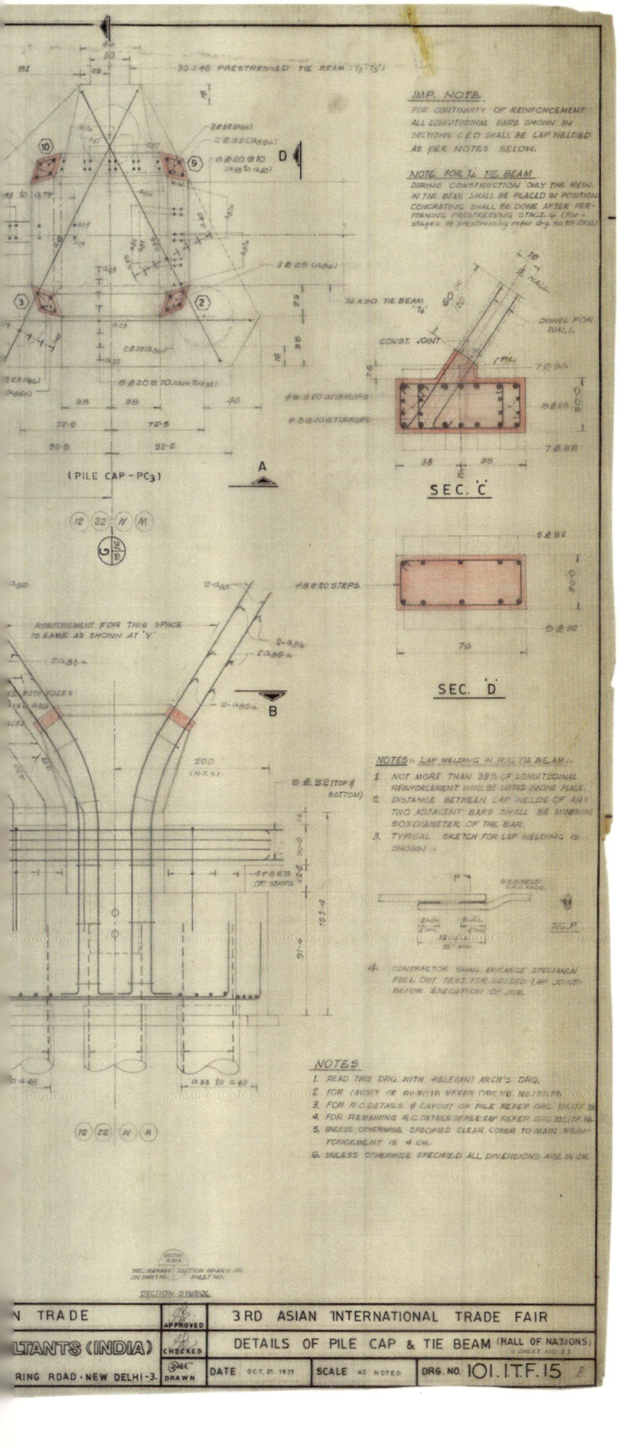

Construction photos of pile caps and the reinforcement being arranged for the space-frame members.

Following pages: Photo capturing the springing point of a set of three nodes emerging from the ground with openings on both sides. The height of the basic pyramid module of the space frame is 3.5m and each member is 4.87m long.

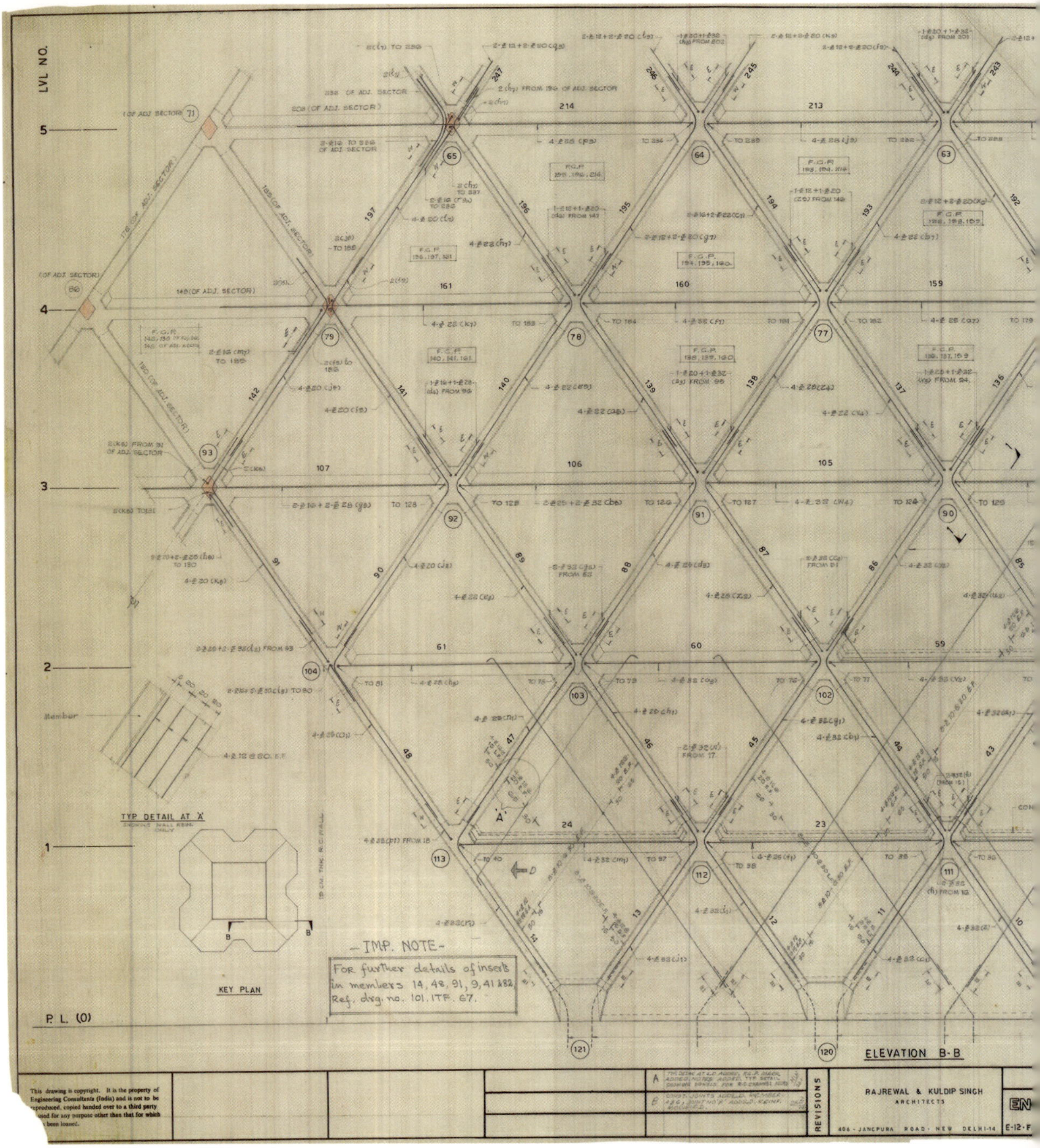

Drawing no. 101.ITF.59CB: (Part) Inner Elevation (Hall of Nations): 16 November 1971

A typical elevational module emerging from three nodes that sprouts into three inverted pyramids. The three inverted pyramids form the base that multiplies three dimensionally to create an elevational module that forms half of each face of the truncated pyramid. The geometry accommodates the triangular openings formed naturally on each side and corners of the elevational module.

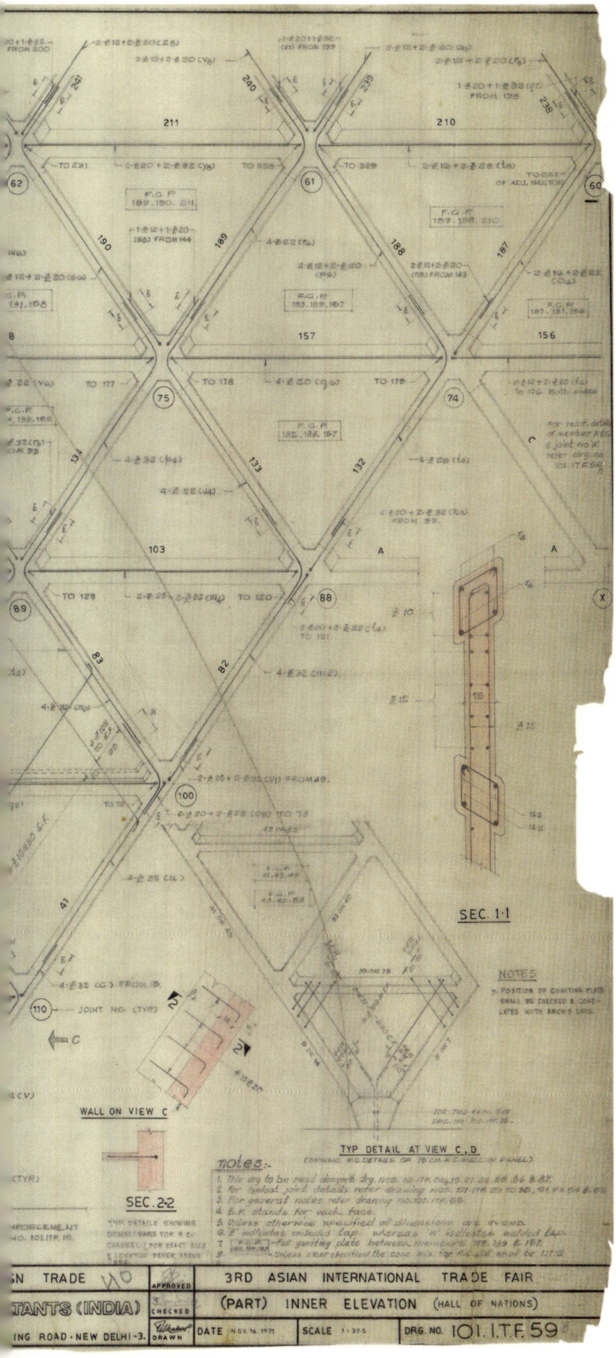

Construction photo of one face of the truncated pyramid form.

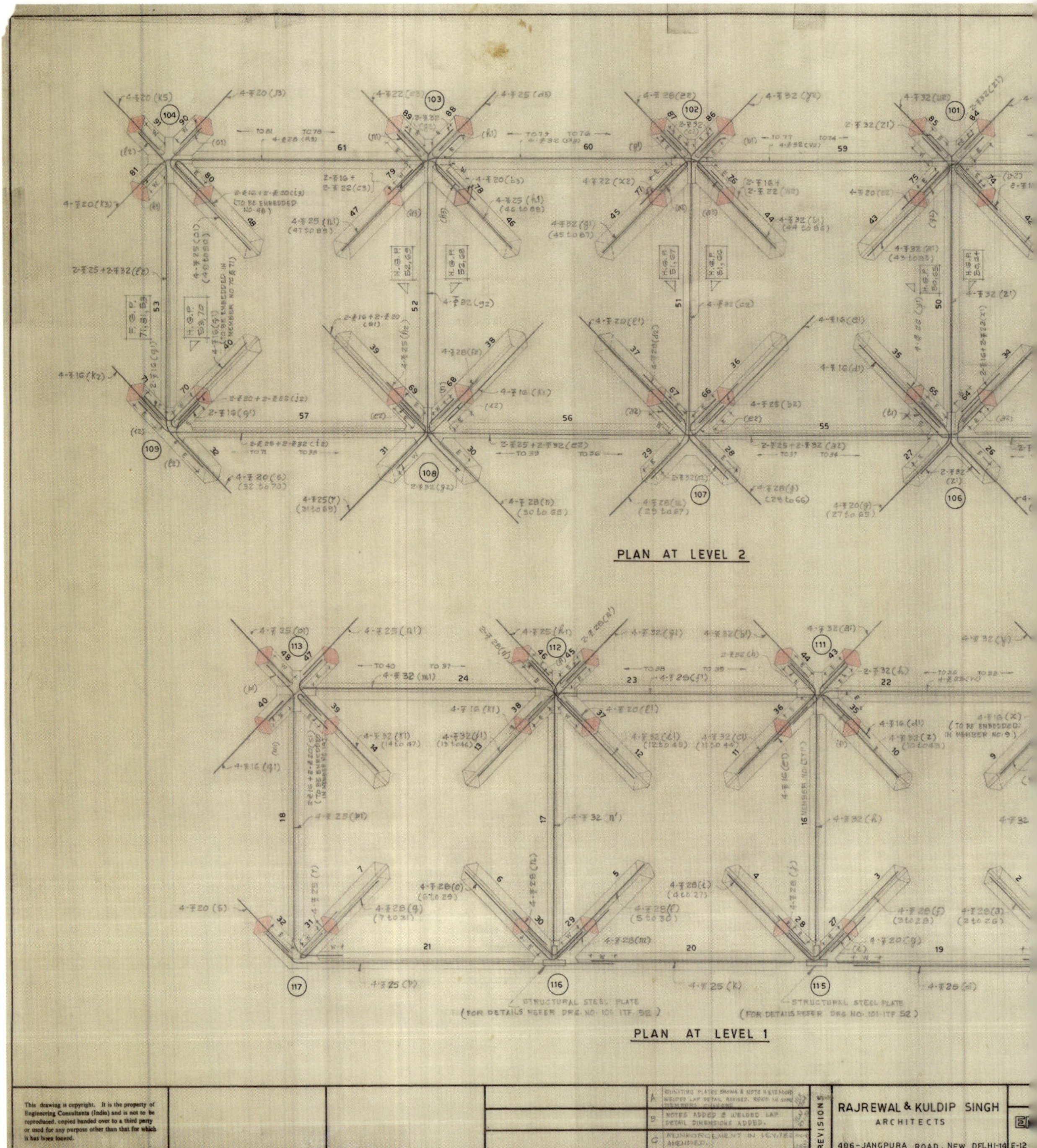

Drawing no. 101.ITF.55C: Reinforcement Layout at Level 1 & 2 (Hall of Nations): 10 November 1971

Plan of level 1 and 2 shows the resolution of one space frame module that emerges from three inverted pyramids. The plans indicate the three-dimensional multiplication on the inner and outer nodes to form a space frame that finally evolves into a truncated pyramid.
The rhombic cross section of the space frame members was an outcome of the search for a joint that would be easier to construct. Evolved from the slopes and intersections of the pyramid members, the geometry of the rhombus was found to be most appropriate versus square, rectangular and circular sections. For the Hall of Nations a rhombic section with chamfered edges and a face dimension of 25cm and cross-sectional area of 585 sq.cm was developed. This section shows Raj's intuitive visualisation and understanding of structures through the rigours of geometry.

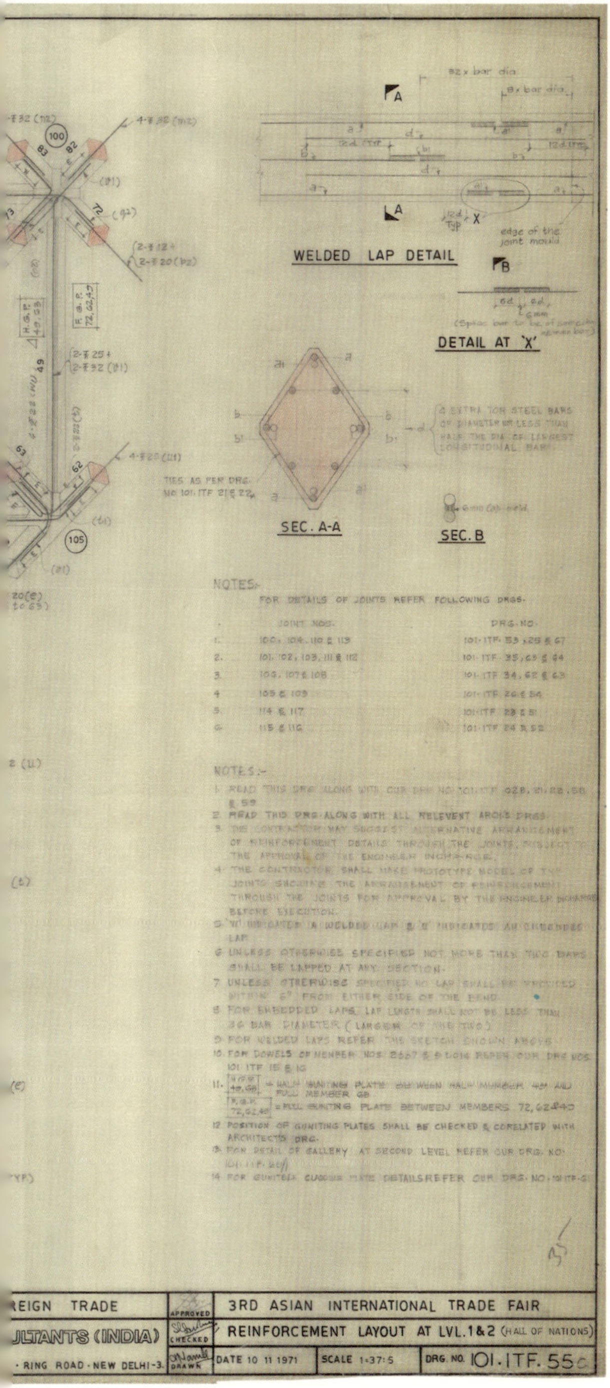

View through a completed section of the space frame during construction.

Following pages: Interior view of three interconnected Halls of Industries.

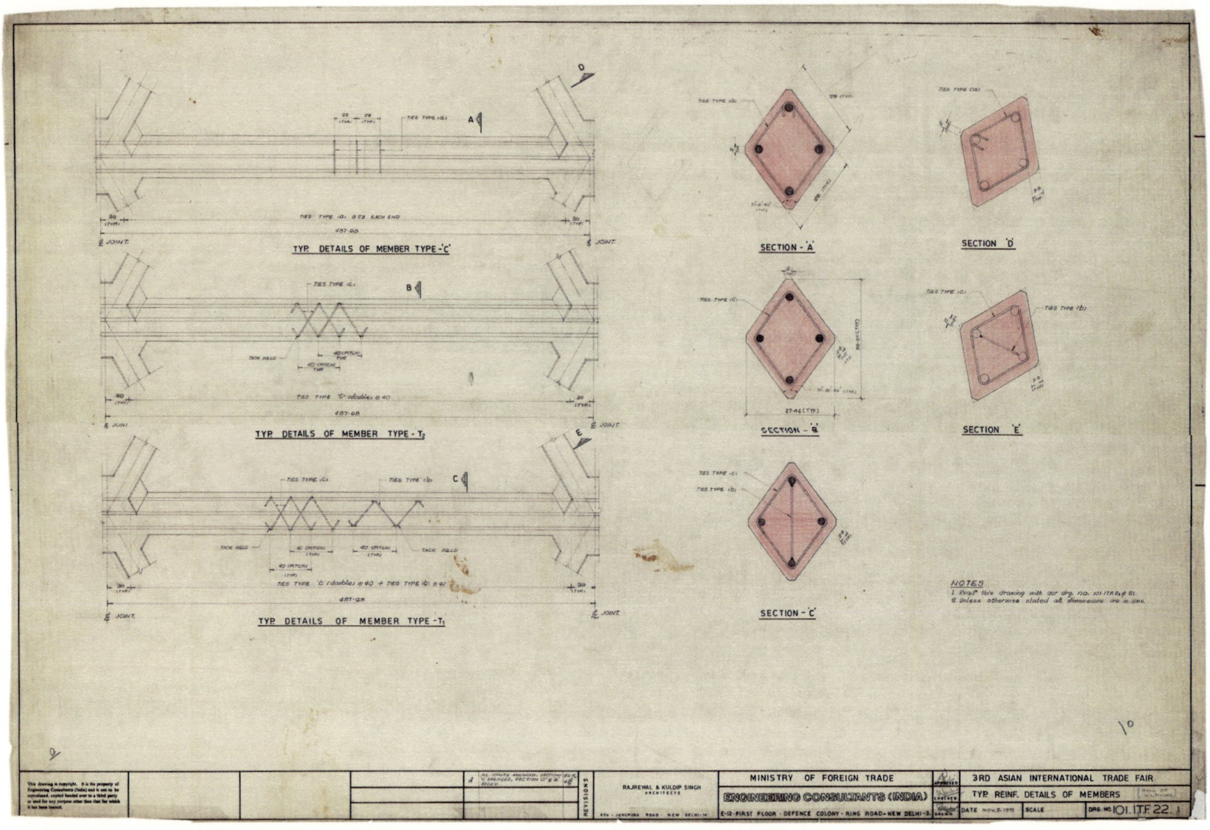
TYP. DETAILS OF MEMBER TYPE-'C'
TYP DETAILS OF MEMBER TYPE - T2
TYP DETAILS OF MEMBER TYPE-T1
SECTION - 'A'
SECTION 'D'
SECTION - 'B'
SECTION 'E'
SECTION - 'C'
NOTES
MINISTRY OF FOREIGN TRADE
ENGINEERING CONSULTANTS (INDIA)
3RD ASIAN INTERNATIONAL TRADE FAIR
TYP. REINF. DETAILS OF MEMBERS
RAJREWAL & KULDIP SINGH
ARCHITECTS
E-12 · FIRST FLOOR · DEFENCE COLONY · RING ROAD · NEW DELHI-3.
DRG. NO. 101.ITF. 22

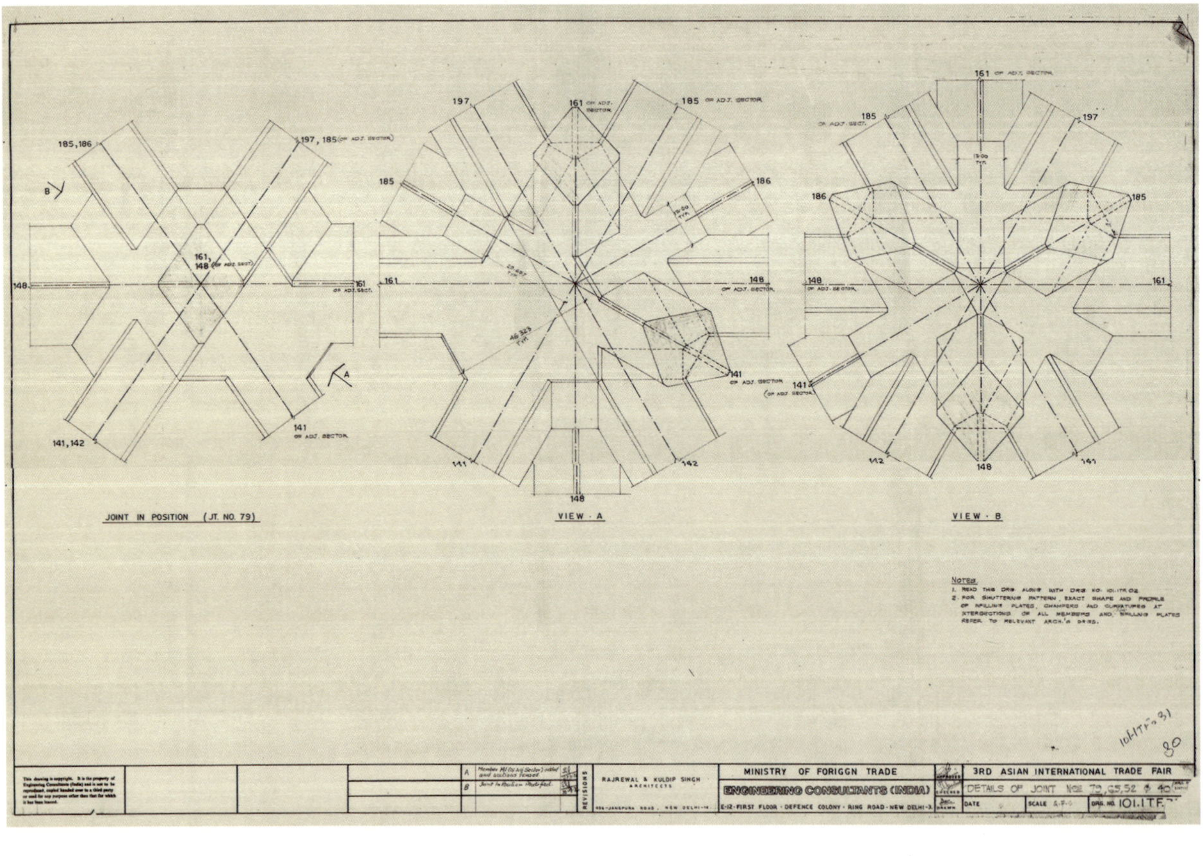
JOINT IN POSITION (JT. NO. 79)
VIEW - A
VIEW - B
MINISTRY OF FORIGGN TRADE
ENGINEERING CONSULTANTS (INDIA)
3RD ASIAN INTERNATIONAL TRADE FAIR
RAJREWAL & KULDIP SINGH
ARCHITECTS
E-12 · FIRST FLOOR · DEFENCE COLONY · RING ROAD · NEW DELHI-3.

Construction photos of the rhombic members and the nodes.

Drawing no. 101.ITF.22.A: Typical Reinforcement Details of Members (Hall of Nations): 5 November 1971

This drawing brings forth another detail responsible for the slenderness of this gigantic concrete space frame: Type T1 and T2 are members in tension with double spiral ties to counter and contain the micro-cracking of the concrete. Type C are members in compression with straight ties.

Drawing no. 101.ITF.31: Detail of Joints Nos. 79, 65, 52 & 40 (Hall of Nations)

Drawing of an eleven-member node visualising the complex intersections of the members with three views.

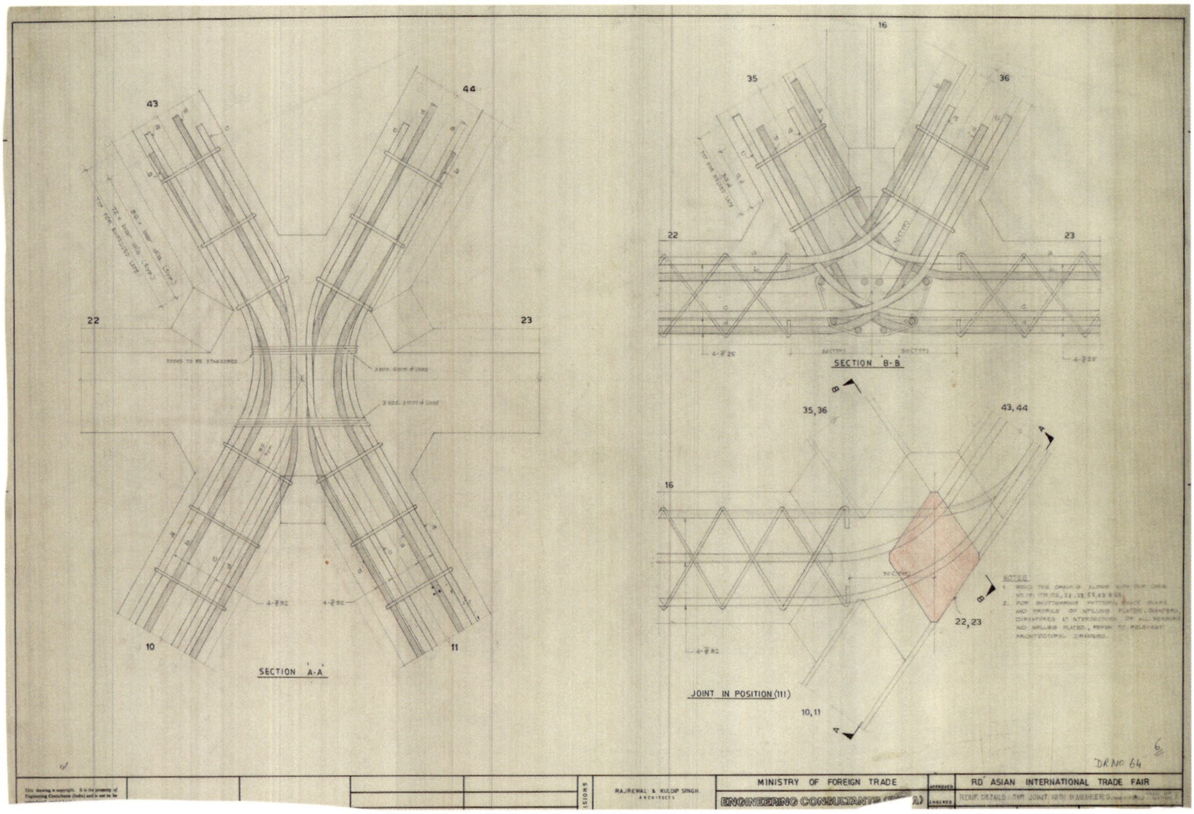
SECTION A-A
SECTION B-B
JOINT IN POSITION (111)
NOTES:
MINISTRY OF FOREIGN TRADE
ENGINEERING CONSULTANTS
RAJREWAL & KULDIP SINGH ARCHITECTS
RD ASIAN INTERNATIONAL TRADE FAIR
DR NO 64

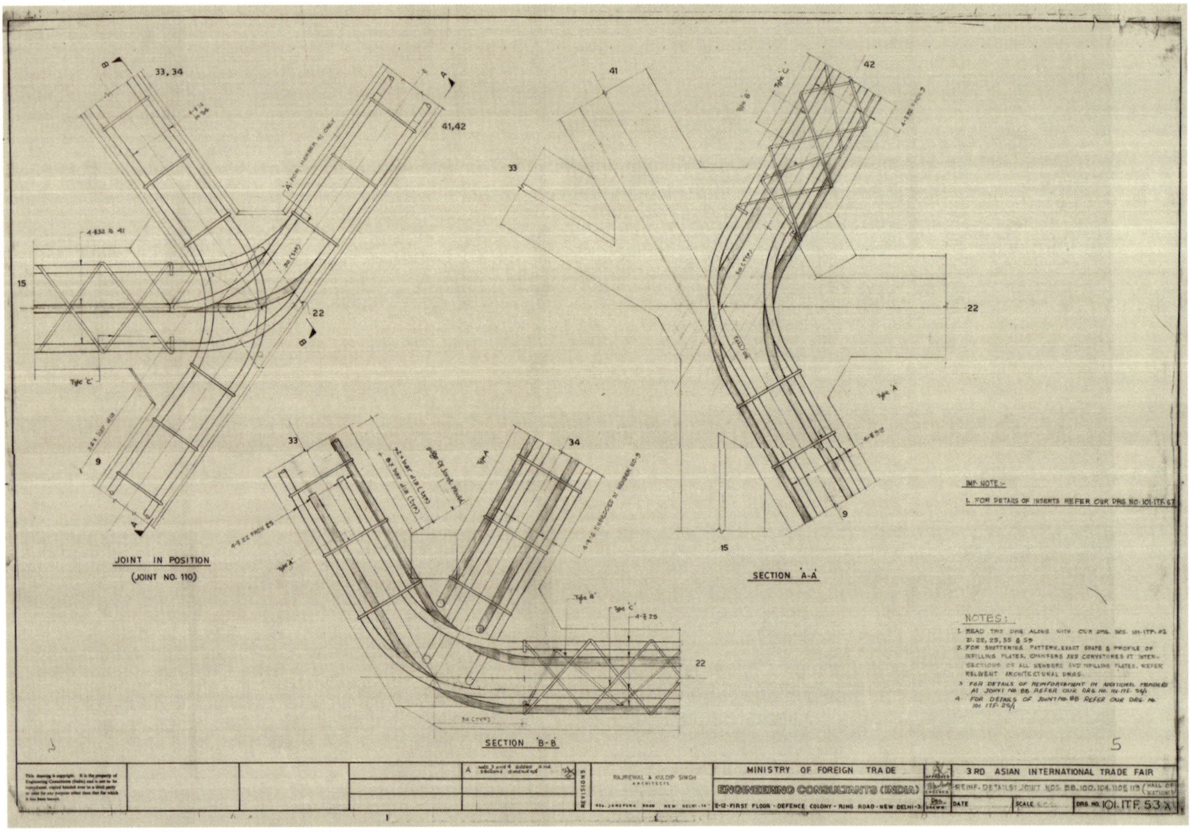
JOINT IN POSITION
(JOINT NO. 110)
SECTION A-A
SECTION B-B
IMP NOTE:-
1. FOR DETAILS OF INSERTS REFER OUR DRG NO-101-ITF-57
NOTES:
MINISTRY OF FOREIGN TRADE
ENGINEERING CONSULTANTS (INDIA)
E-12-FIRST FLOOR - DEFENCE COLONY - RING ROAD - NEW DELHI-3
RAJREWAL & KULDIP SINGH ARCHITECTS
3RD ASIAN INTERNATIONAL TRADE FAIR
REVISIONS
DATE
SCALE
DRG. NO. 101. ITF 53

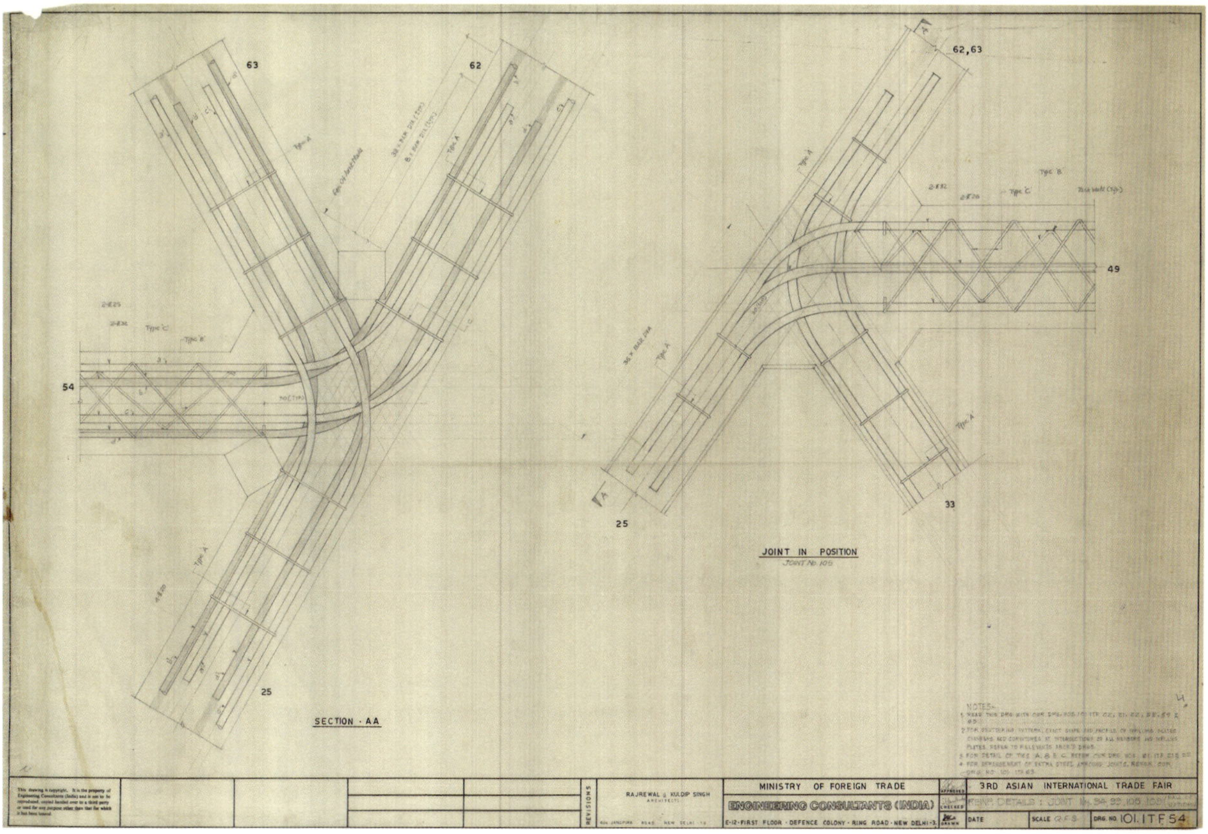

Above: Drawing no. 101.ITF.54: Reinf. Details; Joint Nos. 94, 99, 105 & 109 (Hall of Nations)

Opposite, top and bottom: Drawing no. 101.ITF.64: Reinf. Details; Typical Joint with Nine Members–Inner Face (Hall of Nations)

Drawing no. 101.ITF.53A: Reinf. Details; Joint Nos. 88, 100, 104, 110 & 113 (Hall of Nations)

Details of configurations of lapping and anchorage of reinforcement bars at different nodes. Detailing involved resolution of the congestion of bars at any given joint. On average, nine members met at a node, with four bars per member–an estimated 36 bars would pass each node, some in compression and some in tension. This congestion was artfully reduced by a system of lapping bars from the four lower members with bars of the four upper members accounting for 16 bars and finally, only 20 bars passed through a node.

The nodes were constructed by lap-welding members with straight-length bars to short-length curved and bent bars placed in the joint. The sequence was to cast the straight length of a member from joint to joint, place in position partially pre-assembled formwork of the joints along with placed-in-position short-length curved bars. These bars were welded with straight bars of members, formwork of the members up to the next joint was erected and then the next member and joint concreted. (See construction photos, previous page)

Photo of a node with more than nine members.

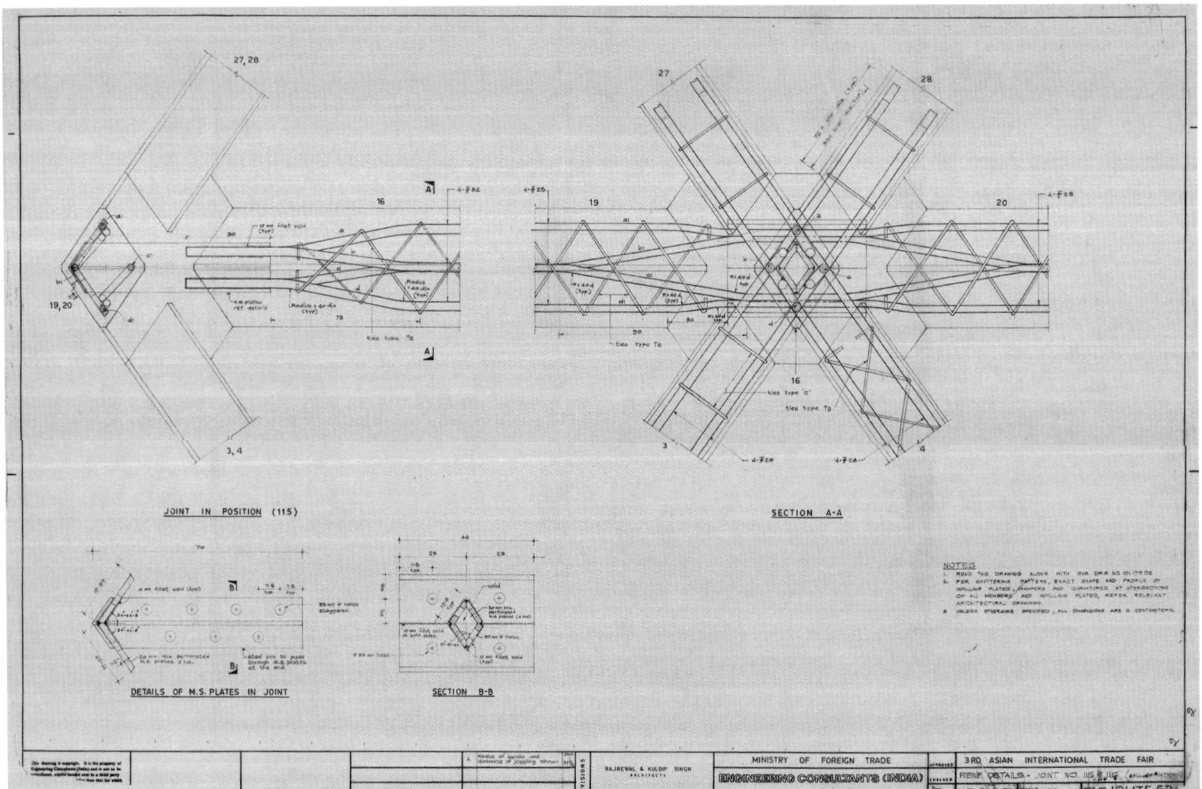
JOINT IN POSITION (115)
SECTION A-A
DETAILS OF M.S. PLATES IN JOINT
SECTION B-B
NOTES
RAJREWAL & KULDIP SINGH
ARCHITECTS
MINISTRY OF FOREIGN TRADE
ENGINEERING CONSULTANTS (INDIA)
E-12 FIRST FLOOR · DEFENCE COLONY · RING ROAD · NEW DELHI-3
3RD ASIAN INTERNATIONAL TRADE FAIR
101-ITF-52

Bird's-eye view of the entire complex.

Opposite: External view of the Hall of Nations.
Drawing no. 101.ITF.52: Reinf. Details; Joint Nos. 115, 116 (Hall of Nations): 6 December 1971

An eleven-member node.

External view of the Halls of Industries.

Internal view of the Hall of Nations.

DCM PAVILION

NEW DELHI | 1972

DCM

The innovative use of folded plate in steel to create a truncated cone with a base diameter of 43m and height of 28m made the temporary exhibition pavilion for the Delhi Cloth Mills (DCM). Three compression rings above and one tension ring at the foundation interconnecting the folds were used to create this structure.

The total structure was divided into three levels and constructed using the cantilever technique. Only the first level was erected with a scaffold. After the compression ring of the first level was put in place, the structure was self-supporting and the second level was erected on it. After the compression ring of the second level was in position, the third level was erected. Finally, after the third compression ring was in place, a fourth inclined compression ring completed the structure with an oval-shaped skylight. The folded plate trusses of the first level were hinged at the base, lending it lightness while at the same time eliminating heavy foundations.

Interior view looking up at the oval skylight formed by the inclined fourth compression ring and roof trusses.

The distinctive truncated cone form of the pavilion.

Previous pages: The DCM exhibition pavilion.

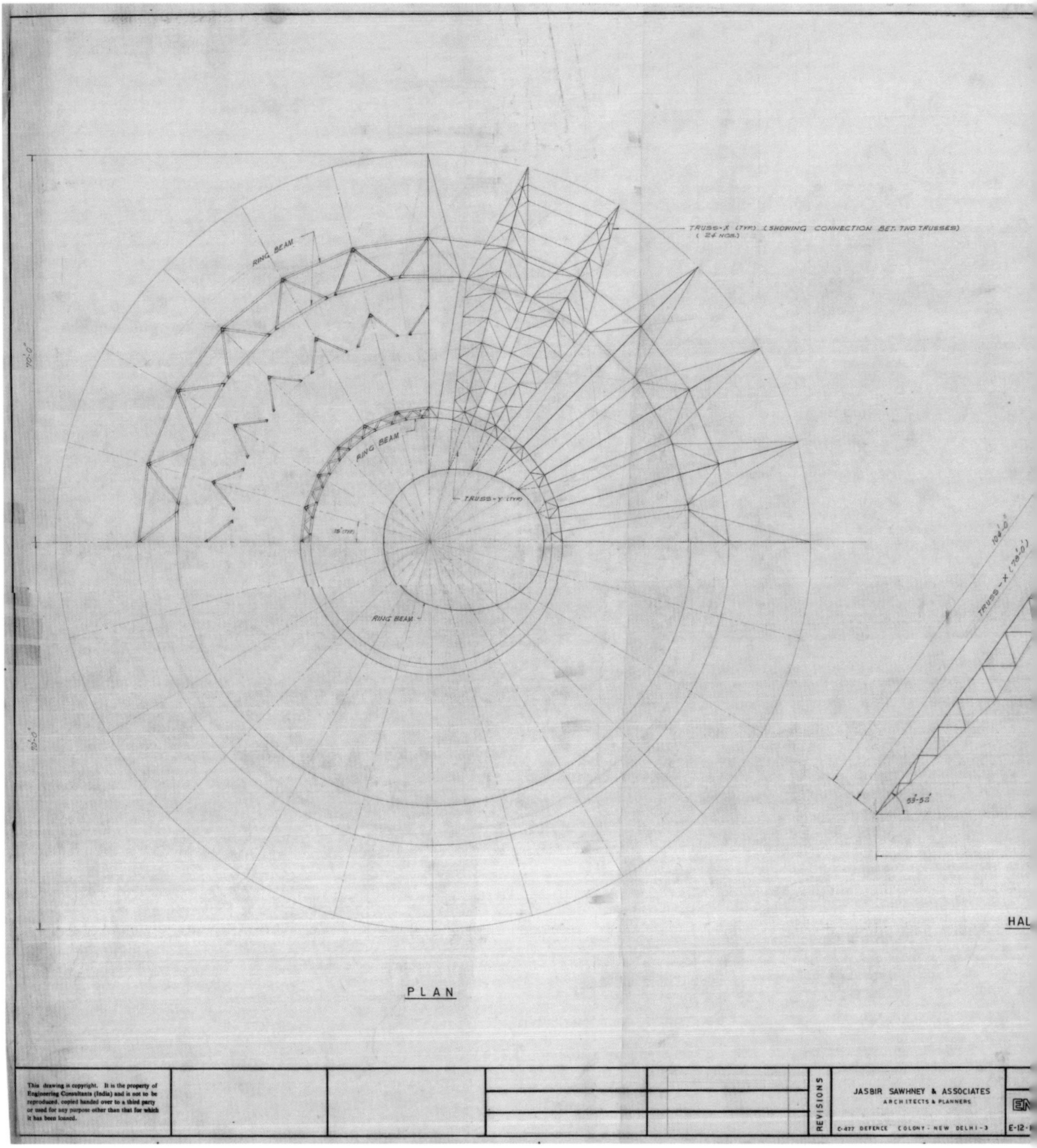
TRUSS-X (TYP.) (SHOWING CONNECTION BET. TWO TRUSSES)
(24 NOS.)
RING BEAM
RING BEAM
TRUSS-Y (TYP.)
RING BEAM
70'-0"
70'-0"
104'-0"
TRUSS-X (70'-0")
HAL
PLAN
This drawing is copyright. It is the property of Engineering Consultants (India) and is not to be reproduced, copied handed over to a third party or used for any purpose other than that for which it has been loaned.
REVISIONS
JASBIR SAWHNEY & ASSOCIATES
ARCHITECTS & PLANNERS
C-477 DEFENCE COLONY - NEW DELHI - 3

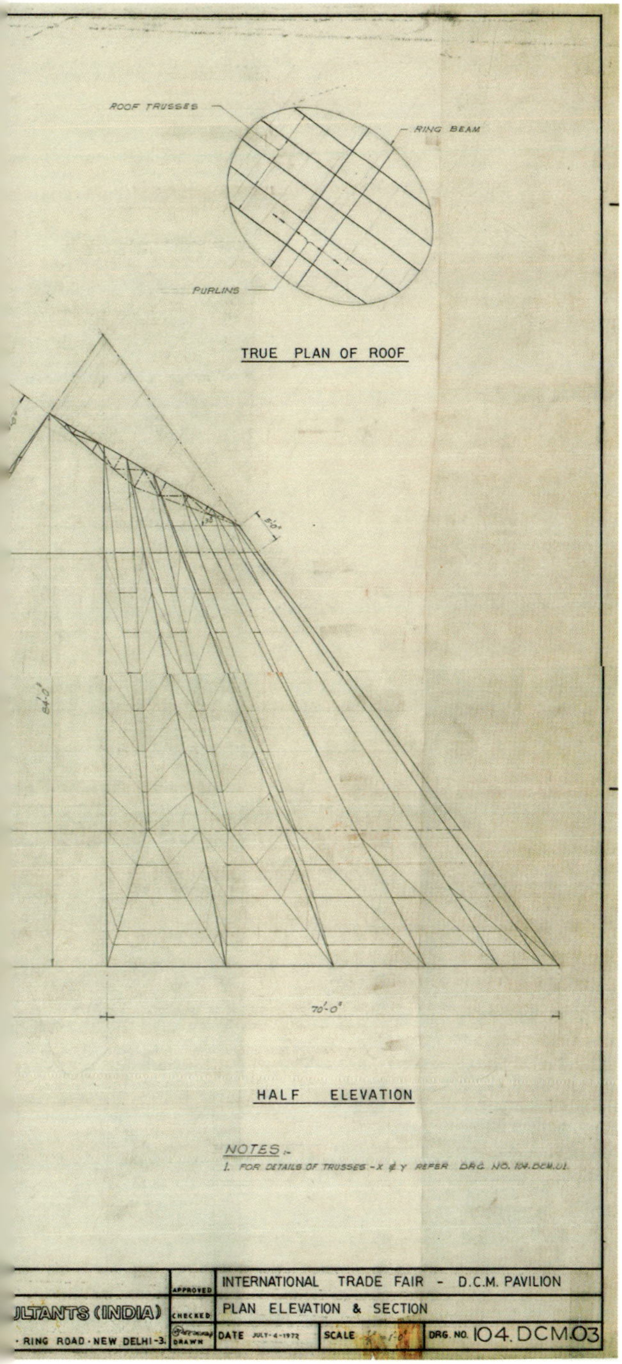

Interior view at the base of the folded plates.

Drawing no. 104.DCM.03: Plan, Elevation & Section: 4 July 1972

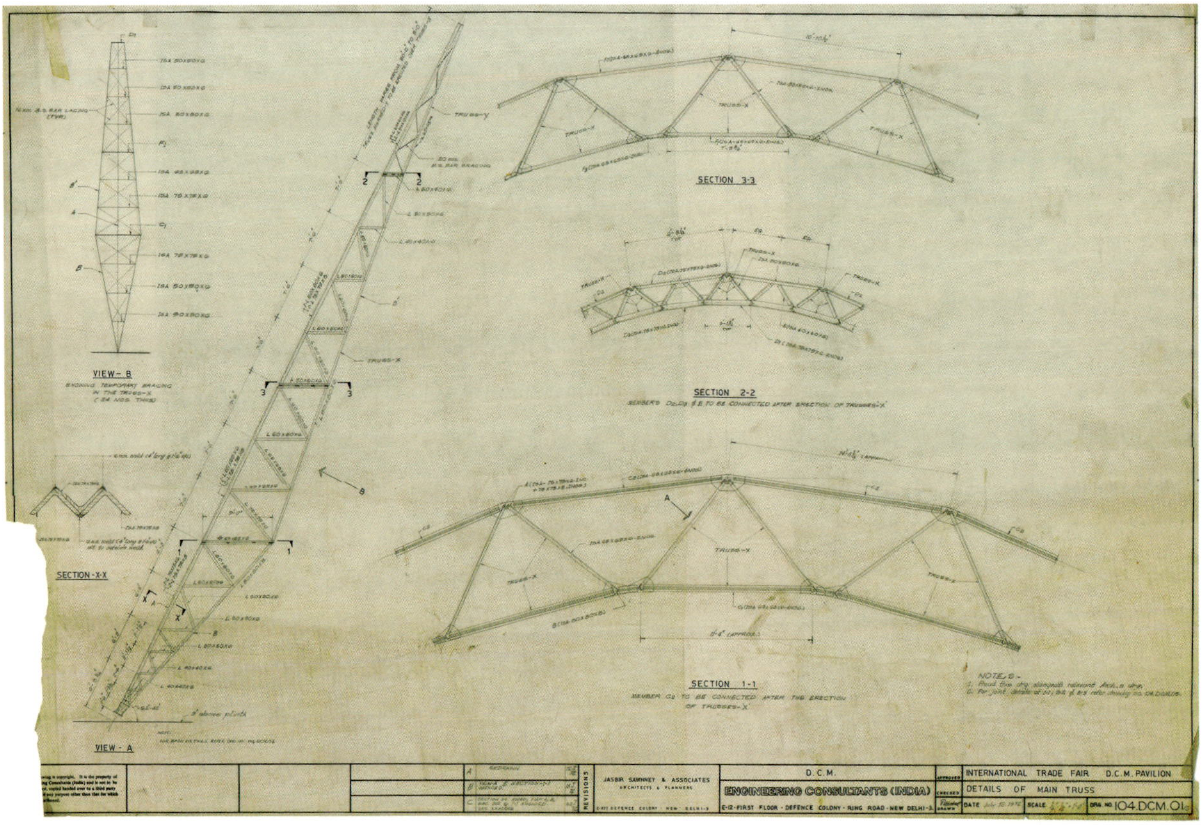

Drawing no. 104.DCM.01: Details of Main Truss: 12 July 1972,
includes details of compression rings and connections.

Below and opposite, bottom:
(Partial) Drawing no. 104.DCM.SK.01: Sketches Showing Scheme of Construction: 5 July 1972
Sketch drawings of an alternative scheme prepared for a different erection system of segments of folded plates fabricated on ground, placed radially around a hoisting pedestal and hoisted in position with a pulley and trolley.

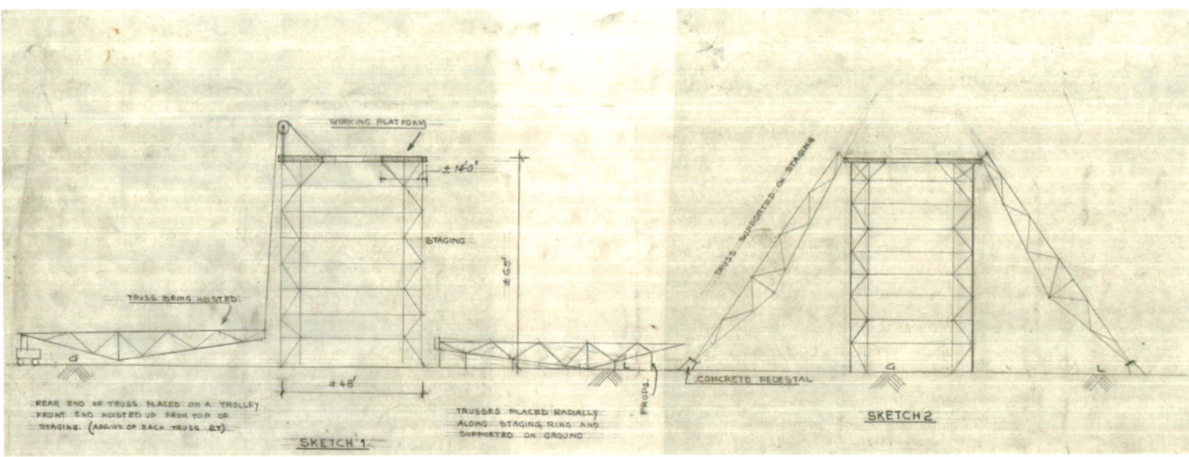

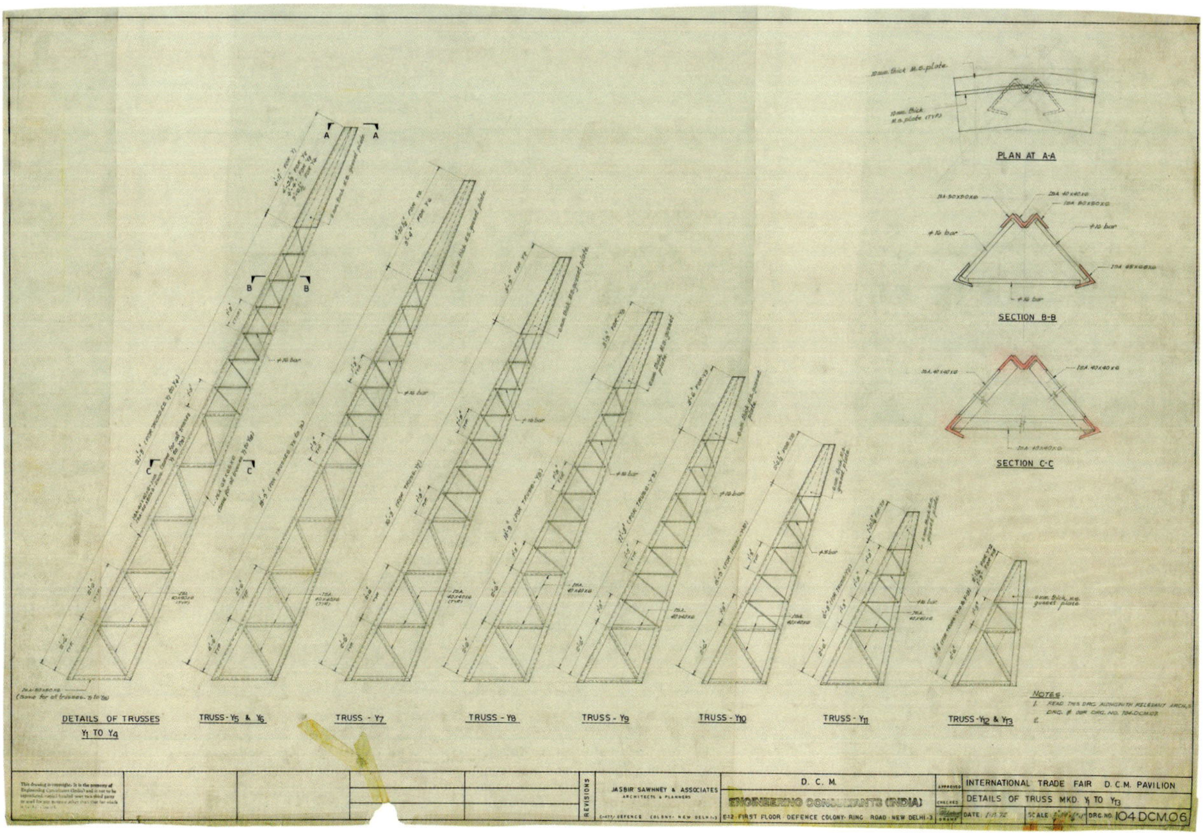

Drawing no. 104.DCM.06: Detail of Trusses Marked Y1-Y13: 1 August 1972, details trusses of varying length above the third compression ring to the inclined fourth ring at the skylight opening.

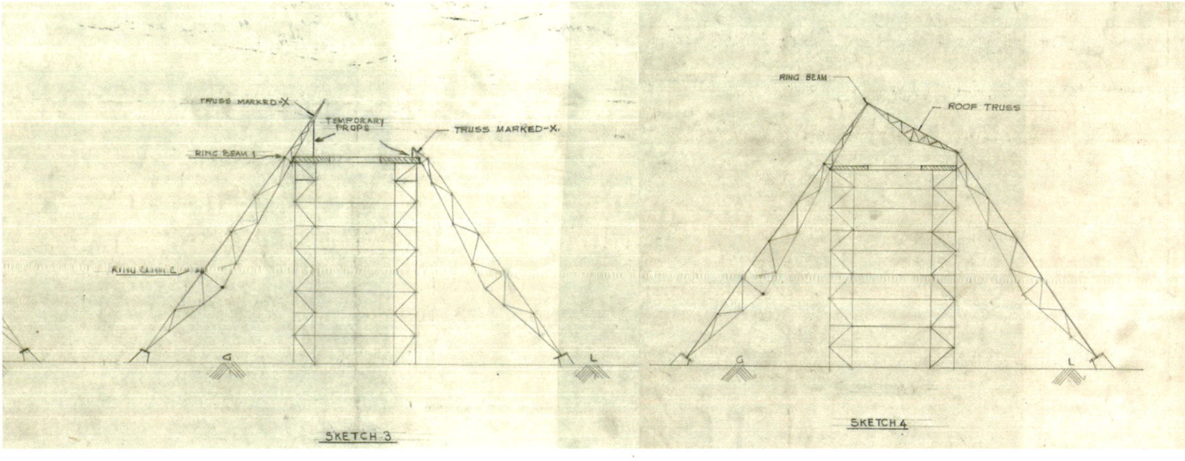

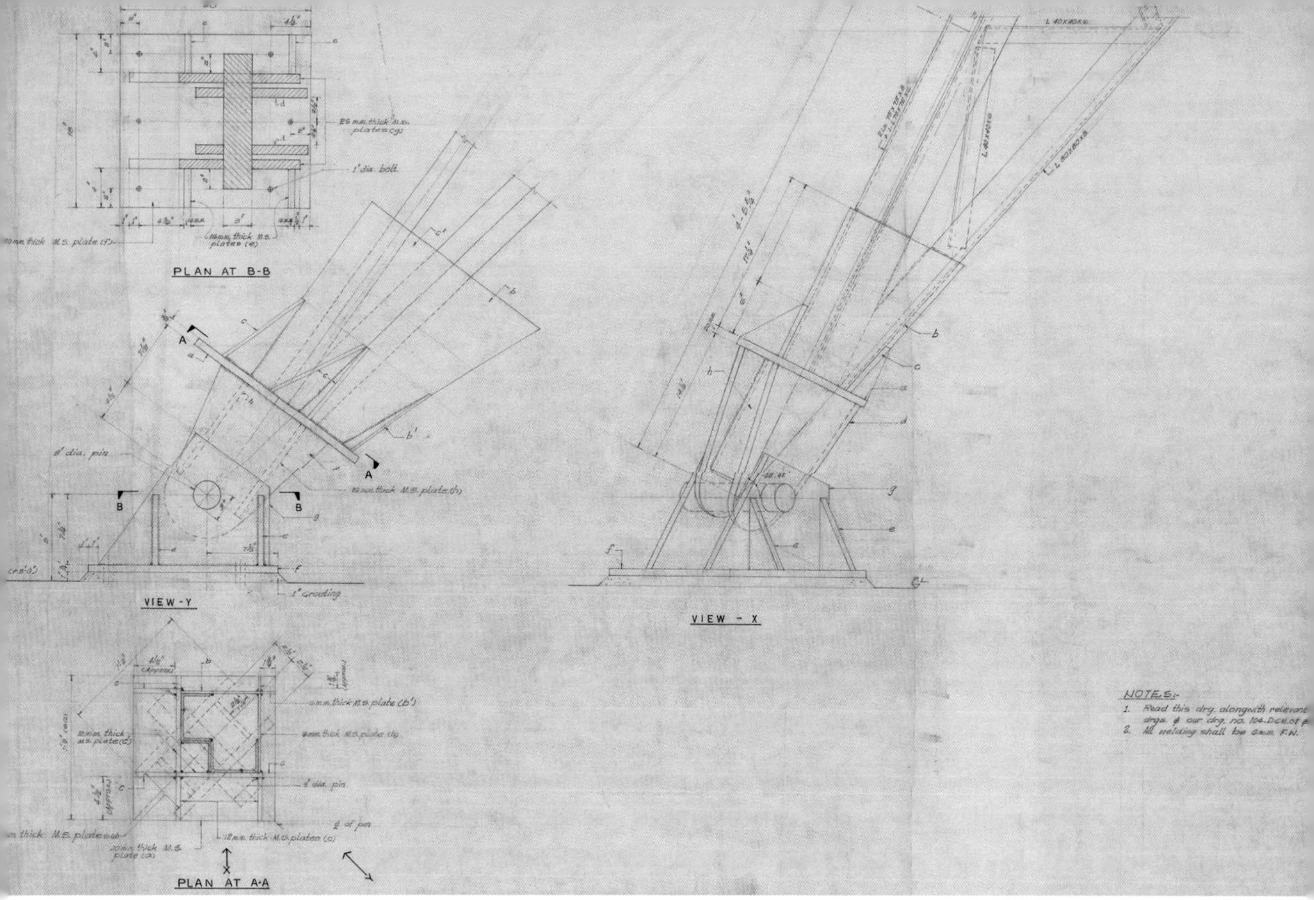
PLAN AT B-B
1" dia. bolt
6" dia. pin
VIEW-Y
1" Grouting
VIEW - X
PLAN AT A-A
NOTES:-
1. Read this drg. alongwith relevant
drgs. & our drg. no. 104.DCM.01
2. All welding shall be 6mm. F.W.

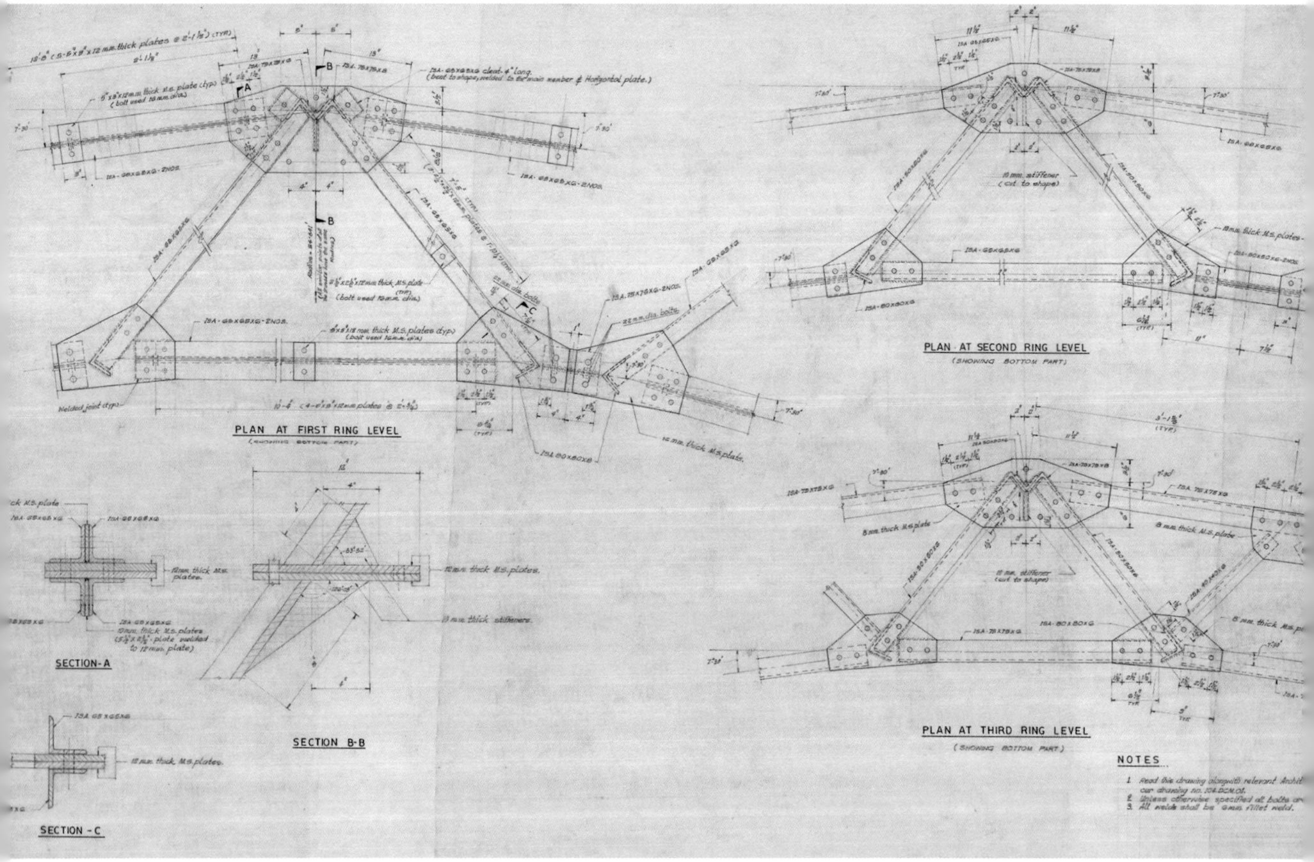
PLAN AT FIRST RING LEVEL
(SHOWING BOTTOM PART)
PLAN AT SECOND RING LEVEL
(SHOWING BOTTOM PART)
SECTION-A
SECTION B-B
PLAN AT THIRD RING LEVEL
(SHOWING BOTTOM PART)
SECTION - C
NOTES
3. All welds shall be 6mm. fillet weld.

Photos of the interior space of the exhibition pavilion lit by triangular openings created between the folded plates at the base.

Opposite, top:
(Partial) Drawing no. 104.DCM.04 A: Base Details of Truss
Drawings of the pivot joint of the folded-plate truss.

Opposite, bottom:
(Partial) Drawing no. 104.DCM.05.A: Ring Connection Details of Trusses: 25 July 1972

NEHRU MEMORIAL PAVILION

NEW DELHI | 1972

This museum, built in memory of India's first Prime Minister, Jawaharlal Nehru, is a modest building, with the bulk of its functional space underground. Its green berm-and-stepped pyramid roof structure is supported on deep frames spanning 24 metres to create a two-level museum space underneath. The controlled dimensions of the structural elements lend a delicate lightness to this building.

Four intersecting portals in reinforced concrete spanning 24m and peripheral retaining walls create the large volume of the exhibition areas.

Previous pages: Interior view of the large uninterrupted and interconnected museum floors, generously sky-lit from the corners and along the stepped sloping roof ribs.

Nehru Pavilion, hidden under a grass mound with a stepped roof structure and plaza above.

HINDON RIVER MILLS

GHAZIABAD | 1973

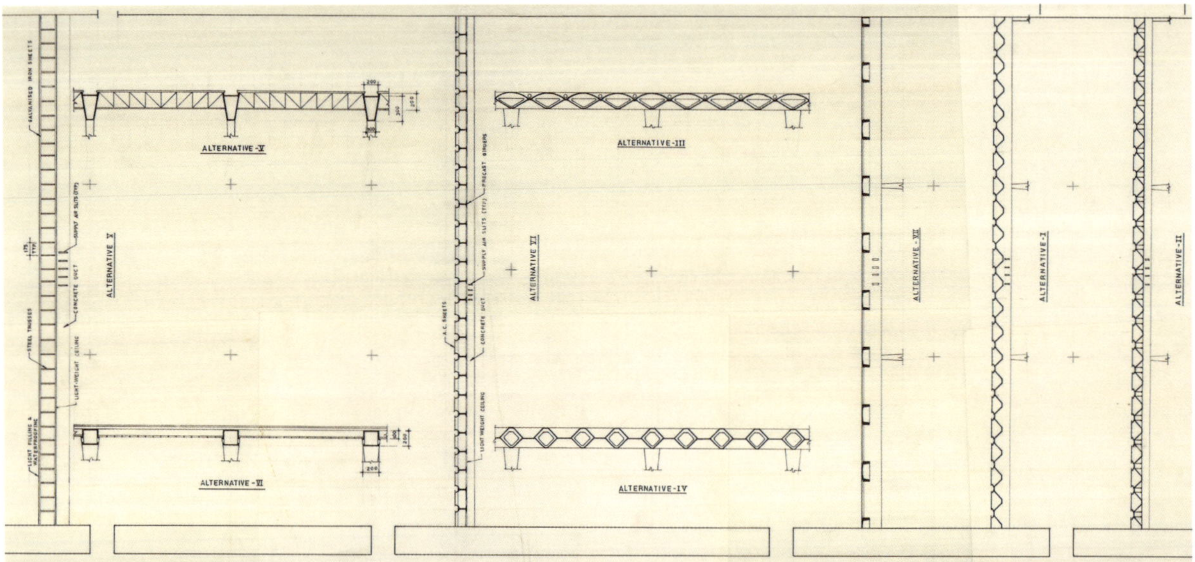

(Partial) Drawing no. 198.D.CM.01: Plan: 24 February 1970
Seven alternatives for 21m x 18m spanning systems. Alternatives I to V envisage 21m-span hollow box girders acting as air ducts and 18m-span roofing elements in different shapes. Alternatives VI and VII envisage 21m-span hollow roofing units placed on 18m-span girders.

Exterior view of the two 48m x 223m long mill sheds.

Previous pages: 9m-high twin arches rising elegantly from the roof of the textile mill building.

This unique textile mill resembles a series of elegant bridges on the horizon and, as with many of Mahendra Raj's designs, presents an aesthetic solution which coordinates a host of functional, structural and constructional requirements.

Textile mills typically require enormous quantities of humidified air in a column-free space of minimum possible enclosed volume to economise on the costs of air-handling units. Seven alternative structural schemes were made for the initial flow diagram and layout which required a 21m x 18m bay. Subsequently, the client revised and enlarged their requirement to two sheds of 49m x 223m each. At this stage Raj conceived of a totally new structure–a twin-tied arch system spanning 49m. This modification allowed a simplified layout with a larger uninterrupted area without interior columns, and reduced the overall span requirement to 48m.

The twin concrete arches span 48m and are placed at 15m centres. A cast-in-situ, hollow post-tensioned girder, 3.5m x 1.5m, carrying humidified air is suspended from the twin arches. High tensile cables are used to post-tension the box girder that also acts as a tie for the arches. Each girder is stressed by 12 cables made of 12–7mm wires,

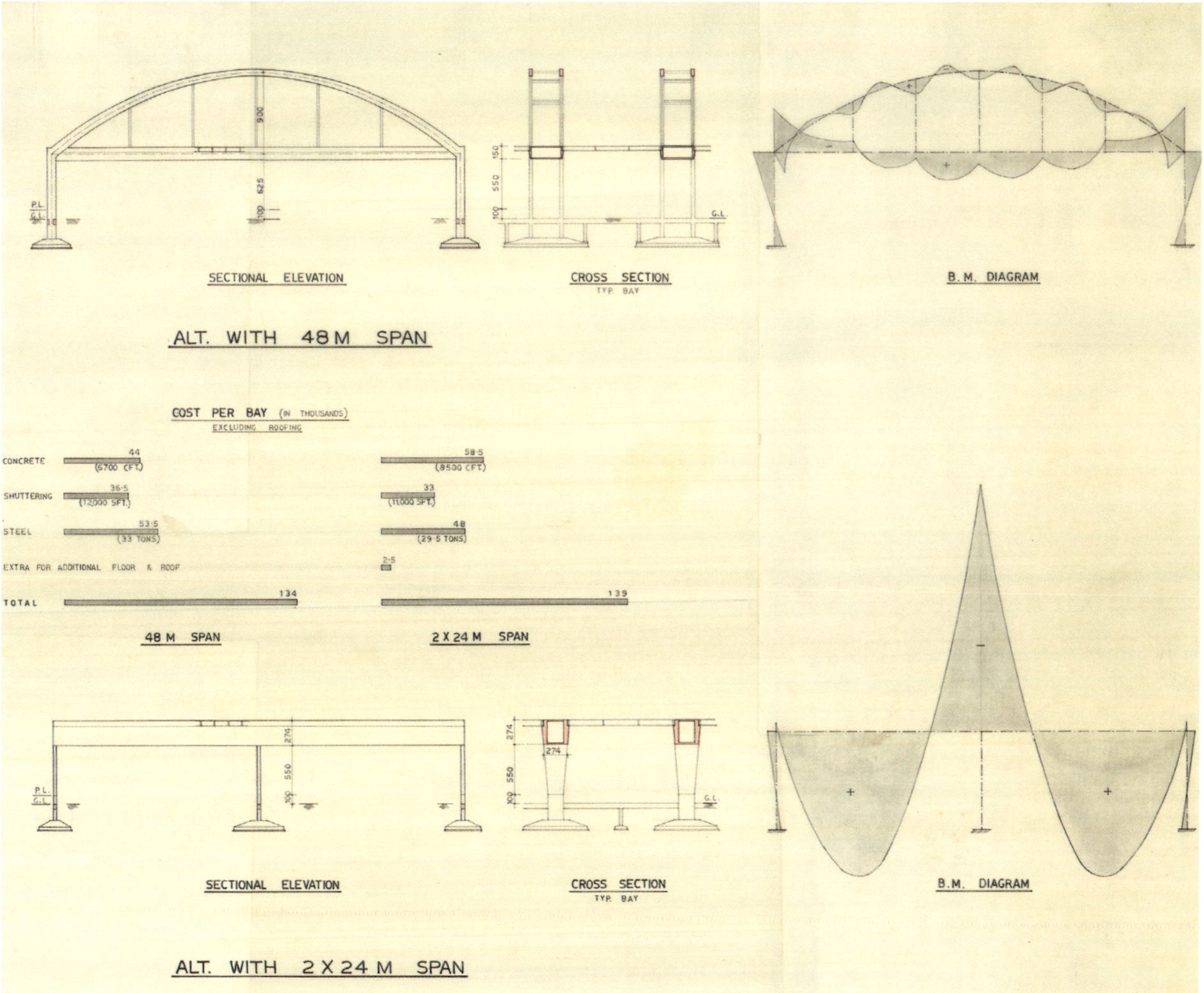

using the Freyssinet system. Pre cast double-T units of 12 tons each are supported on the hollow girders to provide roof covering.

The twin parabolic arches have a 9m rise and a cross section of 38cm x 75cm at the crown, gradually increasing to 38cm x 113cm at the springing point. The arches, 3.5m apart, are interconnected at all five hanger points. Lateral stability for the building is provided by interconnecting the twin columns supporting the two arches with a folded-plate gutter element running along the length of the building. The foundations of the structure consist of spread footings tied together with grade beams. Three different methods of analysis were used to cross-check the design since in the early 1970s the use of computer was very limited.

(Partial) Drawing no. 116-TM-P8: 48m Span Vs 2x24m Span: 29 January 1971

Comparison of the two final options–a 48m-span and a 2m x 24m span structure. The comparison is explained through plans, sections, bending moment diagrams and costs worked out per bay. The analysis shows the 48m twin arch system as a more efficient resolution of structural behaviour and more economical despite the use of higher quantity of steel and scaffolding.

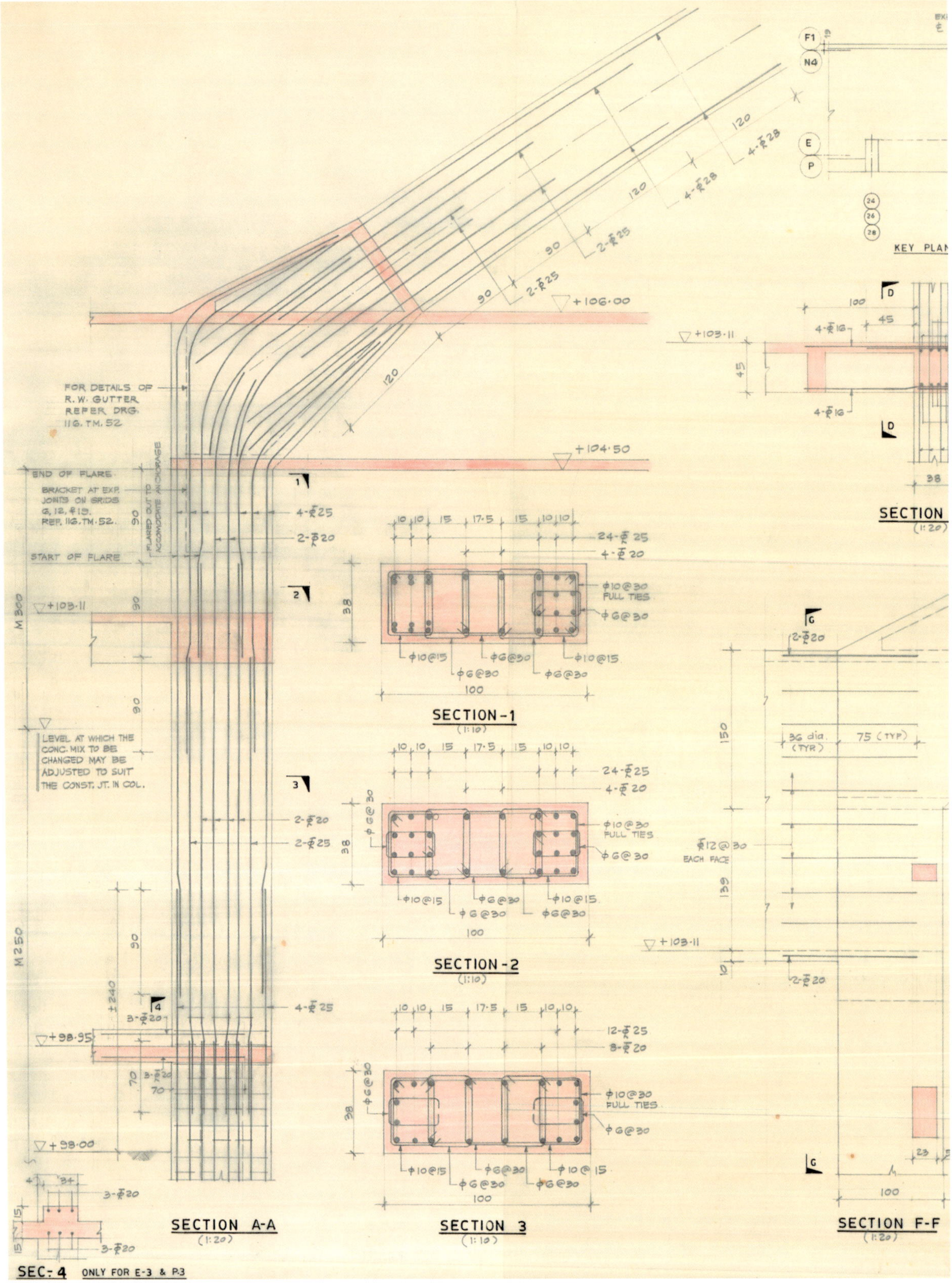

(Partial) Drawing no. 116.TM.15A: Details of Main Columns on Grid E & P: 16 April 1971

Photos of the construction sequence.
The construction sequence for this project was planned in a manner to optimise time and also re-use formwork while minimising use of construction machinery. First the columns were erected, followed by the hollow box girder. After the bottom plate of the hollow box girder was cast, the post-tensioning cables were positioned in place, before casting the vertical plates and then the top plate was cast. Next, 6 of the 12 cables were stressed to carry the self-load of the arches and girders on full span. Then, scaffolding was removed except below the five hanger points and formwork for the arches was erected and arches were cast. After removing the shuttering of the arches, the suspender bars from the arches to the girder were connected with turnbuckles. At this stage, four more cables were stressed to carry the dead load of the roofing units. Next, the pre-cast roofing units were placed and remaining two cables were stressed. Finally, the roof was waterproofed and hangers were encased in concrete.

Following pages (200-201; 202-203): View of the folded-plate gutter system that runs perpendicular to the twin arches providing lateral stability; View of the abutment of the twin arches.

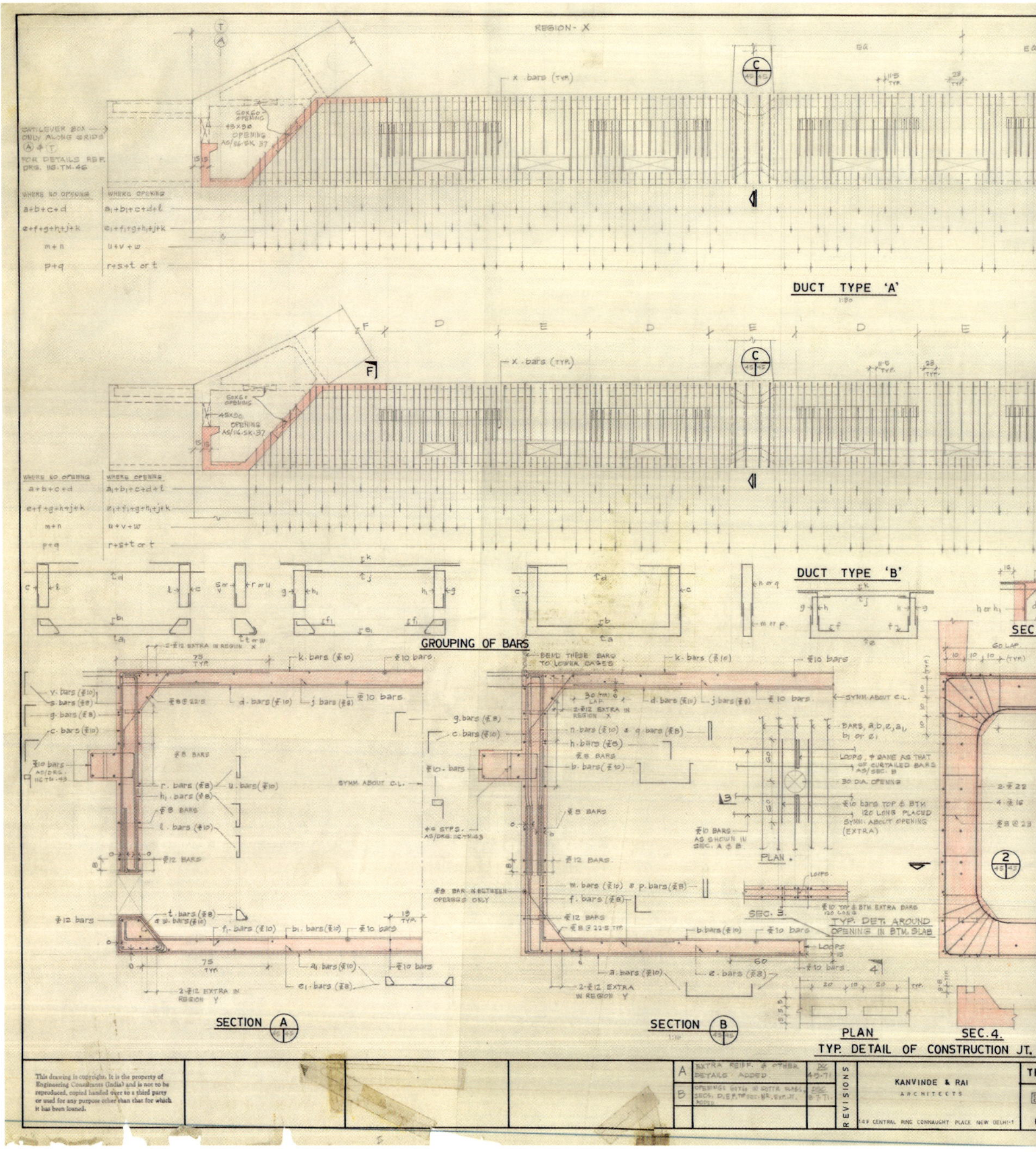
REGION- X
DUCT TYPE 'A'
DUCT TYPE 'B'
GROUPING OF BARS
SECTION A
SECTION B
SEC. 3.
TYP. DET. AROUND OPENING IN BTM. SLAB
PLAN
SEC. 4.
TYP. DETAIL OF CONSTRUCTION JT.
This drawing is copyright. It is the property of Engineering Consultants (India) and is not to be reproduced, copied handed over to a third party or used for any purpose other than that for which it has been loaned.
REVISIONS
KANVINDE & RAI
ARCHITECTS

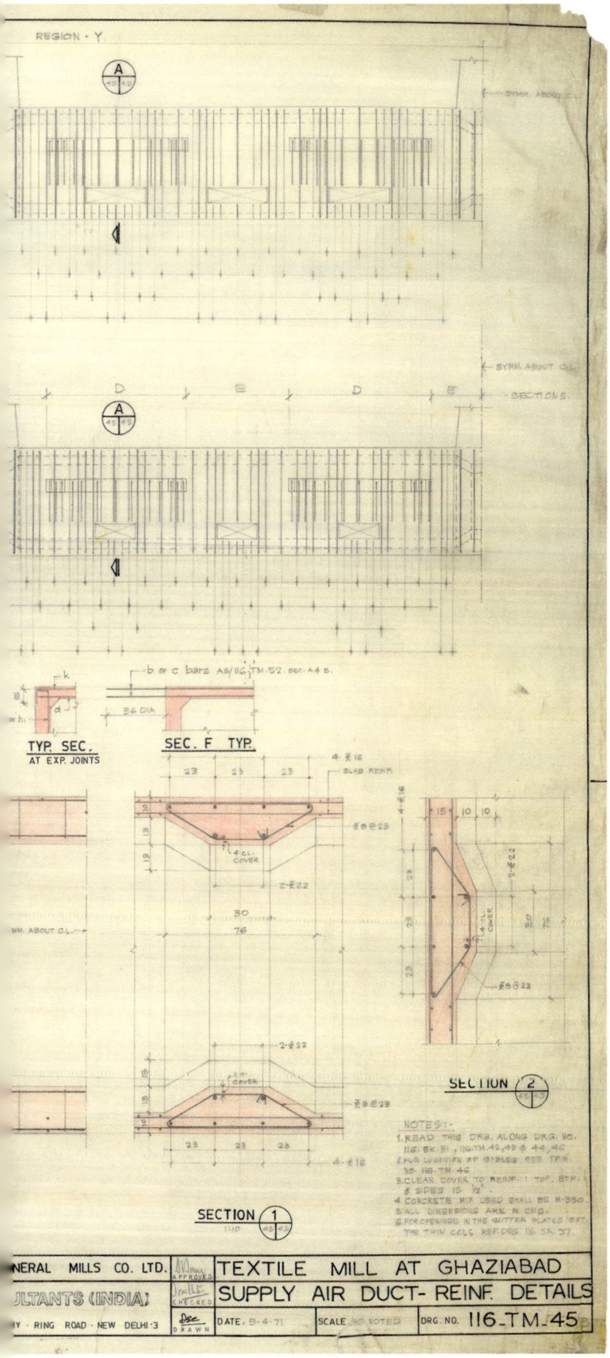

Drawing 116 TM 45 B: Supply Air Duct–Reinforcement Details: 9 April 1971

The twin arches that carry the hollow box girders rise above the roof level and help achieve this immense 223m deep and 48m wide uninterrupted factory floor (shown above), with the optimum volume to conserve the humidified air required.

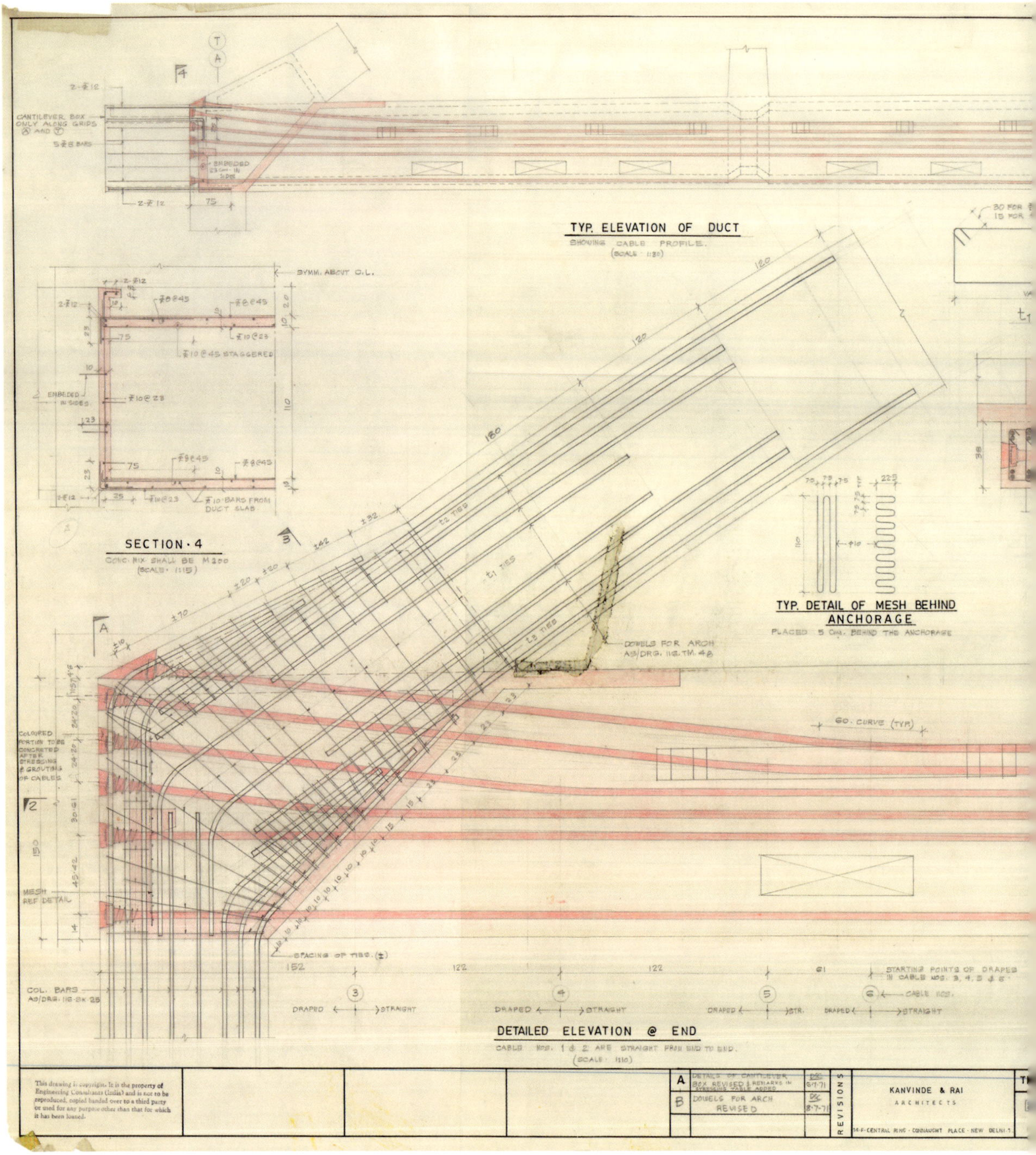

TYP. ELEVATION OF DUCT
SHOWING CABLE PROFILE.
(SCALE 1:20)
SECTION · 4
CONC. MIX SHALL BE M 200
(SCALE 1:15)
TYP. DETAIL OF MESH BEHIND ANCHORAGE
PLACED 5 CM. BEHIND THE ANCHORAGE
DETAILED ELEVATION @ END
CABLE NOS. 1 & 2 ARE STRAIGHT FROM END TO END.
(SCALE 1:10)
CANTILEVER BOX ONLY ALONG GRIDS (A) AND (T)
SYMM. ABOUT C.L.
DOWELS FOR ARCH
COLOURED PORTION TO BE CONCRETED AFTER STRESSING & GROUTING OF CABLES
MESH REF DETAIL
COL. BARS
SPACING OF TIES.
STARTING POINTS OF DRAPES IN CABLE NOS. 3, 4, 5 & 6
DRAPED
STRAIGHT
This drawing is copyright. It is the property of Engineering Consultants (India) and is not to be reproduced, copied handed over to a third party or used for any purpose other than that for which it has been loaned.
DOWELS FOR ARCH REVISED
REVISIONS
KANVINDE & RAI
ARCHITECTS

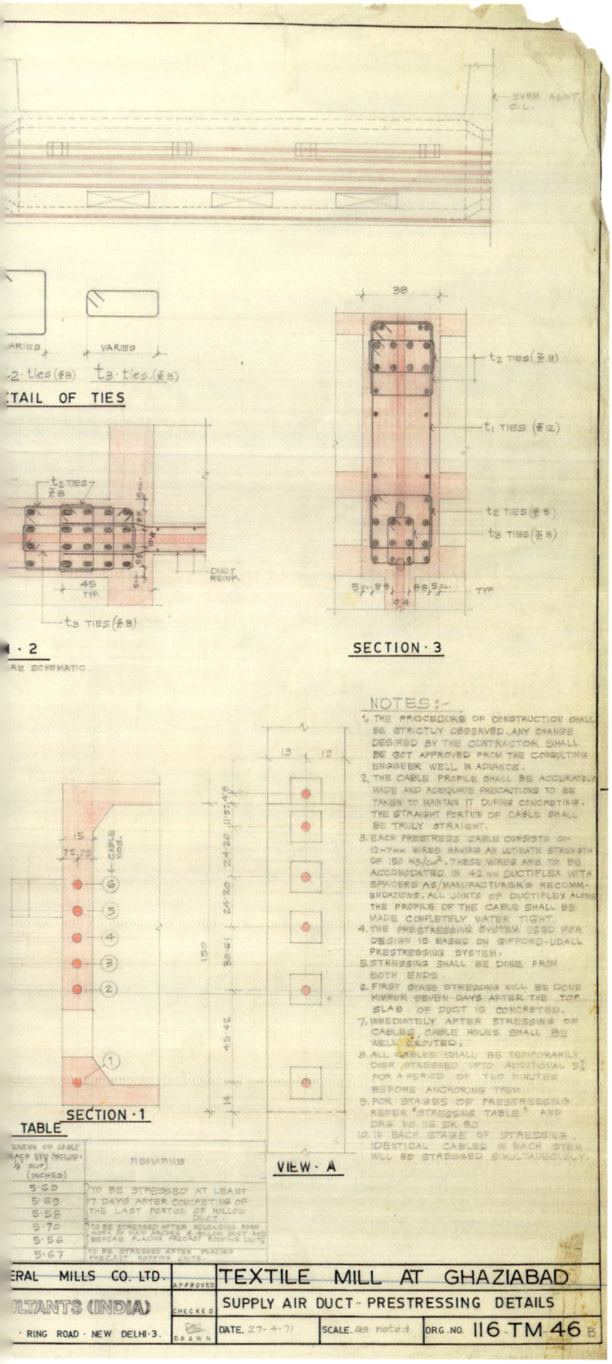

Photo of the exposed cables and reinforcement at the junction of the arch, hollow girder and column during construction.

Drawing 116 TM 46 B: Supply Air Duct–Prestressing Details: 27 April 1971, showing the post-tensioned cables and reinforcement at the junction of the twin arches, the hollow-girder and the columns.

Buckminister Fuller (second from right) visits the Hindon River Mills with Mahendra Raj (extreme left) and DCM personnel.

Landscape of the twin arches as viewed from the roof.

PREMABHAI HALL

AHMEDABAD | 1974

Situated in the old city of Ahmedabad, Premabhai Hall is a dramatic form defined by two end shear walls in exposed concrete, framing the ends of a cantilever that extends onto the street.

'When I first visited Doshi's office in 1960, I saw a very impressive model of a hall,' Mahendra Raj recalls. 'It was a big bowl supported on a few columns and the structural system was very boldly expressed. I later learned that this was the model of Premabhai Hall, a project on which Doshi had collaborated with the Japanese engineer Aoki. Years later in 1970, Doshi asked me to collaborate with him on the same Premabhai Hall since the programme had changed and needed a completely new building. While designing this hall, both Doshi and I had in mind that it should be as impressive as the earlier discarded scheme. The design of this hall emerged very organically [because] halfway through the clients introduced additional requirements. The structure and plan developed from bottom up as Doshi and I worked on it together.'

The roof of the main 55m x 37m hall with a seating capacity of 1000 is supported on four shaft columns connected with deep reinforced concrete girders.

These girders support long-span steel trusses, which replaced the concrete ribs of the original design in order to avoid huge scaffoldings. These trusses cantilever out to carry a hanging library that imparts a sense of tremendous grace and lightness to the building. The seating bowl supported on a 21m-span deep wall-like girder clears the immense space underneath of all its originally proposed columns. The result is a dramatic and volumetrically complex foyer enriched by a structure of deep girders and wafer-thin ribs.

Model of an earlier proposal made with Japanese engineer Aoki.

Above: View of the double shear wall and cantilevered library.

Previous pages: Premabhai Hall's commanding presence in exposed concrete from the street level.

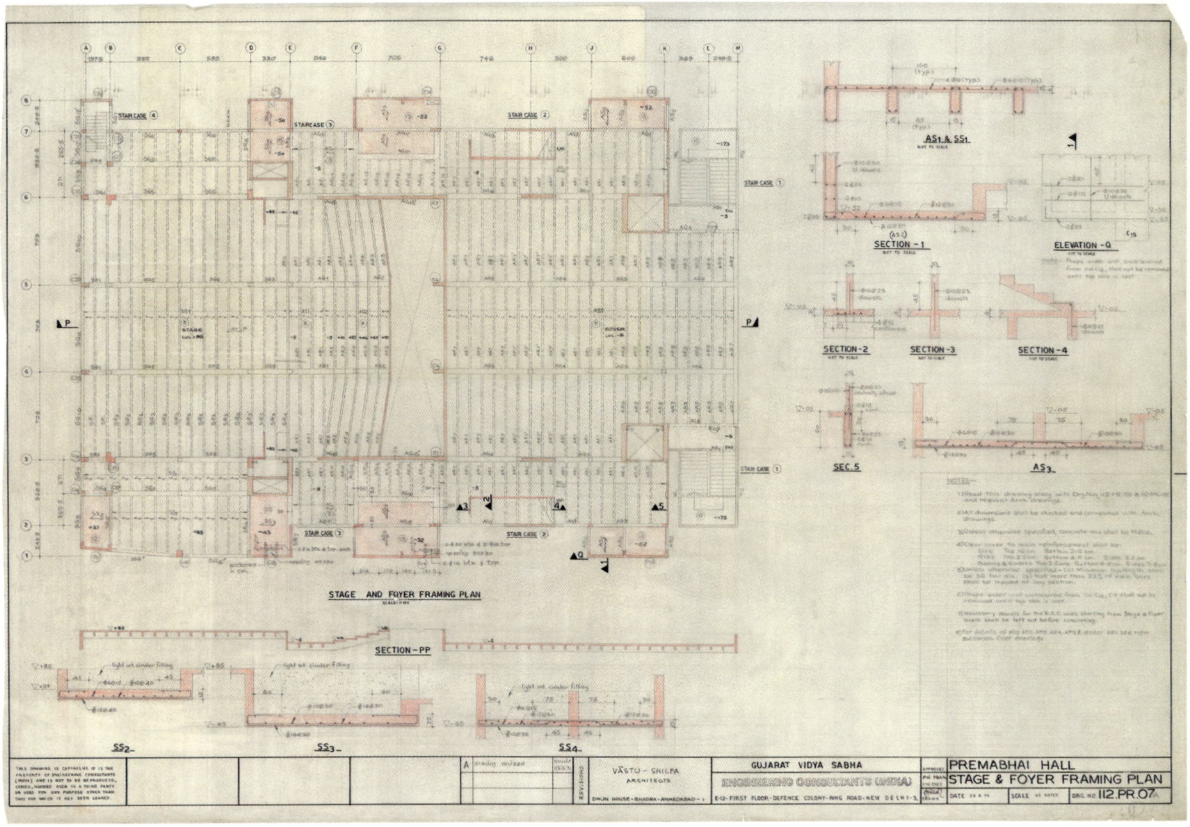

Drawing no. 112.PR.07-A: Stage and Foyer Framing Plan: 28 August 1970

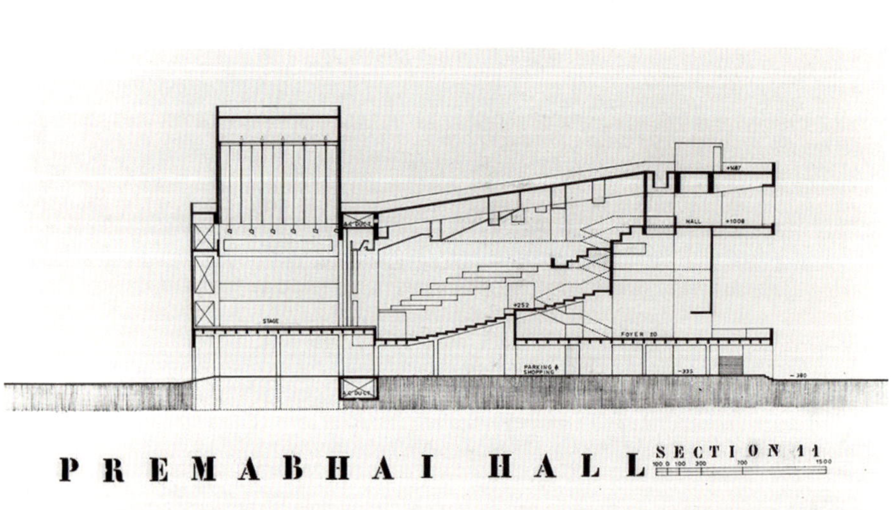

Longitudinal architectural section of the auditorium.

Opposite: Supported on deep girders, tapering ribs springing out at the entrance level free the foyer of columns.

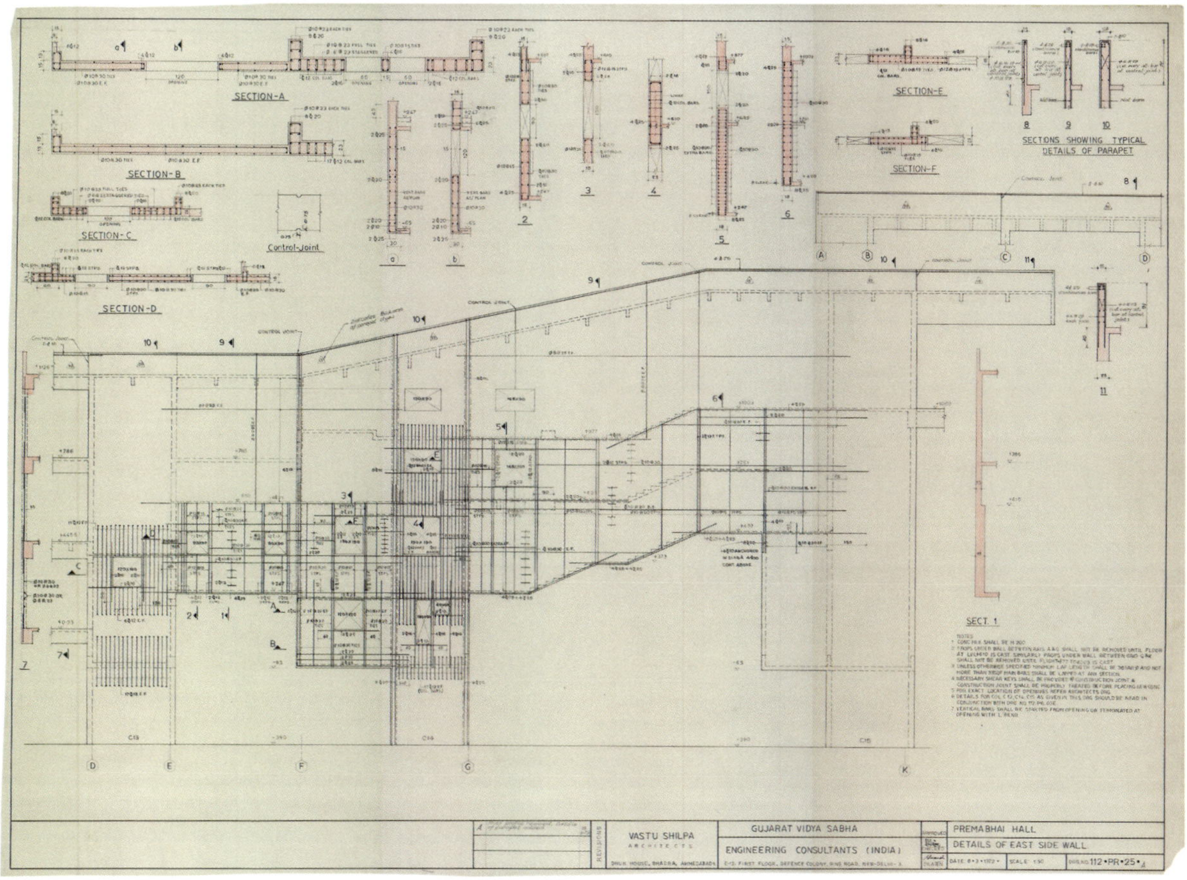

Drawing no. 112.PR.25.A: Details of East Side Wall: 8 March 1972

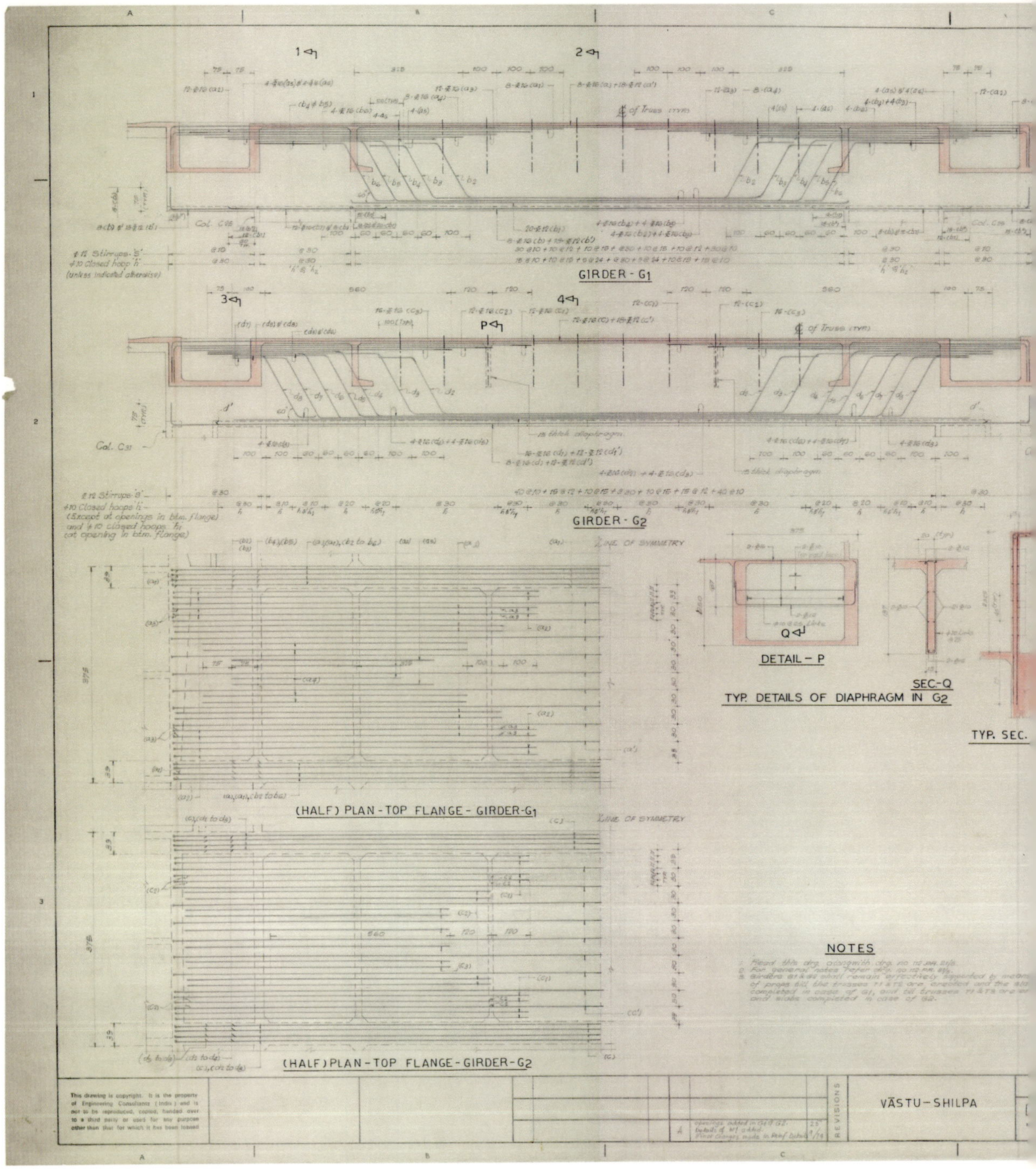
GIRDER - G1
GIRDER - G2
DETAIL - P
SEC.-Q
TYP. DETAILS OF DIAPHRAGM IN G2
TYP. SEC.
LINE OF SYMMETRY
(HALF) PLAN - TOP FLANGE - GIRDER-G1
(HALF) PLAN - TOP FLANGE - GIRDER-G2
NOTES
REVISIONS
VĀSTU-SHILPA

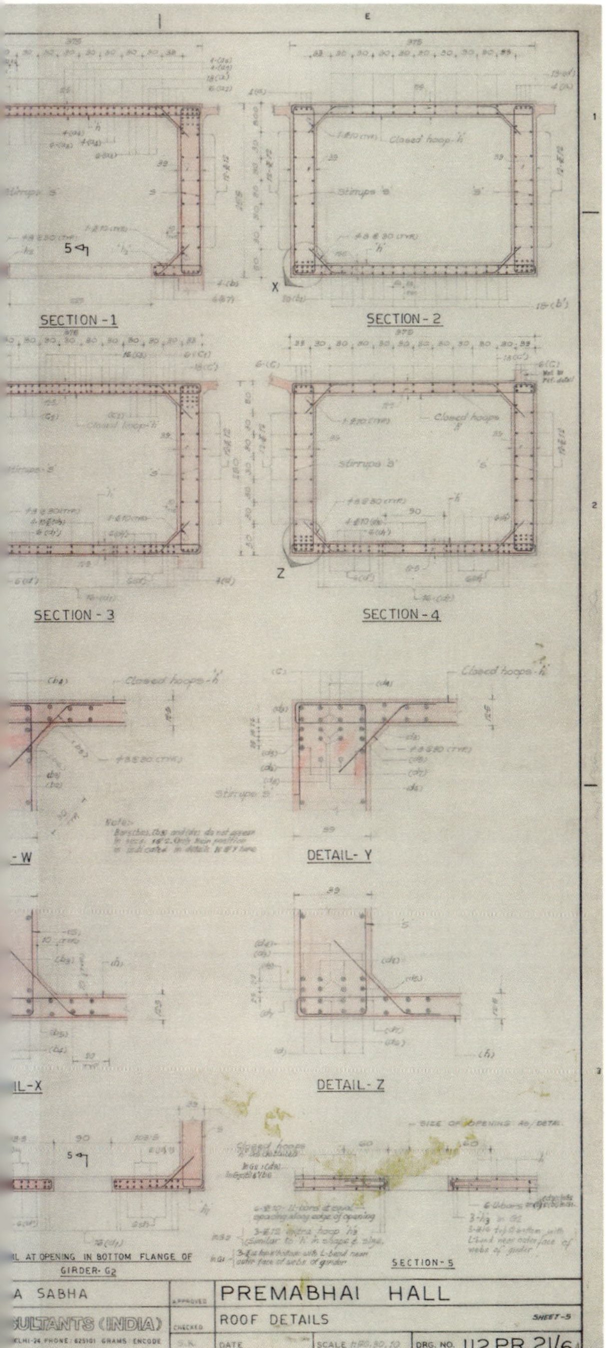

Interior view looking up from the lobby to the roof level. The large column-free space of the lobby, supported on a deep concrete wall-like girder, seen on the left side, carries the exposed ribbed floor of the seating bowl.

Drawing no. 112.PR.21/6A: Roof Details, Sheet 5

Interior view looking down at the lobby below with a deep concrete girder carrying the auditorium balcony on the left side and the end wall of the library on the right side.

Exterior view from the street.

SALVACAO CHURCH

BOMBAY | 1977

The Salvacao Church comprises three dramatic concrete shells on square bases. All the three shells, of varying thicknesses from 10-20cm, have warped surfaces that start as a square and rise to meet inclined circular openings at different heights. The larger shell has a base of 24m x 24m and the two smaller ones are set on 12m x 12m spans. The beams from which these shells spring cater to both vertical and horizontal forces generated at the base, while also functioning as gutters.

To confirm the manual (vs. computer aided) theoretical analysis and design, a model of the largest shell was test loaded at the Indian Institute of Technology, Bombay.

View from the courtyard.

Finished shell with a 24m x 24m base, soon after construction.

Previous pages: Top view of the church's three warped concrete shells.

Construction photos of the shells. Movable top shuttering was used for concreting the shells.

Inside view of the smaller shell spanning 12m x 12m.

Photo during construction with artist MF Hussain (far right) and Monica Correa (centre).

Inside view of the large central shell spanning 24m x 24m. The circular opening was painted by the artist MF Hussain.

ALL THE ENDS OF
THE EARTH HAVE
SEEN THE SALVATION
OF OUR GOD.

The Structural Poetry of Mahendra Raj

AN ENVIABLE CREATIVITY AND FINESSE MARK THE PORTFOLIO OF AN ENGINEER

Interview by Sanjay Prakash

The path of realisation of buildings from concept to completion is fraught with heartbreak. Many a conceptual dream comes down to earth–literally and otherwise–in this process. Rumour in the corridors of the building design profession has it that the structural engineer (as a consultant) is most responsible for diluting the imaginative visions of an architect. From the engineer's point of view, it is said that most historically significant works of structural engineering lie inside the domain of heavy engineering: bridges, dams, harbours, etc. Buildings seem to fall under a separate historical tradition, with their authorship being claimed principally by architects.

In Mahendra Raj, we find at least one happy example of an engineer working primarily in the field of building structures, who refutes the above-mentioned simplistic paradigm. His work, spanning four decades, has shown an uncommon inventiveness and willingness to experiment with new concepts, and he would undoubtedly be one of our most creative engineers alive.

Unlike architects and artists, engineers tend to play down their achievements and do not like to lionise peers. Despite this, Mahendra Raj's work has for years silently inspired at least two generations of engineers and, yes, architects. His portfolio of works is probably far more impressive than any of our architects, something not commonly realised.

As I went over various drawings and files and talked to him and others at his office, I got the impression that he is someone who has deeply understood the nature and behaviour of his materials–the mark of a true poet. In his intense but modest way, he ruminated over the past and speculated over the future in my presence, taking me through a fantastic voyage: into the skewed skeleton of the joints at the Hall of Nations in Delhi, through the tense sinews of the Srinagar Stadium, over the warped geometry of the Salvacao Church, past tree-like transfer girders of the New Great Insurance Building in Bombay and the formidably balanced thrusts of the NCDC building, and on to the state of development of our construction industry. What I have tried to recreate here is that voyage–with much paraphrasing, though not in the same order as I traversed it.

One gets a strong impression that the works of Mahendra Raj represent projects in which neither has the expressive intent of the architect been obliterated by the process of realisation nor has the structure become an anonymous framework to hang architectural elements from. His work seems to proclaim the seldom remembered fact that buildings are created by teams: of architects and engineers, managers, builders, workers and users–and there need be no leaders and followers in this effort for the best results. It is a humbling truth.

At what stage of a project do you get called in for engineering services? Does the stage of the project determine how effective you can be in providing these services?

Raj We are sometimes called upon to take on a project at the conceptual stage, sometimes later. If we are called in at the conceptual stage, then we have an opportunity to interact intensely with the architect and make substantial contributions to the ultimate shape and form of the building as in the Halls of Nations and Industries and the Premabhai Hall. On the other hand, our influences on the overall design may not be inconsequential even if we start out at a later stage. In the Sher-e-Kashmir Stadium and the New Great Insurance Building in Bombay, we were able to change the concept quite dramatically, even though we stepped in quite late. Sometimes the architect comes with complete drawings, but in working out the alternative structural solutions, the ultimate shape and form get substantially affected as in the Akbar Hotel and the Shri Ram Centre.

What work do you personally handle in your office?
Raj Within the organisation I look at all the projects in some detail, particularly at the concept stage, and provide guidance. Nowadays the senior engineers in the organisation largely carry the project through. I get the greatest joy in the concept and analysis stages. I like to be involved in all the tricky analyses and dealings, which I find exciting.

A fact not commonly recognised by outsiders is that all engineering is teamwork. An individual can rarely be the sole author of a structure. So it is important to keep in mind that while we are talking about my work, it is really the work of many teams all essential to that work that is under discussion.

Outside the organisation, what is the kind of interaction that you typically have with professionals on any projects?
Raj The initial meetings in any project are often sketch based and interactive, though sometimes we have only a verbal discussion. In subsequent meetings, we make some sketches and both parties develop the design. Architects understand structures quickly and also respond to suggestions quite well.

Is it because you have educated the architects who work with you, or are you just lucky with your architects that they have a good structural understanding?
Raj That's not for me to say! With every architect, though, we try to understand his concept, his expressive intentions, before suggesting changes. If the architect is responsive, these suggestions are incorporated. Sometimes we also have to respond to the requirements of services and the major aim in all cases is to work together effectively. Only when there is this desire on the part of all the professionals involved can we integrate all the elements properly together. The structure then will be such that it is acceptable to the architect, and he doesn't just want to cover it up. The services are neatly arranged, the users are comfortable, and so on.

Choice of Materials

You are generally considered a master in the art of shaping concrete but not known much for using steel. Is this indicative of some aversion on your part, or is it just a circumstance?
Raj Although I usually know that concrete is the material to choose for non-industrial buildings due to heavy loading in factories, we definitely investigate both steel and concrete, or their combinations–the time and ease of construction becomes far more important there. In the Hindon River Mills, the mandate was for concrete since high humidity was to be maintained.

One reason why we do not use steel more often is because of its market structure: the availability of sections is poor. Also, until recently, steel required heavy fireproofing, before vermiculite came up. Even so, at the Indira Gandhi International Airport in Delhi, we tested the viability of steel because the spans were to be large (with typical bays of 16m x 19m and up to 35m spans); height was available and we had to cut down construction time. The entire airport building is finally in RC and the theme is box girders. But it was studied in four modes: RC, steel, composite (RC and steel) and post-tensioned concrete. We were personally eager to use pre-cast concrete. Steel was not viable as the loads were too large.

As I mentioned earlier, we have gone in for steel work in various projects, mainly industrial, as in the

Gabriel factory at Parwanoo, the DCM foundry at Ropar, the roof structure of Sher-e-Kashmir, Eicher Mitsubishi, DCM Toyota and the DCM Pavilion. The last named is an interesting small structure conceived in the form of a cone which is pretty high, with a large diameter. The construction technique was such that the sections were made separately, and when the first one was done, the ring was put in place and the temporary supports removed. From there on, the next section was erected as a set of cantilevers, another ring, and so on. We did face construction problems because a certain level of precision was involved, and errors came up–the circle was not completing. The solutions had to be reworked from the first principles, and adjustments made, providing for an even distribution of error.

Another project where we planned the use of steel (in combination with concrete) was the prize-winning project for the Indraprastha Indoor Stadium, which used the structural concept of the bicycle wheel, familiar to architects and engineers. While the wheel is normally put on columns or walls, here we put it on another frame with a ring beam tying the entire structure. The forces given by this element integrate the whole configuration and it behaves like a different type of structure.

But in this case were you not proposing HT cables, and not steel per se?

Raj No, this structure was not planned to be built using cable technology since that cannot be easily put up in India. We planned it in steel because that technology is readily available here. We have wanted to use cables, but have not really had the opportunity of doing so as much as we would like to.

Exploring Alternatives

You mentioned the study of alternatives. There is a feeling that engineers don't really explore many alternatives for any structure…

Raj In most types of structures, it becomes necessary to do a number of studies. I am sure many other structural engineers do similar exercises, so we are not unique in that respect, and I am not aware of the feeling you mention. The number of studies needed depends upon the importance and the size of the project. Sometimes the process of trying out one or another system is a little painful, even more so because other professionals have to go along with you. But I think of it as an essential, not a waste of office time. There is no point in rushing into a decision and suffering later, so we try, by and large, to consider a number of systems. Looking back, I think that most of our projects have worked well because we have explored all alternatives before selecting the optimum solution in terms of strength and economy.

The Aesthetics of Engineering Design

Another feeling is that engineers somehow find it more difficult than architects to visualise non-conventional forms, and thus tend to simplify a concept till it no longer retains any aesthetic or symbolic value.

Raj Lack of communication between the architect and the engineer may result in what you are saying. But I don't think engineers, as a group, are poorer at visualisation as compared to architects at all! I can only describe it as an individual problem, not of the profession as a whole.

But engineering in India hasn't given us any unconventional solutions…

Raj Unconventional structures are definitely designed by engineers in India! The past 10–15 years have given a lot of opportunities to many good engineers. Engineering capabilities develop with opportunities. Today, a lot of engineering firms have been exposed to complicated structures, tall structures, nuclear plants, offshore rigs and the like. About 15 years ago, this was not the case.

To bounce off a contrary feeling amongst at least some engineers, it is felt that engineers can find more satisfaction in working on 'functional' structures, such as dams or bridges, where an architect is not present to dictate the aesthetic terms.

Raj I don't agree with you here. I wouldn't say that

designing such structures is any more or less satisfying. Every profession has a role to play, and, in different degrees, that role can be played without being a hindrance.

Yet many purely engineering forms can be quite pleasing–such as hyperbolic paraboloid tanks. But this has nothing to do with the presence or absence of an architect or any other professional. The elegance of any structure is related to correct proportioning, which is also the basis of strength requirements. Something very heavy jars; something very slender also gives one an uneasy feeling.

And foundations?
Raj Foundations are designed solely on the basis of functional requirements, and all of them are quite challenging, though unfortunately the effort never shows. But I wouldn't spend my time in seriously discussing the aesthetics of foundation design!

In this context, would you give me a brief appraisal of some of the engineering structures of the last few decades that have also been known for their unusual form, for example, the Sydney Opera House, or the TWA Terminal by Saarinen?
Raj In the Sydney Opera House the proportioning was well conceived, but that alone is not enough. A sector of a dome by itself is only a sector of a dome. If many of these sectors were to be placed together, then it is to be seen whether the resulting structures would still behave as a shell. In Sydney, it is probable that when the architect was developing the roof shapes, he thought he was developing a structural form, but the analysis required a great effort to make it physically possible.

The TWA Terminal by Eero Saarinen was planned to be a shell, but at that time, computer analysis was not available. Manual analysis being limited, it turned out to be a pretty heavy structure, but it was analysed, designed and built the way it was because Saarinen insisted upon that very form (though he did make some changes upon the recommendation of the engineers). Subsequently, similar concepts have been tried, which, with minor modification to the shapes and forms, come much closer to true shell behaviour, and the development of analytical methods by a computer has also made this possible. But generally, a shell-like structure is difficult to realise.

On Computers

Yet another opinion is that engineering can be reduced to computation or algorithms and the work in engineering firms will eventually be taken over by computing machines.
Raj Engineering is not an exact science. As a structural engineer, you try to idealise the structure, understand its behaviour and then make some assumptions on your own. You can never be sure that the properties will, in fact, follow these assumptions, and you are not sure of the loads, the placement of materials and the strength; yet you are required to design a building and put reinforcement in it–very accurately. Analysis and design have a large number of grey and dark areas, but you can't say that you won't design something or the other for this reason; you have to decide upon some strength and materials on the basis of assumptions and experience.

Computers are required for structures like the Srinagar Stadium, but cannot be used at the conceptual stages. We first try to realise the behaviour and evolve some approximate techniques before final computer analysis can begin. We have to first fix the size, depths, and so on, by manual computations. And even after computer analysis, we have to read the output critically and see to it that the forces and stresses computed were compatible with the behaviour as envisaged. If we are not able to understand the computations, there are same conceptual problems. Sometimes it becomes necessary to have multiple runs when certain unanticipated behaviour of the structure becomes evident. On the basis of the behaviour, you modify your concept, which sometimes happens after computer analysis.

Conversely, does the use of computers promote a kind of laxity amongst engineers?
Raj Electronic computation was not available at the beginning of my career. Even now, computational

facilities in India are scarce, and the computer is used essentially as a tool for those analyses which are manually difficult. So, the younger set of engineers here has not become conceptually lax. But abroad, with computers so popular, and programmes so available, this laxity is appearing and it's not very desirable at all. Sometimes I feel that the young engineer who starts off his career depending on computer analysis, designs and drawings, will not turn out to be a very good engineer. One has to develop on one's own, a capability of understanding or trying to predict the behaviour of a structural form, and analyse it manually to an extent where one is confident that the final analysis will be close to the initial one.

Early Years

You started your career when our country had just got its independence from colonial rule. Can you tell me about those days in India, and your work in the USA?

Raj In the Fifties, except for Bombay and Calcutta, there were few places with constructions of more than three storeys. The Chandigarh High Court was a very big challenge for us then. I was working as an assistant engineer on this project and the calculations were done by me under the guidance of my executive engineer, a very capable man, the late Mr Gulzar Singh. We had to think deeply about every element, with nobody to guide us. We had some interaction with Corbusier who always insisted on getting exactly what he wanted. If we said that something was structurally difficult, he would say that in that case he would get it done in France–so we had to do it! A couple of times he listened to our suggestions, but his initial reaction was always that what he'd done was final.

I said earlier that engineering capabilities develop with opportunities. My own interest in engineering was stimulated because of exposure and opportunities, first in Chandigarh, and then in America in the late Fifties. Otherwise, I might have remained in the PWD. Anyway, in the US, I mainly designed special, large-span structures, like hangars and stadia, which force you to attack them with rigour because they are so complex. It was in the US that I had the opportunity to first appreciate the importance of working in a team. I value the open discussions we used to have amongst engineers, and strive to inculcate that feeling amongst our younger engineers even today.

After coming back in 1959, I had various alternatives facing me, including my government job, various offers in the US, and starting on my own. I stayed back and started my practice from Bombay, even though at that time I still held my Green Card. It was a new age in India and not many of us opted to remain abroad.

You have worked with Corbusier and Kahn, amongst the architectural giants of our century. But, talking of an engineering giant, even though you haven't worked with him, am I correct in seeing a strong influence of Nervi, especially in your early work?

Raj Every engineer must be inspired by Nervi! The likes of him are too rare for me to even try to compare my work with his!

But let's take some specific examples–the columns at the Tagore Memorial Hall, or the roof of the Talkatora Stadium?

Raj The Tagore Memorial Hall columns are sculptural. But then it was Doshi, the architect, who made the final selection of the form after I had provided some alternatives. Again, in the Indoor Stadium at Talkatora, the only particular inspiration from Nervi may be that the supports are triangulated. The project was constructed by a government department that redesigned the roof dome, and the corrugated roof suggested earlier was changed to a simple surface.

Technology in India

Would you say that we have achieved in India a satisfactory level of technological development with respect to structural engineering?

Raj We have as much engineering capability as anywhere else in the world. To some extent, the use of computer techniques has to be developed here. Modelling has to be improved. It is after modelling that you have to analyse. In all this, we are not that far behind Western technology. Broadly speaking, what can be idealised or analysed there can be analysed here. We are also rapidly developing our plant engineering design skills now and I feel we are on the correct path in this area.

Where we are behind is in our construction techniques. The construction industry is beset with a lack of professionalism and has lagged behind. Actually, the construction industry has two parts: the heavy construction industry and the general building industry. In the case of heavy construction, advancement is taking place out of necessity. Contractors have been adopting the latest methods and techniques because they often have to directly compete with foreign companies. Our recent development has been out of necessity in the areas of sports structures, harbour structures and other large civil engineering structures.

However, in the building construction industry, we are still using age-old techniques. We still mix concrete at site, carry concrete in *taslas* or buckets, vibrate by hand or use *ballis* for scaffolding and use methods of measurement which are extremely dubious, for example, while measuring ingredients for a concrete mix. Of course, there has been availability of cheap labour and lack of availability of machinery. There has also not been a necessity to change. We have shown a great ingenuity for deploying human labour to do any task, whether suitable to human effort or not. The major tall buildings of Bombay, which came up about two decades ago, were all built without any cranes and with very few hoists–we had human ladders and human hoists!

And what of the future?

Raj Personally, I feel there will be major changes in the next twenty years. For a start, we are producing better equipment indigenously such as cranes, hoists, loaders and excavators. Then, labour is not so cheap anymore and skilled labour is scarce. Also, builders are being exposed much more today to new techniques, and the earlier generation of contractors, who largely proceeded on a subcontracting basis with very low engineering know-how, very little planning, no shop drawings, no bar-bending schedules, are now giving way to a professional breed. Contractors will now have to apply themselves to better management, by designing proper scaffolding, formwork, hiring professionals, etc. Now, we will never reach that state of mechanisation that eliminates labour content. But we do have to develop tools to save time, achieve greater human (and machine) productivity, obtain greater safety, and involve lesser effort from the better utilisation of repetitive formwork, systems building management, reliable concrete mixing, placing, curing, and vibrating methods, and so on.

Research

Do you have to rely on a huge volume of research information?

Raj In our organisation, we collect and use a large quantity of data, updated over the years. We also develop and use indices such as concreting or formwork per unit area. We ourselves have not been writing research papers or articles although we would like to, and we could write about so many of our projects which would make interesting papers, but there is no time. We have our own extensive library, and frequently refer to papers and documents published by others.

What should India's priorities for research be?

Raj Exploration of low-cost materials and replacement of scarce materials come first. The Structural Engineering Research Centre (SERC) has been a bit more into theory, and it has essentially been the task of the Central Building Research Institute (CBRI) to do the applied research. Although CBRI has been around for quite some time, there has not been as much of this as one might wish. One of their main contributions was pozzolanic cement that found application during the days of the cement shortage. They have also done some work on fly-ash, hollow block work, some pre-cast battens and joists. The National Building Organisation (NBO) is also making an effort, but so far their work has found application only in EWS (economically weaker sections) and LIG (lower income group) situations, and most practising engineers are unaware of it.

They also give grants for adopting experimental techniques, such as the rubble block masonry. Building research institutes need to do a lot more solid work. They don't interact effectively with practitioners. In the next 20 years, we shall probably see the emergence of private R&D establishments also.

What importance would you give stabilised soil as a construction material of the future?
Raj After the Partition, they had used this technique for housing for refugees in various areas of Punjab, due to a shortage of cement, coal, etc. The main reasons for its failure seem to be that the level of precision and technology required is very high, and the fineness of control required to obtain *pucca* houses from only clay, silt and sand is not always possible. The interest in the field has been there since the early Forties. Did you know that before the Partition there was a regular department with an attached research laboratory on soil stabilisation in Punjab–the chief aim being to apply it to road work? But, to date, we don't have a clay research institute in the country, and we have been asking the NBO to initiate one.

And what about diametrically opposite, large-panel prefabrication systems of the future?
Raj Attempts in large-panel prefabrication and factory-based construction systems have been made in India, mainly by the Hindustan Housing Factory (HHF), but they have not been successful. I have my doubts whether these attempts will eventually succeed. The centralisation of facilities and the large outlay of capital required need assured continuity of work, and, in addition, the system has to be competitive. Even though the HHF had a captive market since the Fifties, it was not able to compete. I do wish some types of systems building designs would become popular, but I don't think it needs to be large-panel pre-fabrication. Even in the West, a whole lot of systems have been used, but they have been changing and adapting themselves to better systems like steam-curing large dismantlable formwork. Planned systems, in general, are good for India, and buildings should be shaped and moulded to accept these systems.

Then what are the likely solutions to our massive urban housing problems?
Raj I do not think we will ever be able to resolve our urbanisation problems with any simple answers. We can't really go in for high capital outlay plans, which, in principle, require unlimited resources. That approach may deliver very fast, but will be a drain on resources. We want something fast and cheap. We must have our own approaches. Some elements of planning and systems building, some new materials–a mix is necessary. Some good examples of technologies in recent years have been sheet steel door frames, rice-husk boards, dry gypsum walls (they will have to come in one day), cellular concrete and hollow blocks. The problem of shelter is basic and acute, but heavy industrialisation is not the answer.

Building Codes

Considering your office is known for a pioneering attitude towards design, how often do you find that the IS codes are a constraint?
Raj We have to follow Indian Standard codes; that's the law of the land. But where there aren't any codes, we use our discretion. We have to decide in the face of uncertainty and take risks all the time. You don't need codes for some minor methods or extremely simple designs. Codes do not really specify how to set up your analysis of the overall configuration–they only guide element design.

And how well do the new limit state design based codes function in practice?
Raj By and large, the new codes based on limit state design give similar section sizes as the earlier codes. I started work in 1946 and codes have become more complicated since then. I am often asked what I think of the new codes and I say that I have designed many buildings before these codes, and if you were to check them today, some parts may be considered unsafe, but then they are still standing, so there is no reason to pull them down! The dialogue between the practical engineer and code-maker is important, but basically code-making is a very time-consuming task best suited

to professors, the IIT, SERC, etc. Off and on I have also been involved in code-making, but I avoid doing so.

Competitions

Most large architectural competitions do not have a structural engineer as one of the jurors. The assumption seems to be that the structure of an otherwise good scheme can always be made workable, and, if workable, economical.

Raj I think it is very unfair that in architectural competition juries, the structural design is not normally evaluated. I have very strong opinions on this. Competitions are organised in a certain fashion because the Council or other associations of architecture who have framed the rules exclude engineers, which is absolutely wrong. Juries *must* have a strong representation of engineers. The architects cannot go into the structural or cost aspects in a very detailed way, but when you are evaluating housing in an EWS scheme or a structurally sophisticated building, assessing without engineering support at the jury level is incorrect. Engineering representation must come not at the advisory level, but from the composition of the jury. At least one member must be a strong engineer.

The Future

What are the kinds of projects you'd like to do or like to do more of? And how would you evaluate the contribution of your organisation to the field?

Raj As mentioned earlier, we have been wanting–or should I say, waiting?–to use cables but not really had the opportunity of doing so, so far. Again, although we have worked on some tall buildings such as Usha Kiran and the NDMC (civic centre) we'd like to do more of these. But in India, we do not truly have a need for buildings taller than 40–43 storeys (the Bombay Stock Exchange being about the tallest that we need to go), and possibly we don't have the construction technology.

What our organisation has contributed is not really for me to say, except to mention that a project becomes interesting to us if we succeed in making a structural statement, and in contributing something, somewhere, dramatic or otherwise.

View of the NCDC office building, New Delhi.

Originally published in A+D, *November-December, 1986*

1980s

NCDC OFFICE

NEW DELHI | 1980

Set along an important arterial road in Delhi, the National Cooperative Development Corporation (NCDC) office building is a fine example of exposed reinforced cement concrete work not only in the overall structure but also in all of its elements, details and finishes. The building is formed by two inclined wings of nine-storeys each, which mirror each other along a central core and atrium. What distinguishes this structure are the thin end walls, zigzagging and tapering from ground up.

Says Mahendra Raj, 'Kuldip [Singh] showed me his idea of the meandering walls for a competition entry and I told him that it would be doable. The jury liked the design but questioned him on the feasibility of the structure. He took my name, so I had to go and convince the jury members that it was possible to construct the building. After he got the job, we worked out the structural system.'

Interior view of the open office space and central corridor.

Previous pages: Looking up at the exposed concrete office building's meandering form, the slender edge shear wall and receding floors spanned by one-way ribs.

The structural resolution of this original idea with shifting load transfers was resolved by the very slim shear walls at the outer edges of the building and two bays of parallel zigzagging columns in the centre. These columns carry the box girders that tie the

two inclined office wings across the building and also form the central connecting corridor. With a column-free floor of 11m x 25m, this unusual office uses one-way ribs to span between the zigzagging shear walls and columns. The form created by these columns generates high tensile and compressive forces on alternate floors, making the connecting corridor either a tie or a strut. Post-tensioned cables running through the box girders of the corridor contain the tensile forces.

Every part of the building is interconnected and hence is resolved at various stages of construction before it all ties together. 'If we had ignored these interdependencies, this structure would not have been possible,' says Raj.

The post-tensioned cables were stressed in stages as the building went up. The number of post-tensioned cables required also reduced with height and is expressed on the building facade. The end shear walls are connected at the top for earthquake requirements, making it a portal. Another interesting feature of the building is its staircase with steps that cantilever from a circular core wall.

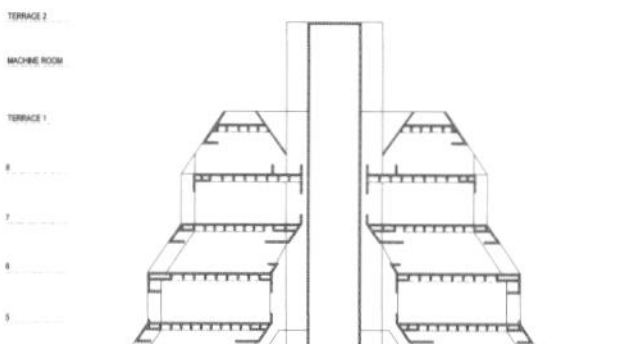

A section of the structure.

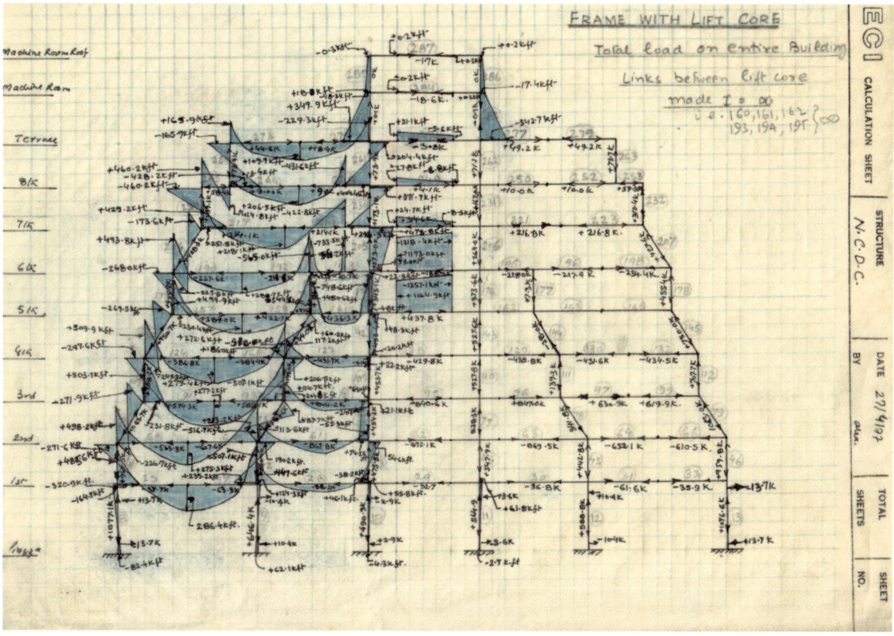

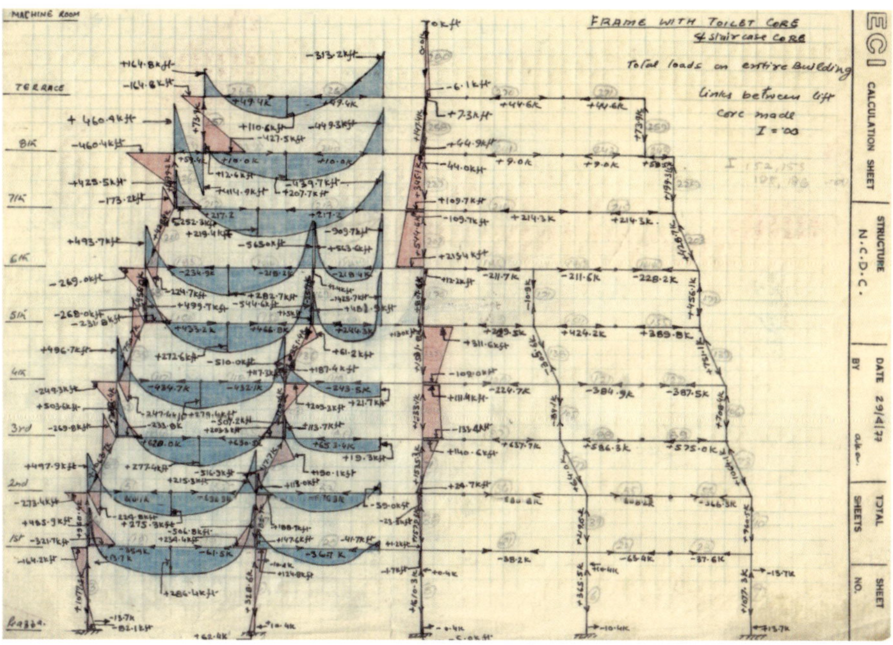

Sample calculation sheets analysing moments and the total loads on the frame at the lift and toilet cores.

Exterior photo during the final stages of construction showing the two inclined office blocks with central service cores supported on meandering shear walls on the edges.

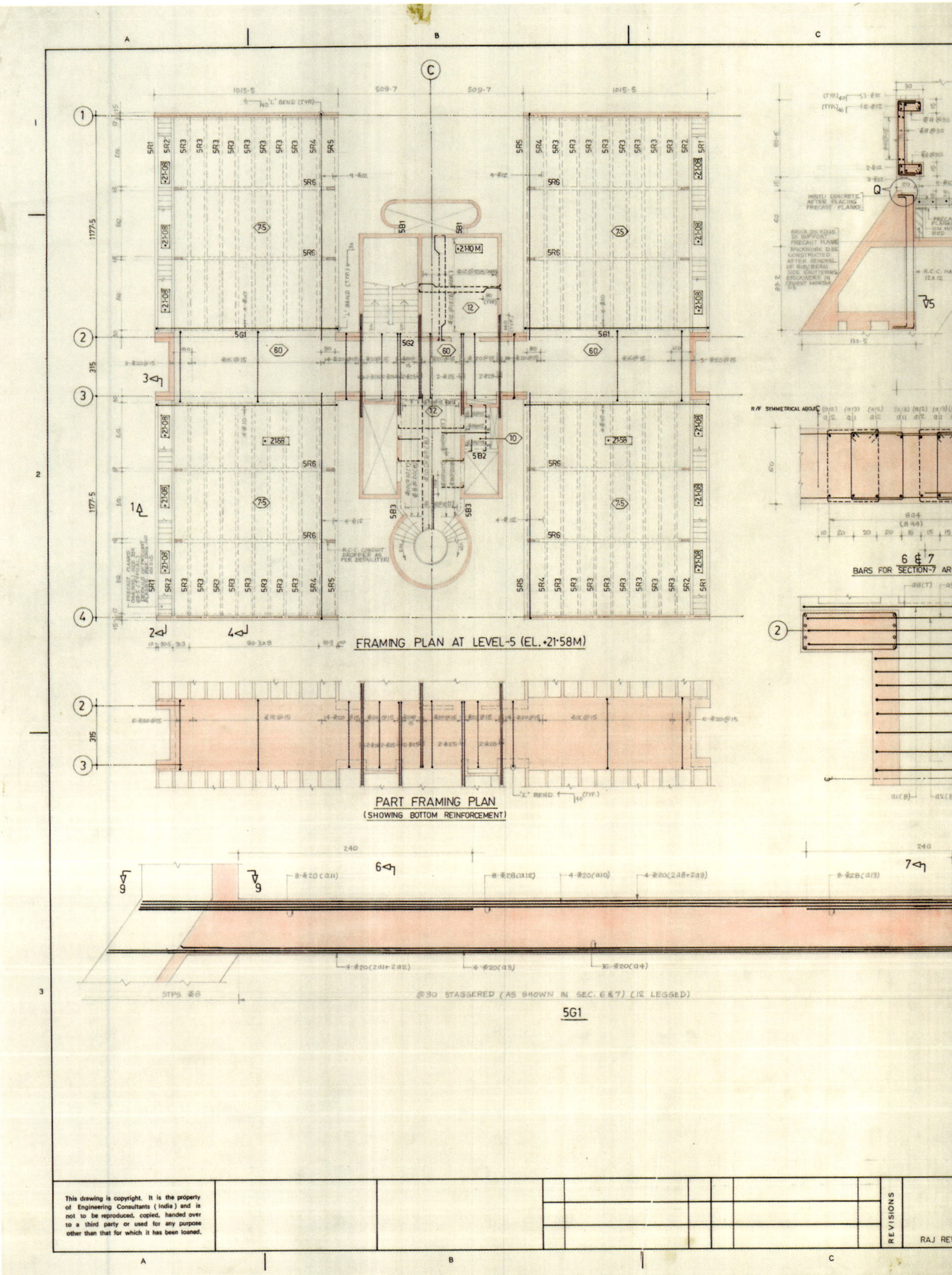
FRAMING PLAN AT LEVEL-5 (EL. +21·58M)
PART FRAMING PLAN
(SHOWING BOTTOM REINFORCEMENT)
5G1
BARS FOR SECTION-7 ARE
This drawing is copyright. It is the property of Engineering Consultants (India) and is not to be reproduced, copied, handed over to a third party or used for any purpose other than that for which it has been loaned.
REVISIONS

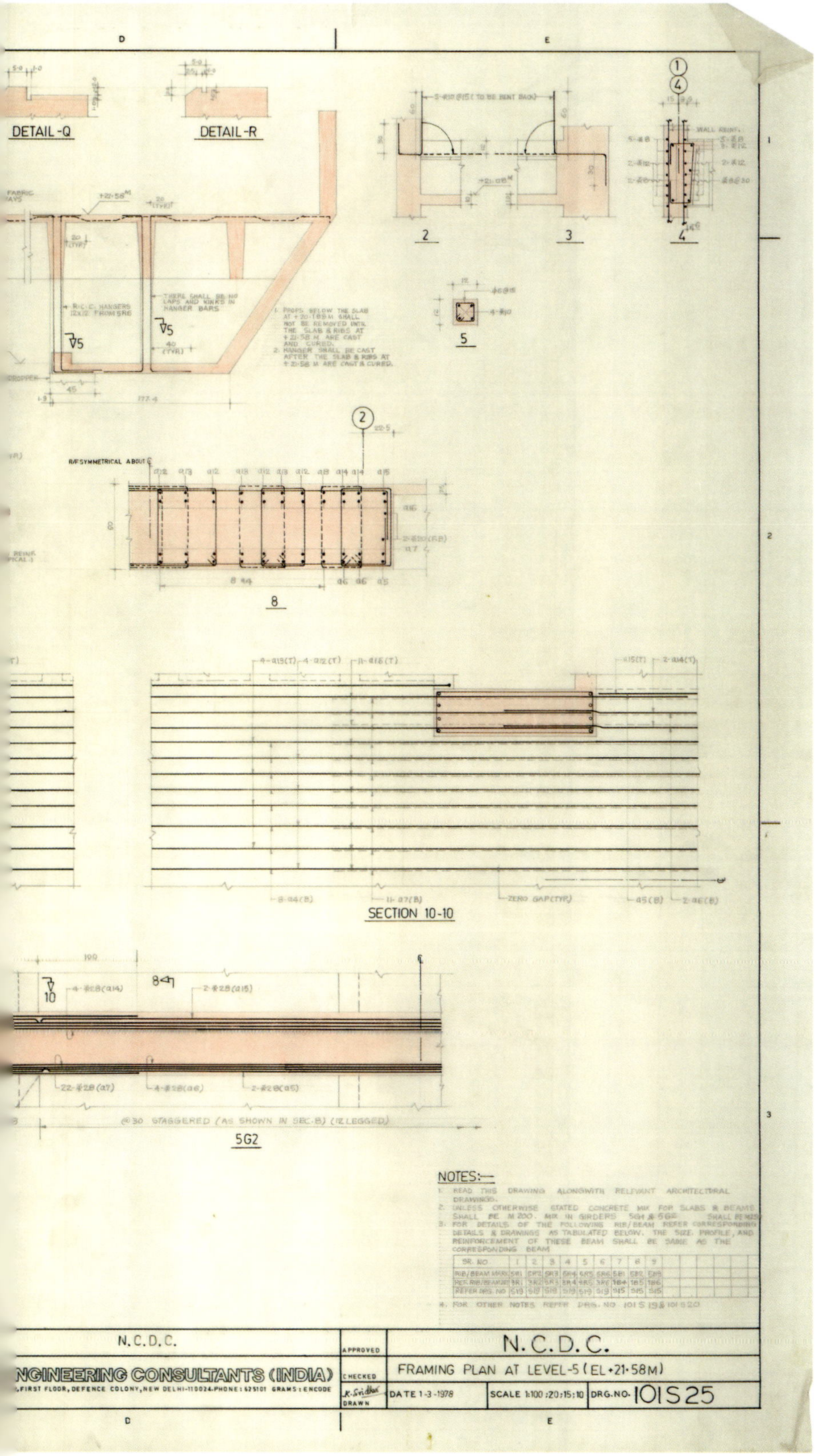

Drawing no. 101S25: Framing Plan at Level 5 (El +21.58m): 1 March 1978
Framing plan of level 5 with sections at the edges of the building and reinforcement details of the central corridor box girder in compression.

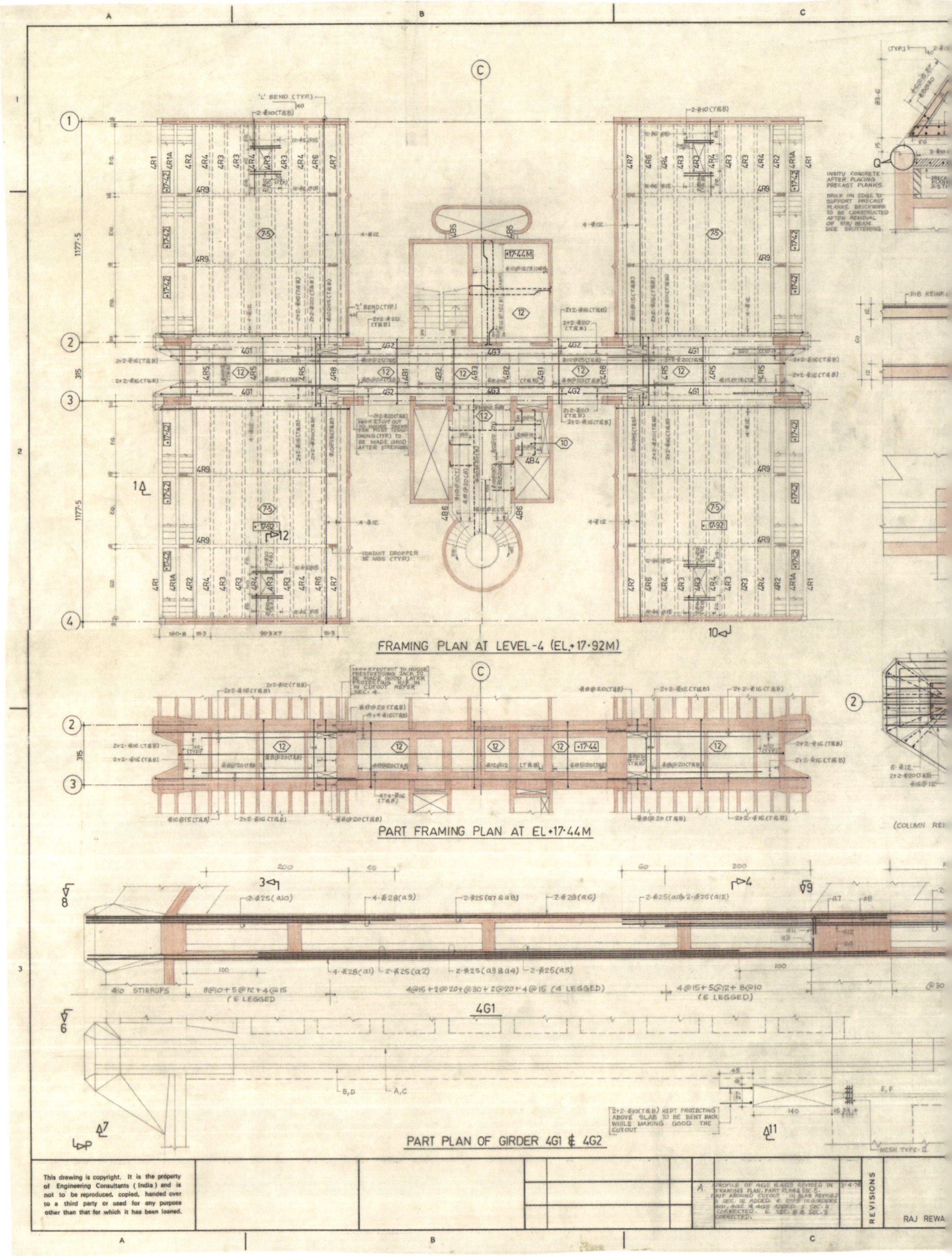
FRAMING PLAN AT LEVEL-4 (EL.+17·92M)
PART FRAMING PLAN AT EL+17·44M
PART PLAN OF GIRDER 4G1 & 4G2
4G1
INSITU CONCRETE AFTER PLACING PRECAST PLANKS.
BRICK ON EDGE TO SUPPORT PRECAST PLANKS. BRICKWORK TO BE CONSTRUCTED AFTER REMOVAL OF RIB/ BEAM SIDE SHUTTERING.
CONDUIT DROPPER 16 NOS (TYP)
'L' BEND (TYP)
4ϕ10 STIRRUPS
(6 LEGGED)
(4 LEGGED)
(COLUMN REI
RIB REINF
MESH TYPE-II
REVISIONS
RAJ REWA
This drawing is copyright. It is the property of Engineering Consultants (India) and is not to be reproduced, copied, handed over to a third party or used for any purpose other than that for which it has been loaned.

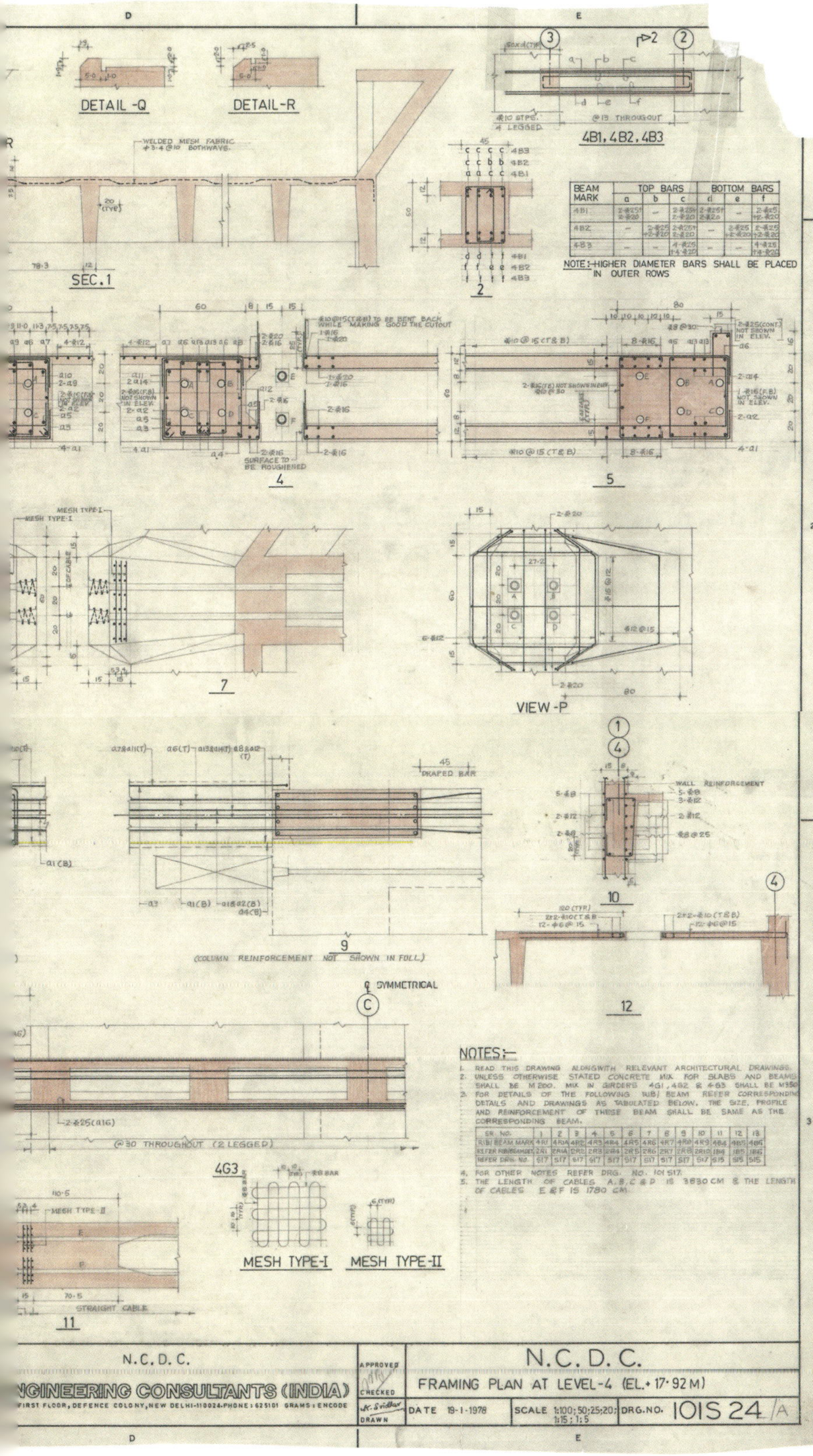

Drawing no. 101S 24/A: Framing Plan at Level 4 (El +17.92m): 19 January 1978

Framing plan of level 4, with sections at the edges of the building, reinforcement details of the central corridor hollow box girder in tension, and the post-tensioning cable details.

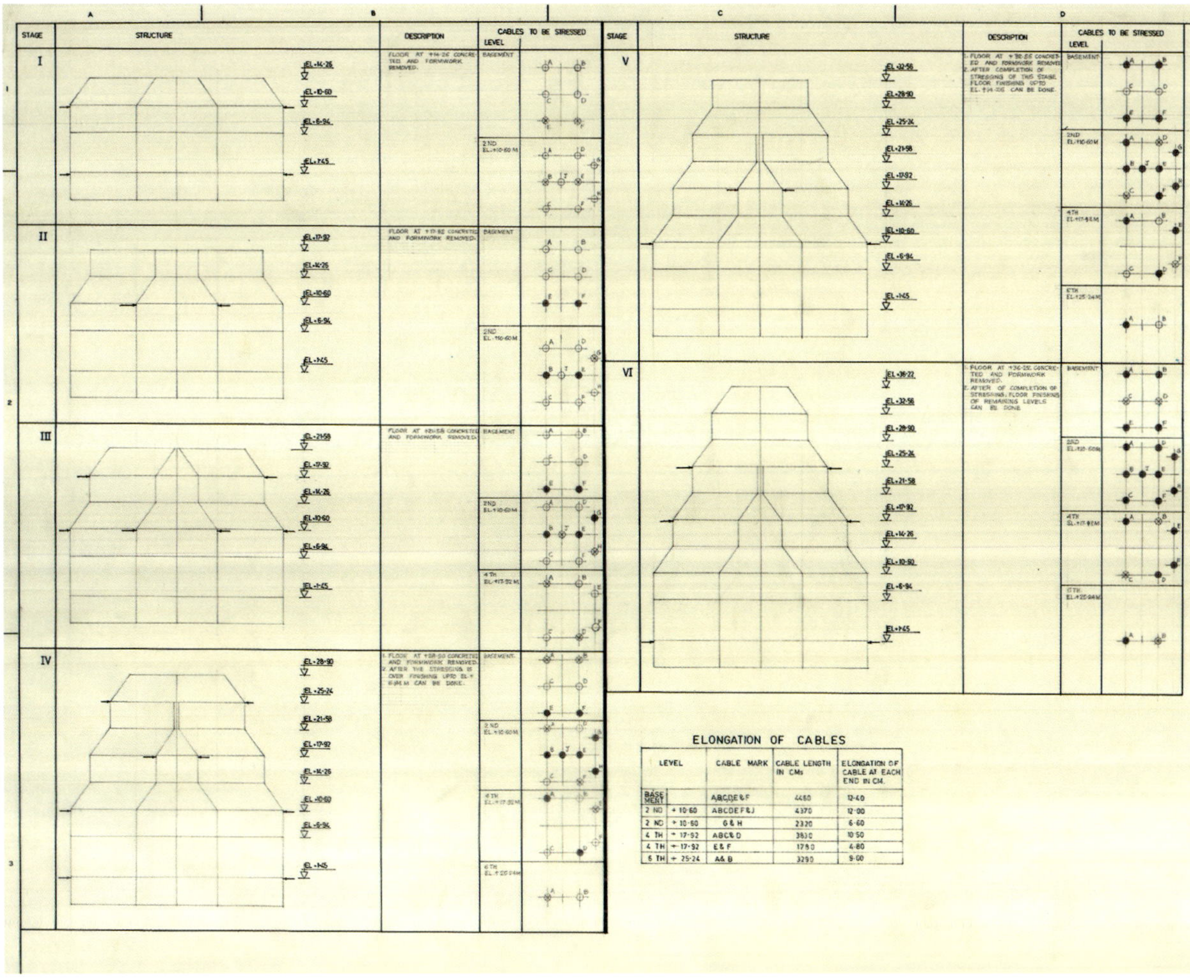

Drawing No. 101S31: Sequence of Stressing of Cables: 15 June 1978
Drawing detailing the sequence of stressing of cables and its relationship to concreting and formwork removal at different levels. This integrated and precise approach to structural design and the construction of buildings defines the finesse and economy of Raj's structures.

View of the ground floor lobby.

Photo of the exposed post-tensioned cable ends expressed on alternate floors of the facade. Notice the number of post-tensioned cables reduce as the building rises.

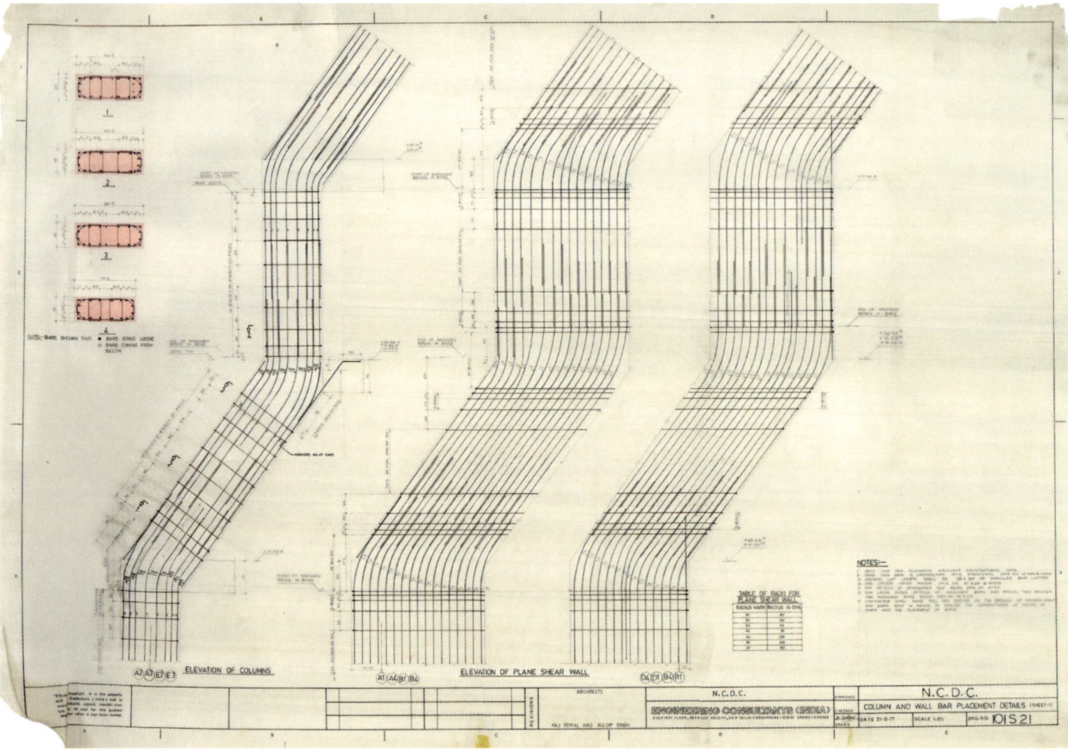

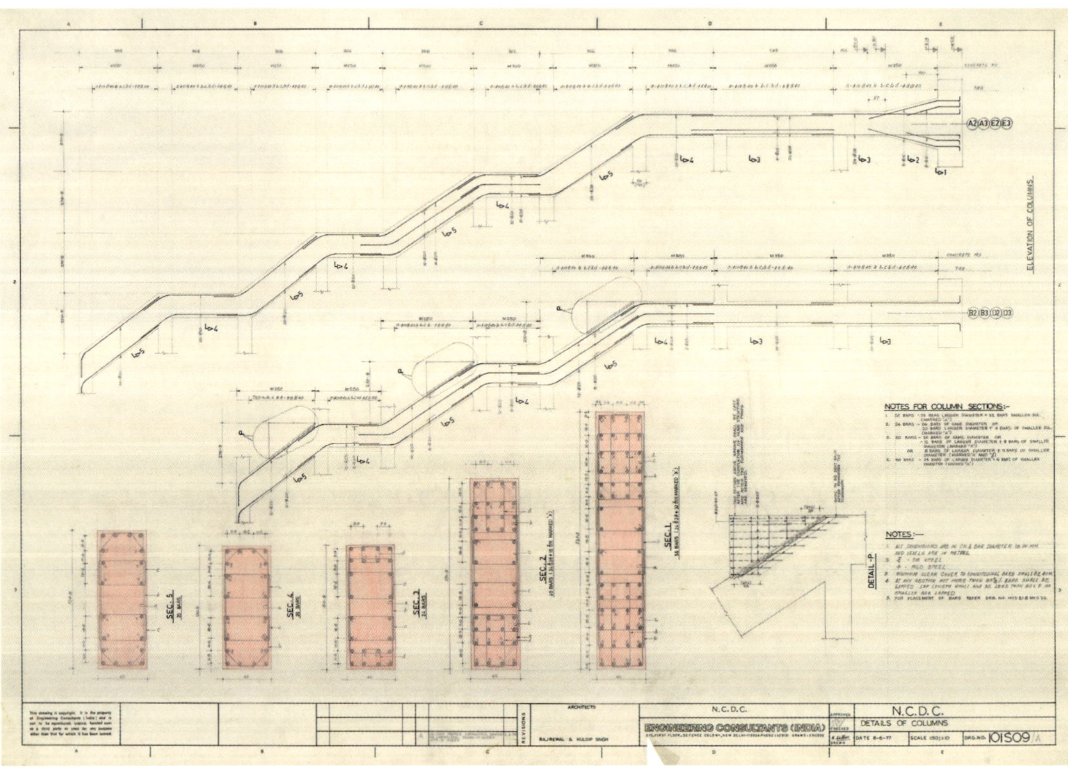

Drawing no. 101S 09/A: Detail of Columns: 8 June 1977

Above: Drawing no. 101S 21: Column and Wall Bar Placement Details (Sheet 1): 21 December 1977

View of the central atrium looking at the corridors, the zigzagging columns and the service core.

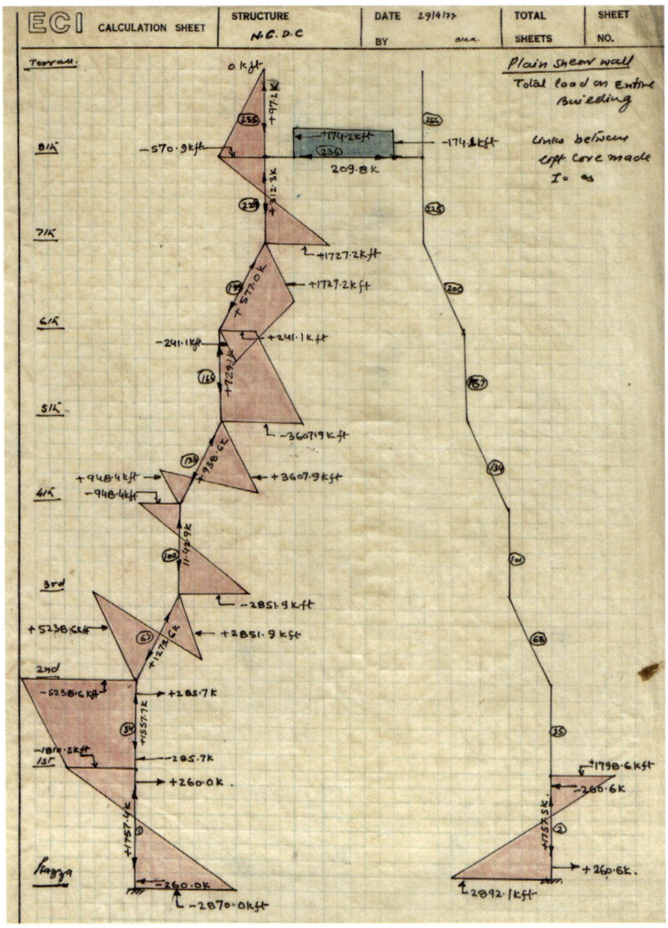

A sample calculation sheet showing moments and loads on the shear wall.

View of the meandering shear walls connecting at the top and the central corridor that ties the two wings and service core together.

INDOOR SPORTS STADIUM

SRINAGAR | 1982

With a maximum cross-length of 74m, the Indoor Sports Stadium, with a seating capacity of 5000, was originally designed by Kanvinde & Rai as a cross plan with central pylons to help span the roofs. Mahendra Raj suggested eliminating the pylons and instead proposed a structural system that radically altered the aesthetic quality of the building, while still adhering to the original intent and basic cross plan: 'It made some sense, at least architecturally, and was better than any other box-like shape conceived for seismic factors. Our effort was to evolve a structure which would create the entire enclosure, while completely transforming it volumetrically,' explains Raj.

Reminiscent at first glance of the Hall of Nations, the stadium is based on an entirely different structural concept. Both have triangular grids, but where the Hall uses a double-layered concrete space frame cast-in-situ, the stadium uses single-layered triangulated plates, which taper outwards from the cross-shaped base to form a continuous ridge on which rests the seating system. Although not visibly part of the enclosure structure, the seating integrates the stadium's triangular geometrical complexity with ease. The space between the two lower inclines of the seating and lower enclosure plates is used as a foyer.

A reverse taper going up from the central ridge closes the structure and supports the roof, also made of triangular steel plates. Twenty-eight plates, inclined in different directions, come together, balancing all forces. Since each incline is at a different angle, each edge is unique and complex. Raj can remember how meticulously he and his engineer, Krishna Hegde, had to work on these complex edges and junctions, drawing them manually one by one. He adds, 'In such a structure every joint/junction is very important as it is systematically collecting and transferring forces to the ground.'

Post-tensioning was an important tool to contain high tensile forces in some of the members. Four diagonal members forming a dramatic incline hold the post-tensioned cables running through the inclined plates. The roof is supported by a post-tensioned edge member that is also supported by the diagonals. The pile foundation is also an interconnected triangulated system containing heavy, staged post-tensioning. The high seismic requirements and strong horizontal forces of the structure necessitated such foundations.

View of the almost completed structure: the roof and enclosure follow the same structural systems albeit in different materials to create an integrated large-span structure that required no expansion joints. Seen here, the three different triangulated concrete plates 1, 2 and 3, and steel plates 4 and 5, their integration through the post-tensioned tie members and the numerous lugget joints in each plate. See drawing on next page.

Previous pages: Exterior view of the Indoor Sports Stadium.

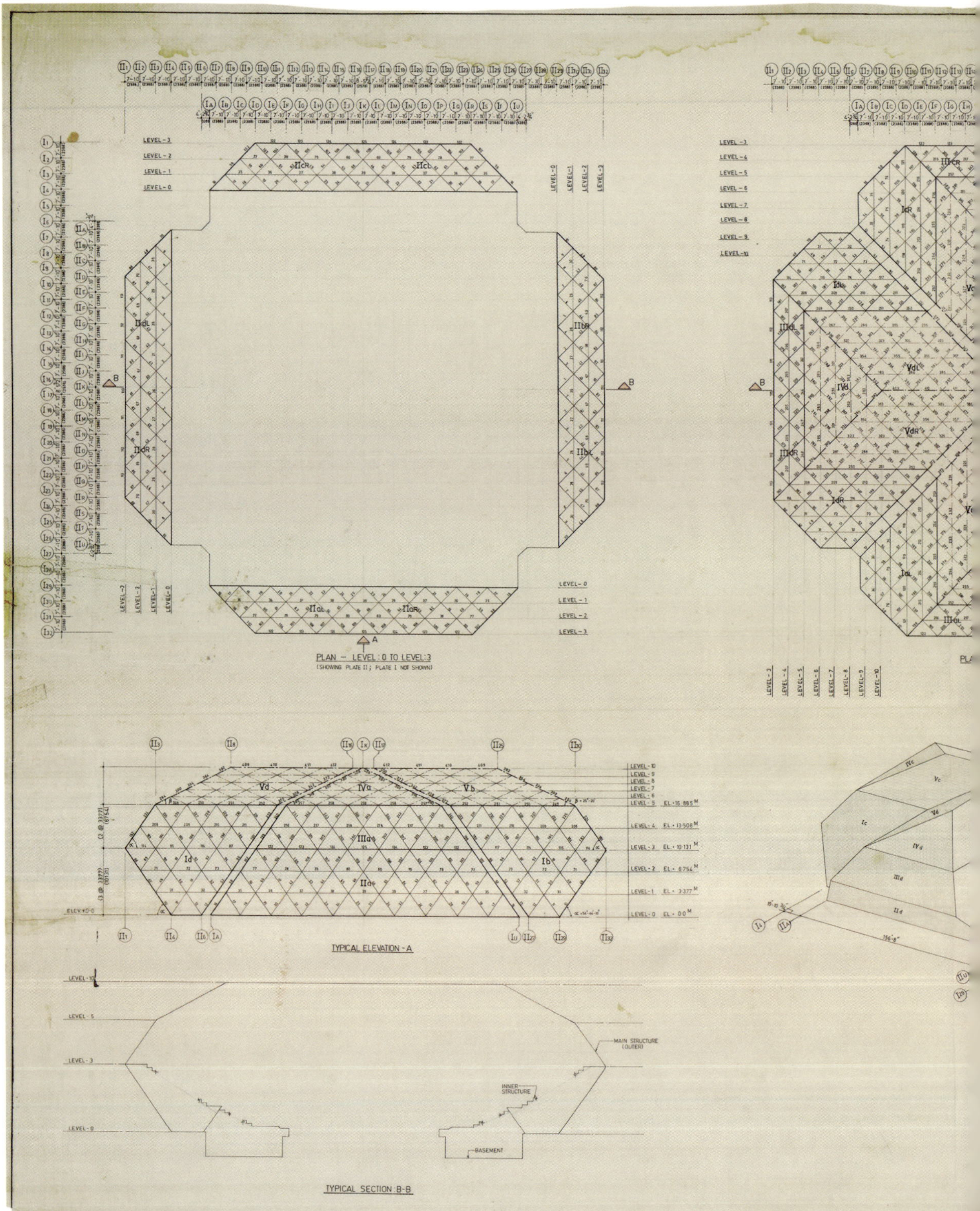

Drawing no. 107-SS-S01: Main Structure, Grid System, Members, Plates and their Numbers: 12 May 1980

Drawing explaining the main structure, grid system of the structure and a view showing the plates with their numbers, used as reference through the entire drawing set. The building is divided into four identical units, each an arm of the cross plan converging at the centre. There are seven typical plates in each unit and 28 plates in the entire structure that geometrically balance each other.

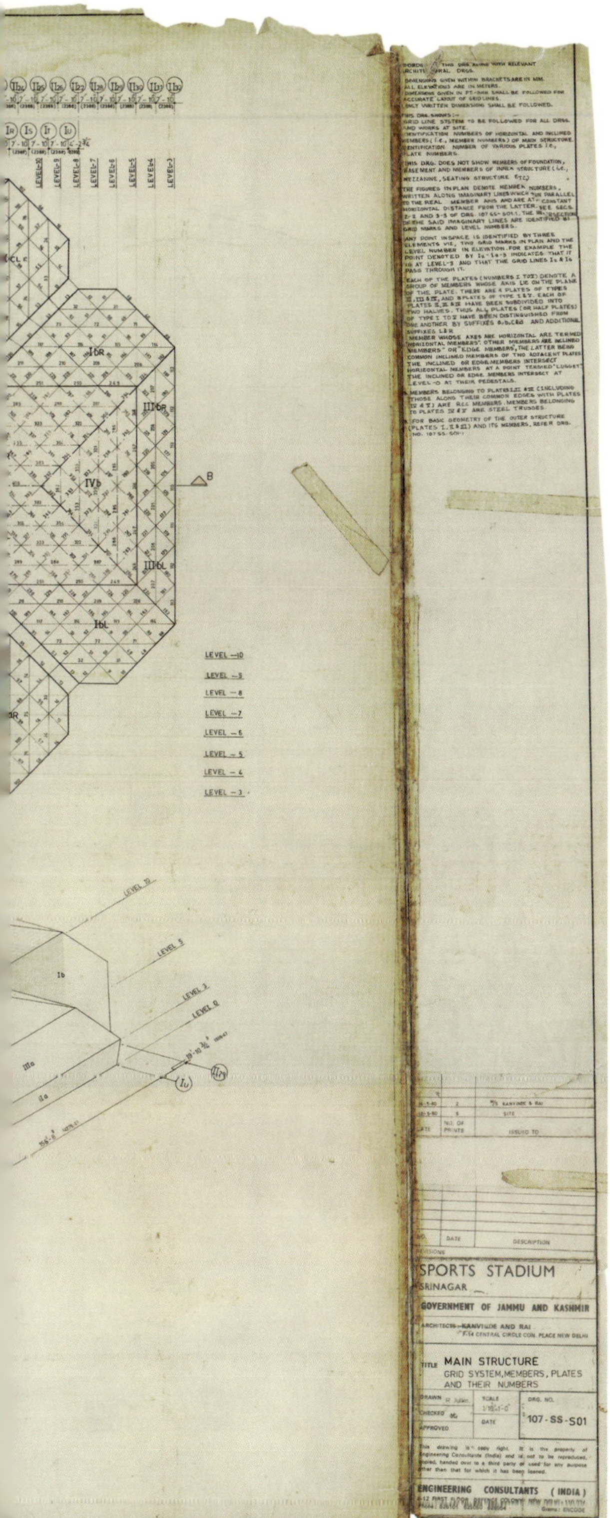

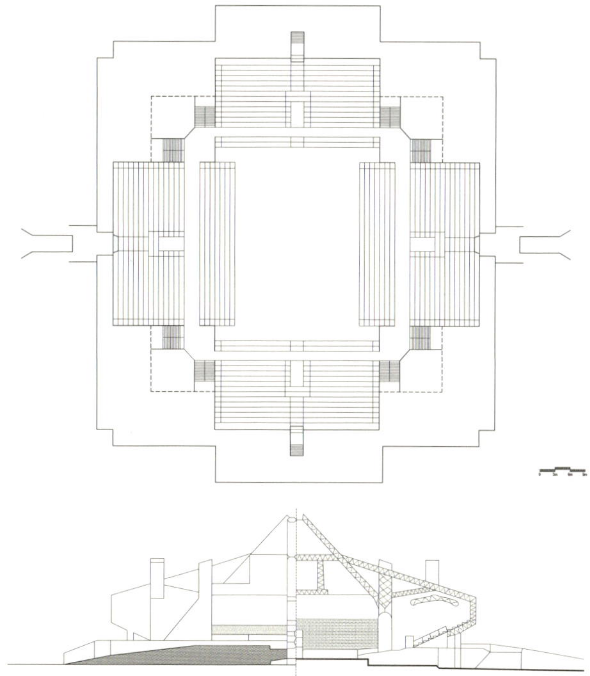

The original architectural idea of the stadium was a cross plan supported with pylons (above). The final geometry of the stadium structure evolved with Raj's suggestion to the architect of using the cross-shaped plan but removing the pylons and introducing instead an inclined triangulated plate structure.

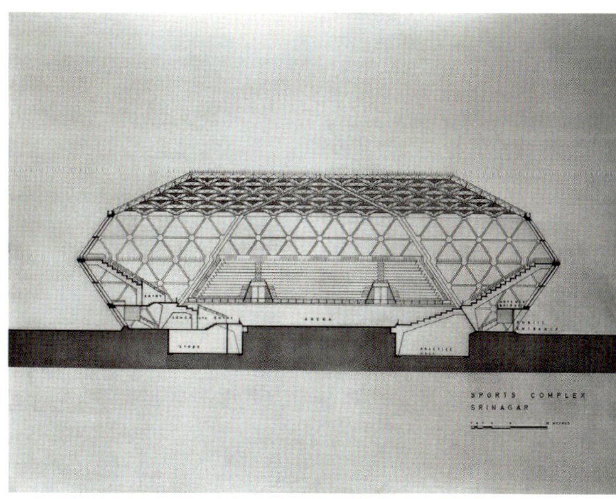

Architectural section of the stadium revealing the integrated structural concept of the inclined plates, roof truss and seating bowl.

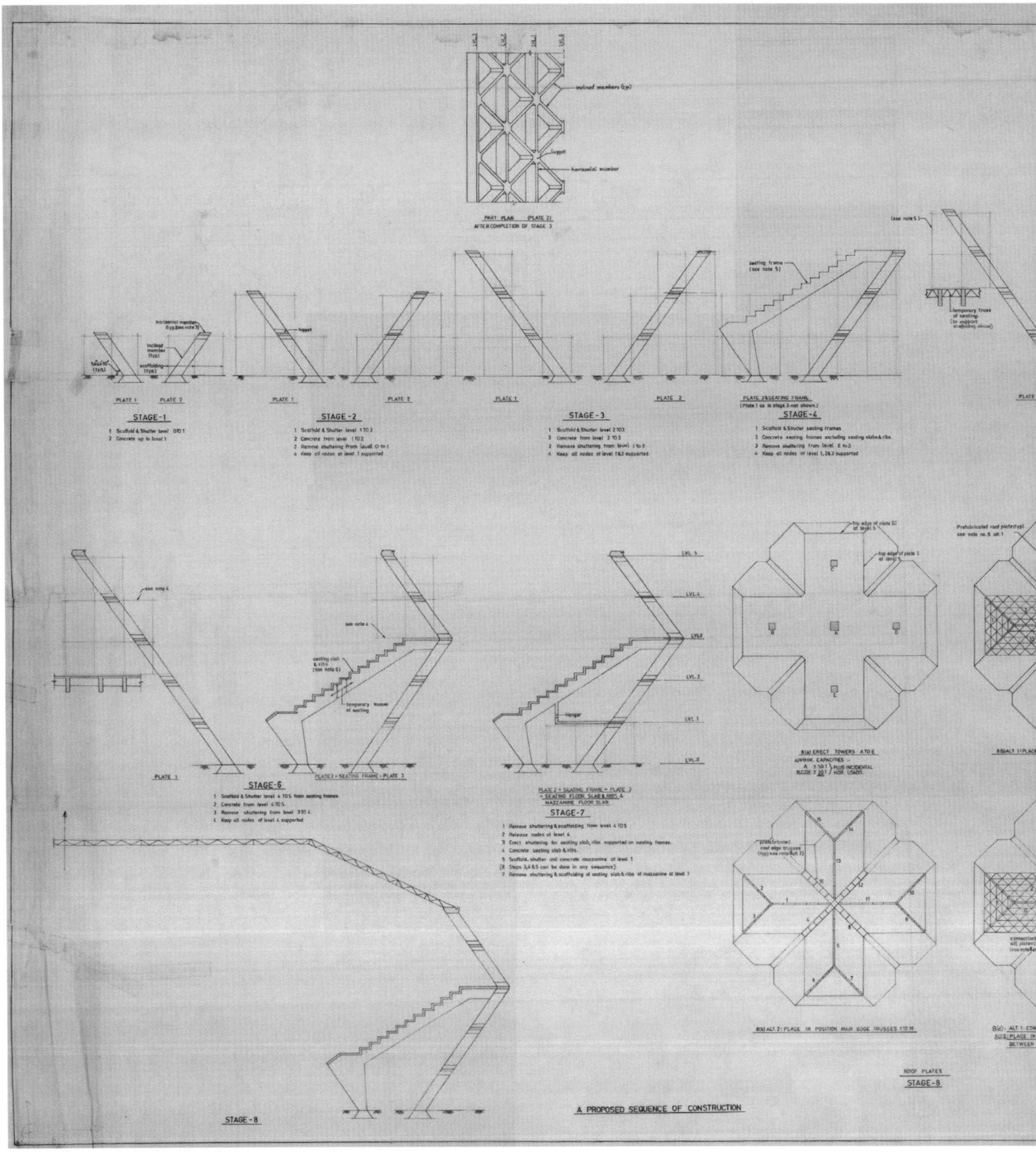

PART PLAN (PLATE 2)
AFTER COMPLETION OF STAGE 3
STAGE-1
STAGE-2
STAGE-3
STAGE-4
STAGE-6
STAGE-7
STAGE-8
ROOF PLATES
STAGE-8
A PROPOSED SEQUENCE OF CONSTRUCTION

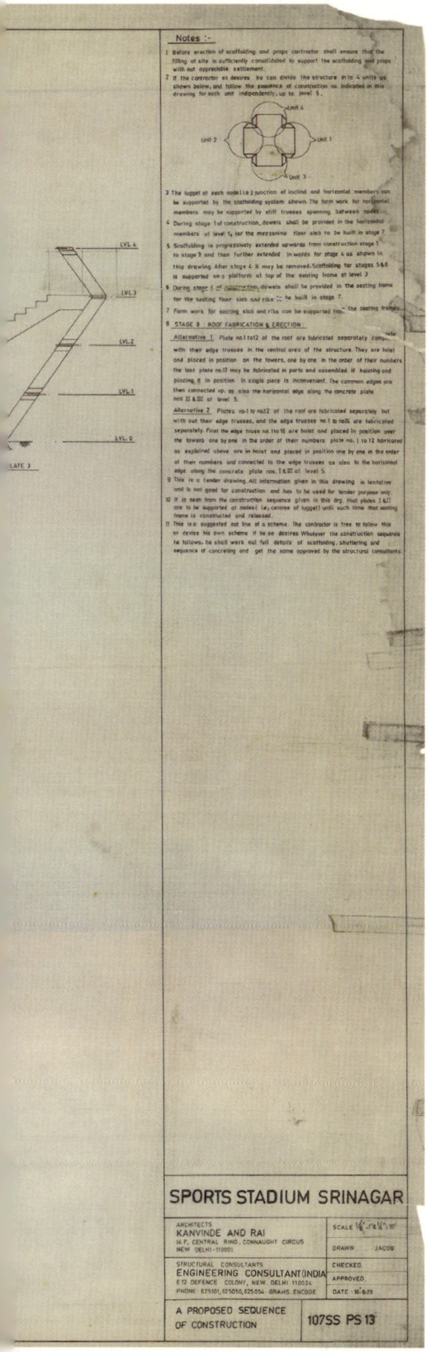

Drawing no. 107-SS-PS-13: A Proposed Sequence of Construction: 16 August 1979

This drawing proposes that the four units of the cross plan can be constructed independently till level 5 and details out two alternatives for roof erection to address possible technological and economic constraints during construction.

Photos showing different stages of construction: the main structure emerges from foundation pile caps into the seating ribs and bars of level 0-1; scaffolding for levels 1-3 is progressively extended upto Stage 3 then extended to Stage 4; after Stage 4 scaffolding is removed with supports provided at the main nodes only; scaffolding for levels 4-5 is set on a platform on top of seating frame at level 3; at level 5 work on the final tie beam is in progress; and finally the roof plates being installed.

Following pages: View of the internal space of the stadium. Notice the strong inward inclined diagnol members articulated by exposed post-tensioned cable ends. These are two of the four diagonally inclined members that hold together all the plates of the structure and are subject to large forces.

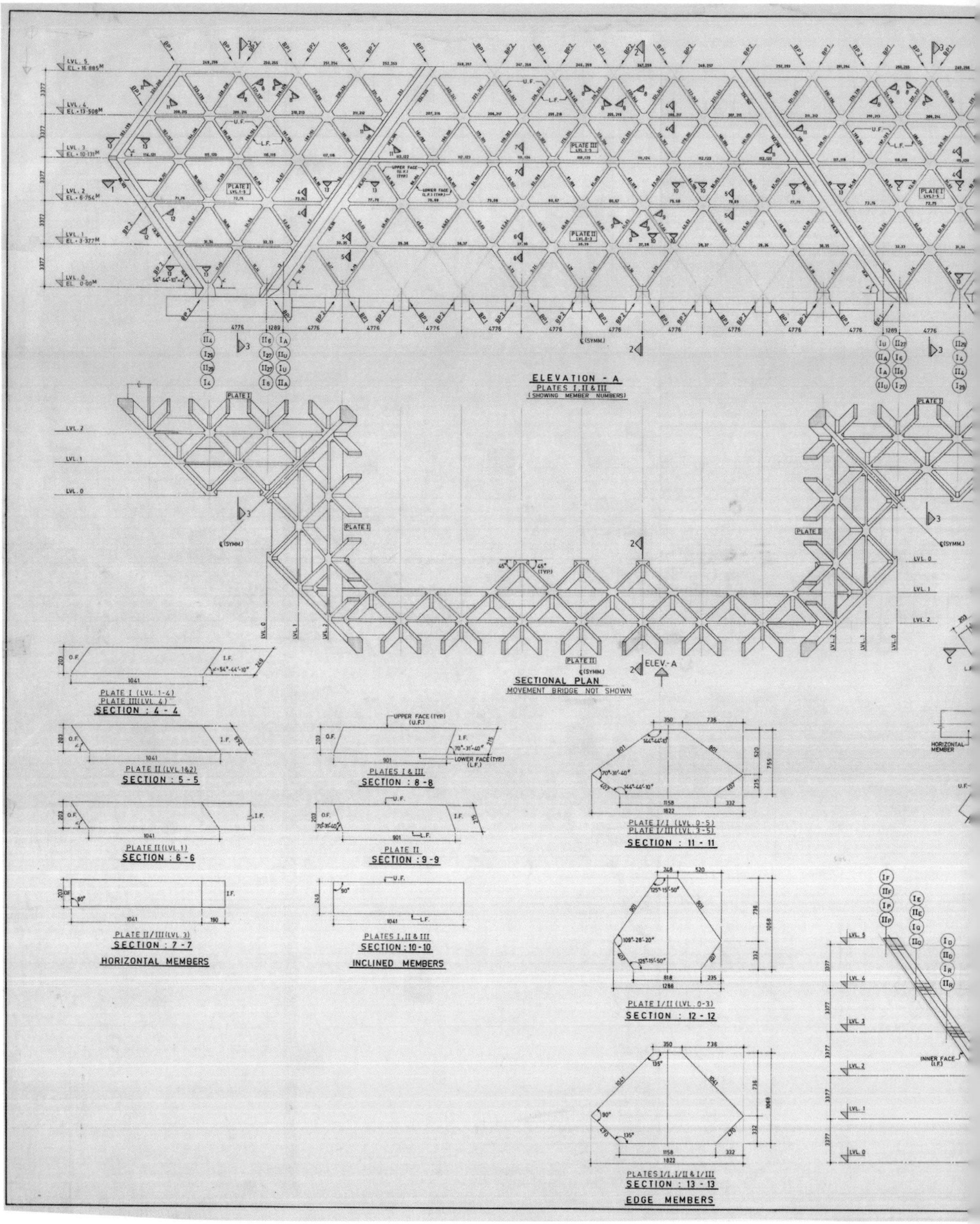
ELEVATION - A
PLATES I, II & III
(SHOWING MEMBER NUMBERS)
SECTIONAL PLAN
MOVEMENT BRIDGE NOT SHOWN
PLATE I (LVL. 1-4)
PLATE III (LVL. 4)
SECTION : 4 - 4
PLATE II (LVL 1&2)
SECTION : 5 - 5
PLATE II (LVL. 1)
SECTION : 6 - 6
PLATE II/III (LVL. 3)
SECTION : 7 - 7
HORIZONTAL MEMBERS
PLATES I & III
SECTION : 8 - 8
PLATE II
SECTION : 9 - 9
PLATES I, II & III
SECTION : 10 - 10
INCLINED MEMBERS
PLATE I/I (LVL. 0-5)
PLATE I/III (LVL. 3-5)
SECTION : 11 - 11
PLATE I/II (LVL. 0-3)
SECTION : 12 - 12
PLATES I/I, I/II & I/III
SECTION : 13 - 13
EDGE MEMBERS

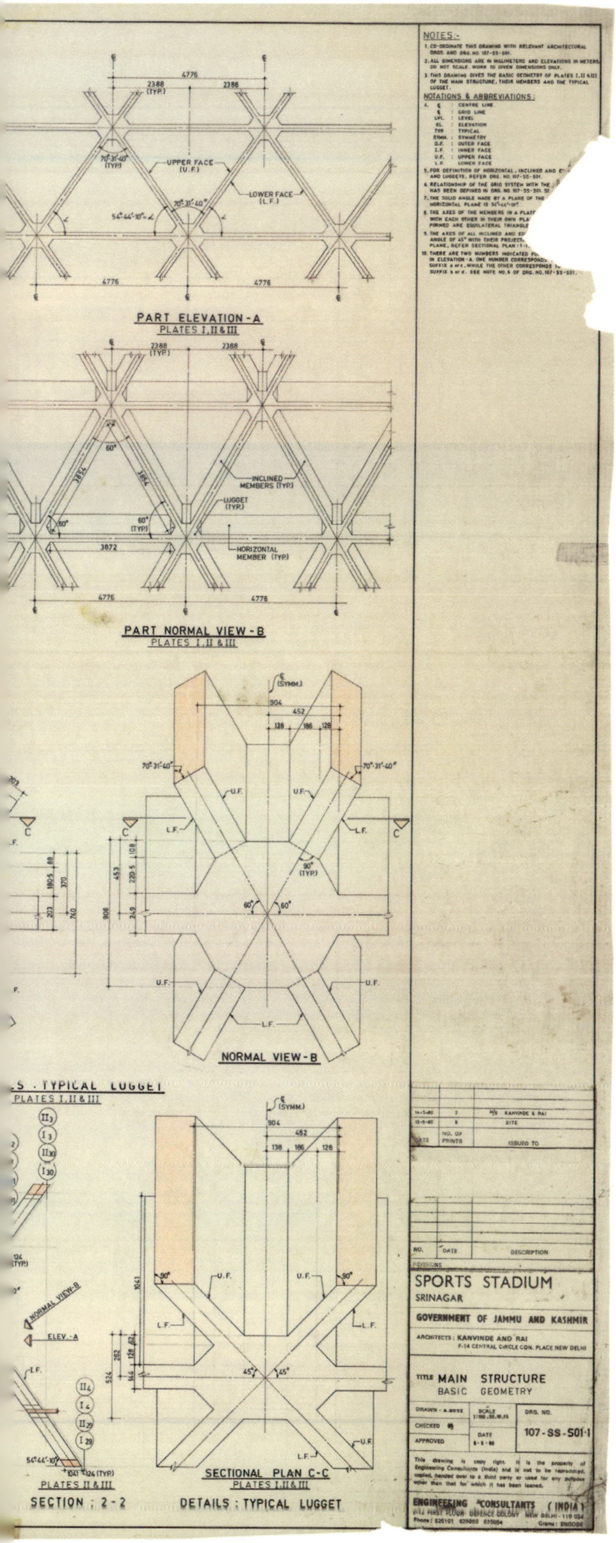

Interior and exterior photos of the joints, plates and members of the stadium structure.

Drawing no. 107-SS-S01-1: Main Structure- Basic Geometry: 8 May 1980
Drawing laying out the basic geometry of the main structure and all its members, not including the roof. The building is divided into four identical units and this drawing details one such typical unit.
Each unit has 3 typical plates–two of plate I and one each of plate II and III. All members are noted in a sectional plan and Elevation A; their inclines and correlating numbers in section through the 5 levels is indicated in sections 1-2 and 2-2. The 'Normal View' is used to visualise and detail each plate and its nodes in the pure geometric form. The solid angle made by all inclined plates with the horizontal plane is always 54°44'10".
This drawing also includes typical geometries of the main edge; inclined and horizontal members in plates I, II and III; and sectional plan, view and elevation of a typical 'Lugget' joint where the horizontal, inclined and edge members intersect.

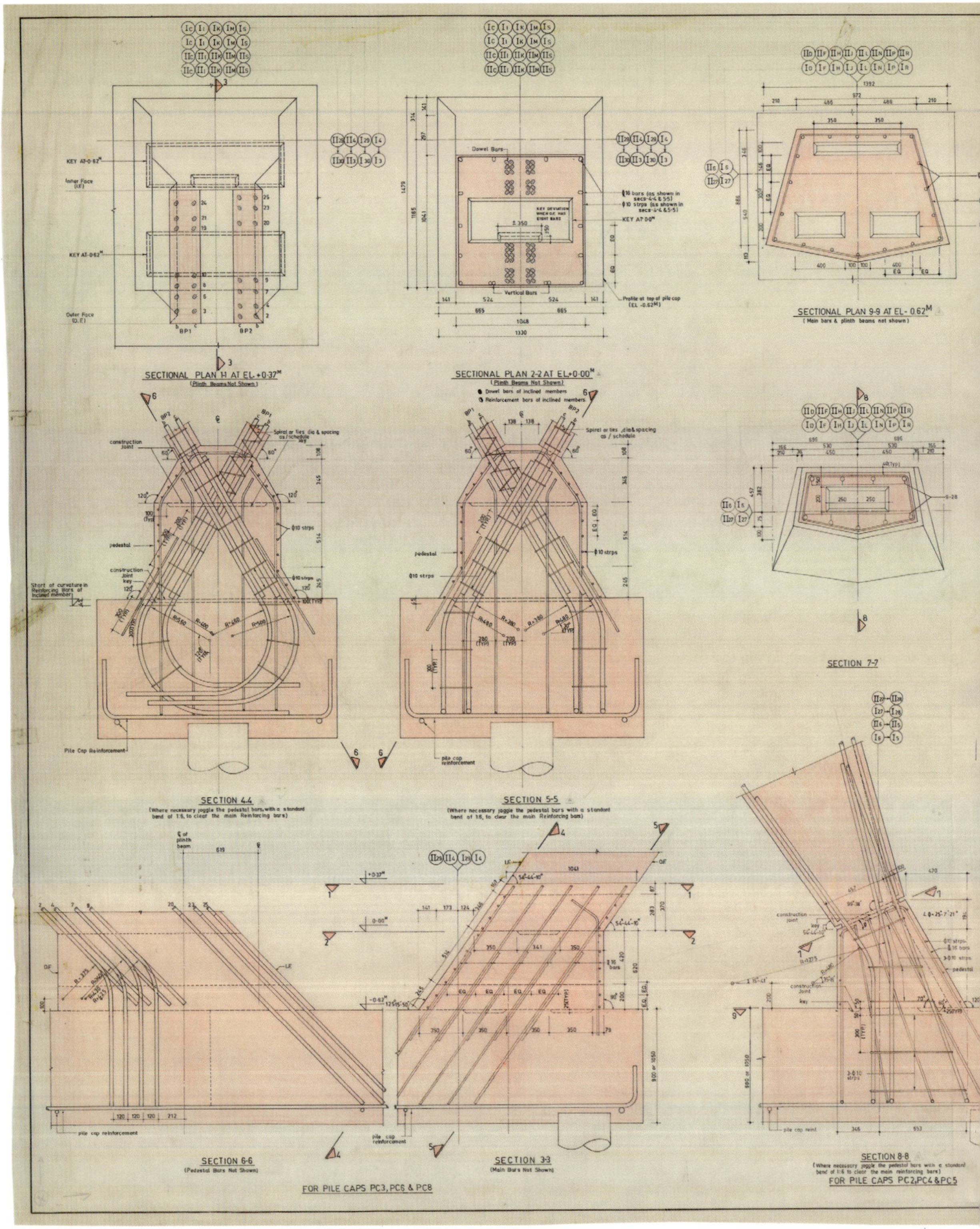

Drawing no. 107-SS-S08/A: Details of Pedestals (Sheet1) All Pile Caps except PC1: 12 April 1980

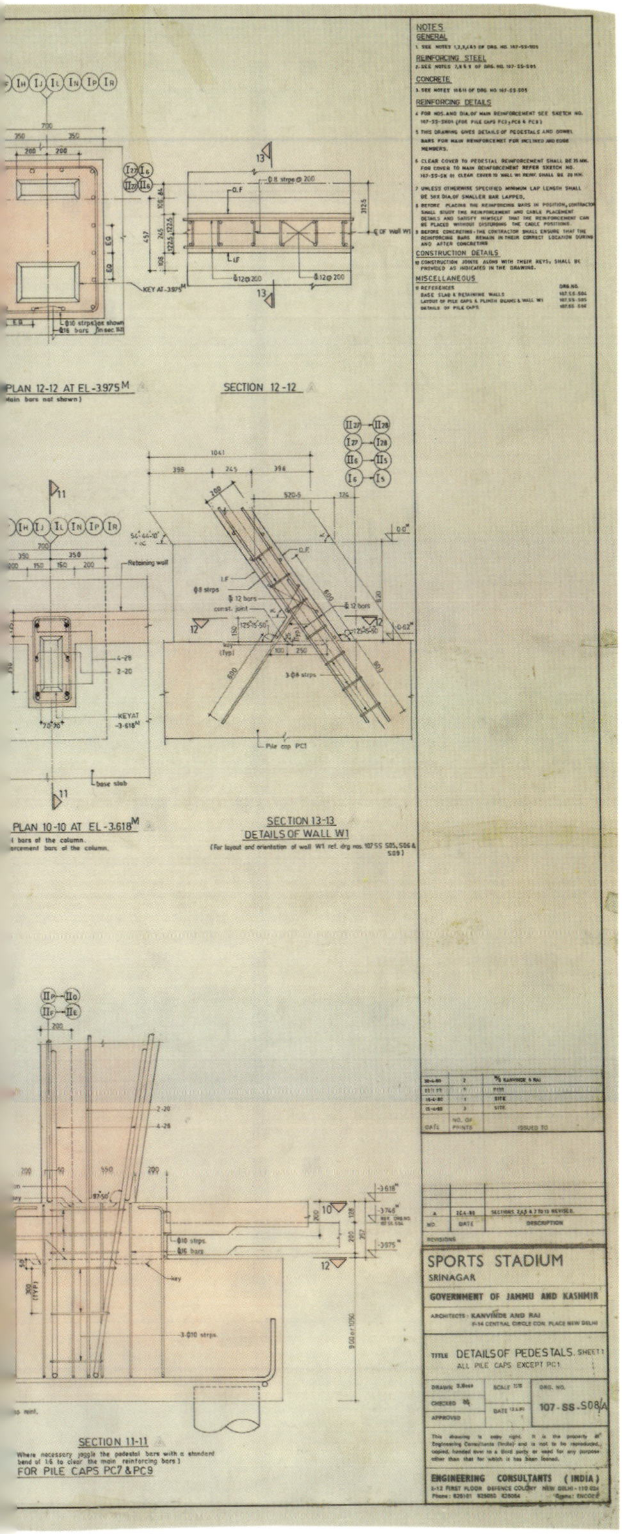

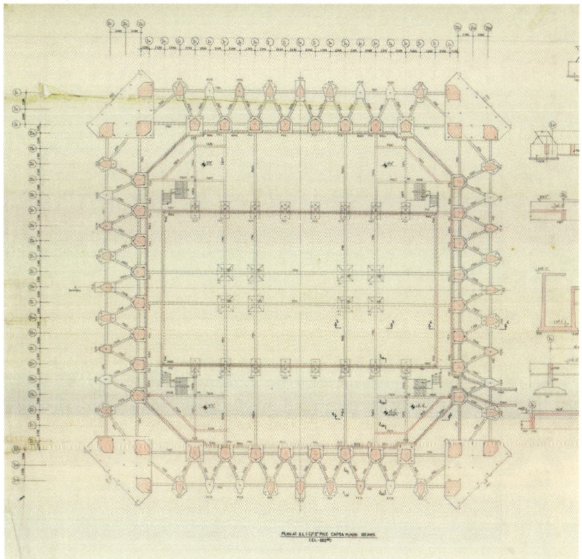

Construction photo of foundation pile caps and pedestals of the external inclined plate.

(Partial) Drawing no. 107-SS-PS-01: Foundation Plan–Pile Caps, Plinth Beams and Basement: 28 August 1979

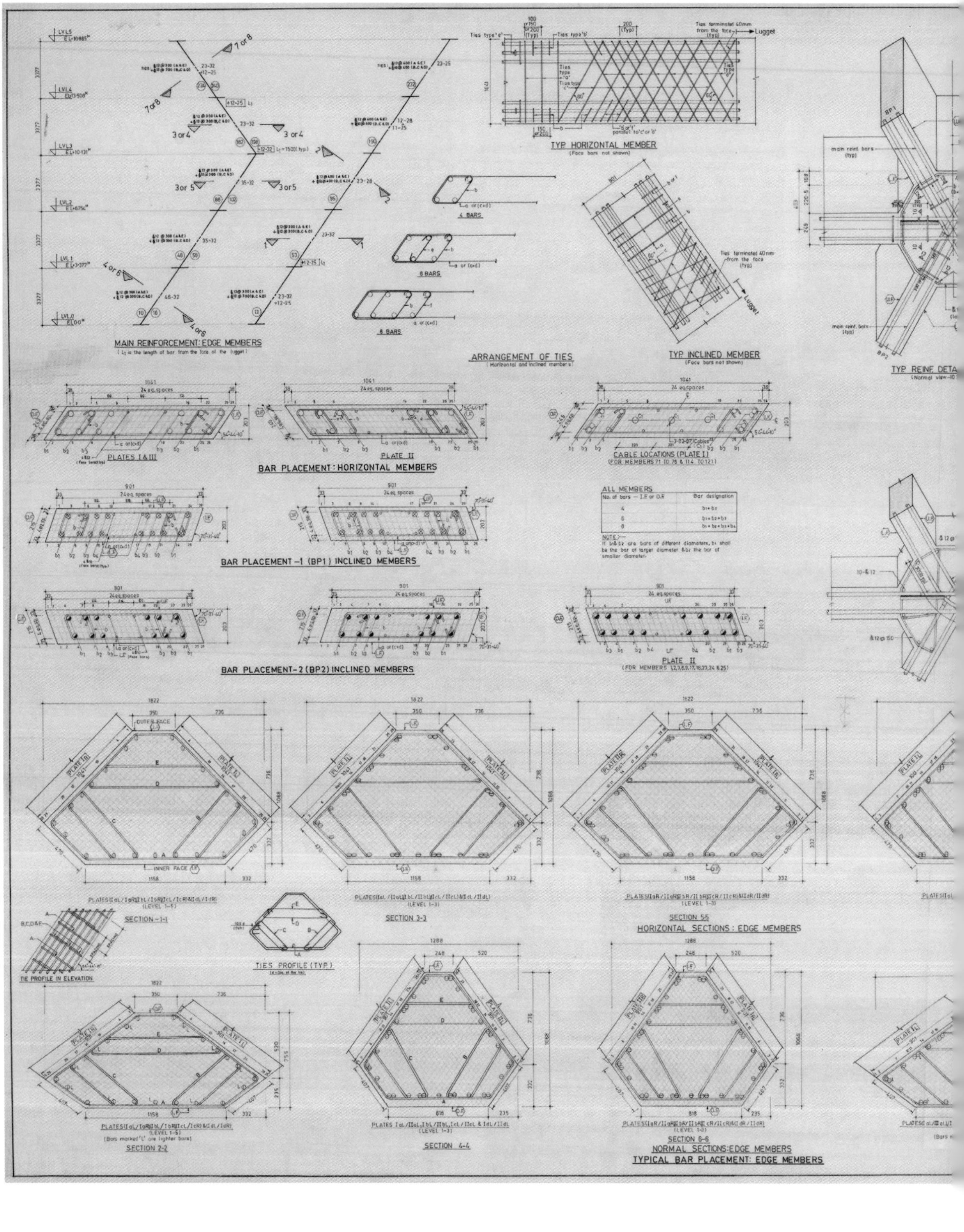
MAIN REINFORCEMENT: EDGE MEMBERS
ARRANGEMENT OF TIES
TYP HORIZONTAL MEMBER
TYP INCLINED MEMBER
BAR PLACEMENT: HORIZONTAL MEMBERS
PLATES I & III
PLATE II
CABLE LOCATIONS (PLATE I)
BAR PLACEMENT-1 (BP1) INCLINED MEMBERS
BAR PLACEMENT-2 (BP2) INCLINED MEMBERS
ALL MEMBERS
SECTION 1-1
TIES PROFILE (TYP.)
TIE PROFILE IN ELEVATION
SECTION 3-3
SECTION 5-5
HORIZONTAL SECTIONS: EDGE MEMBERS
SECTION 2-2
SECTION 4-4
SECTION 6-6
NORMAL SECTIONS: EDGE MEMBERS
TYPICAL BAR PLACEMENT: EDGE MEMBERS

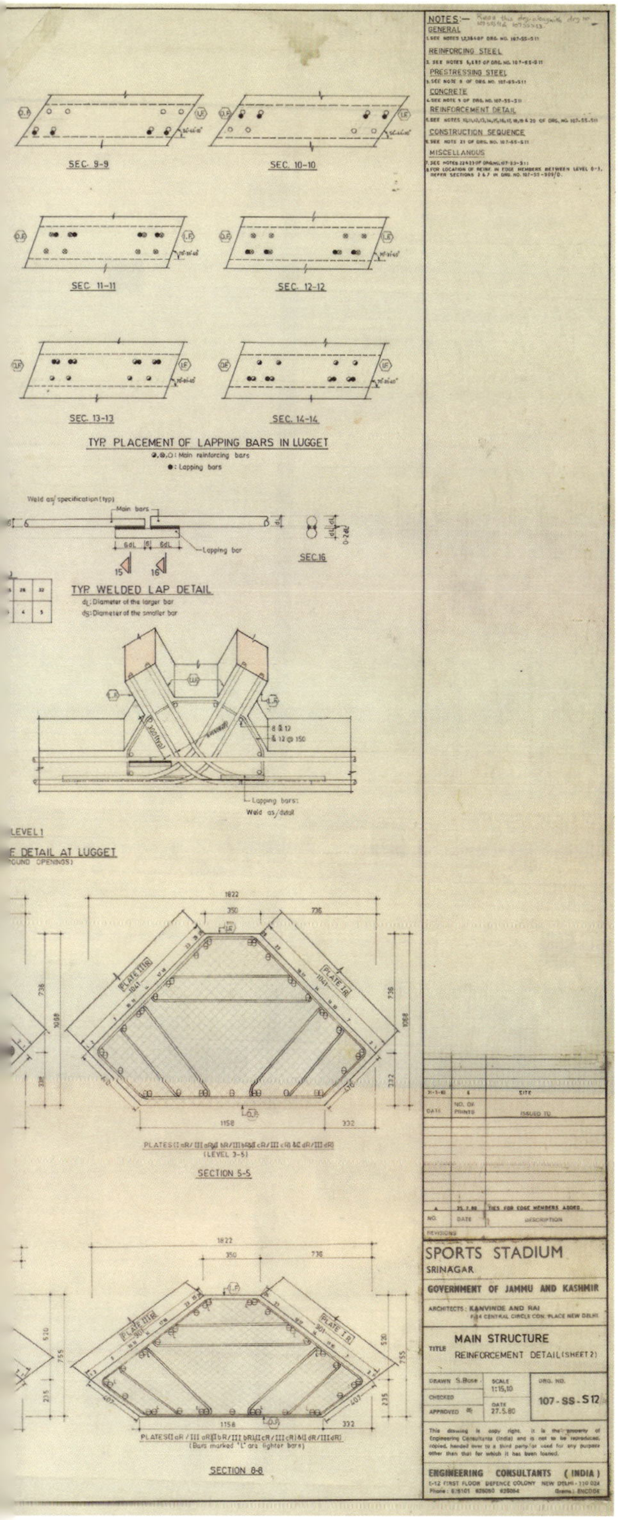

Construction photo showing the reinforcement in place and pre-cast members being inserted into the joint.

Drawing no. 107-SS-S12/A: Main Structure–Reinforcement Details: 29 May 1980

Reinforcement bar placement details, for edge members and pre-cast inclined and horizontal members and their intersection at the lugget joint. Also seen here, details of arrangement and sizes of ties, laps and even the electrode sizes used for welding.

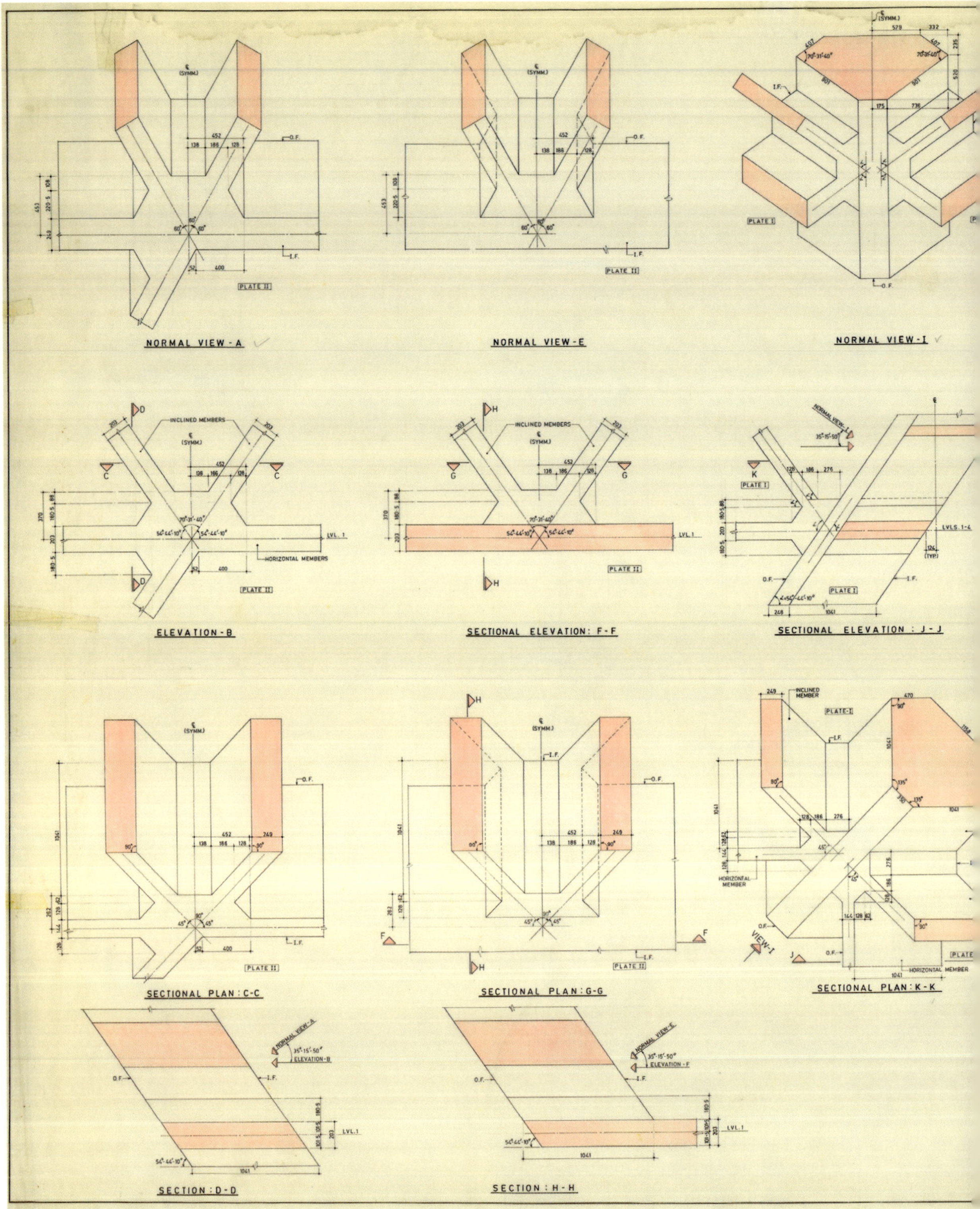
NORMAL VIEW - A
NORMAL VIEW - E
NORMAL VIEW - I
ELEVATION - B
SECTIONAL ELEVATION: F - F
SECTIONAL ELEVATION : J - J
SECTIONAL PLAN : C - C
SECTIONAL PLAN : G - G
SECTIONAL PLAN : K - K
SECTION : D - D
SECTION : H - H

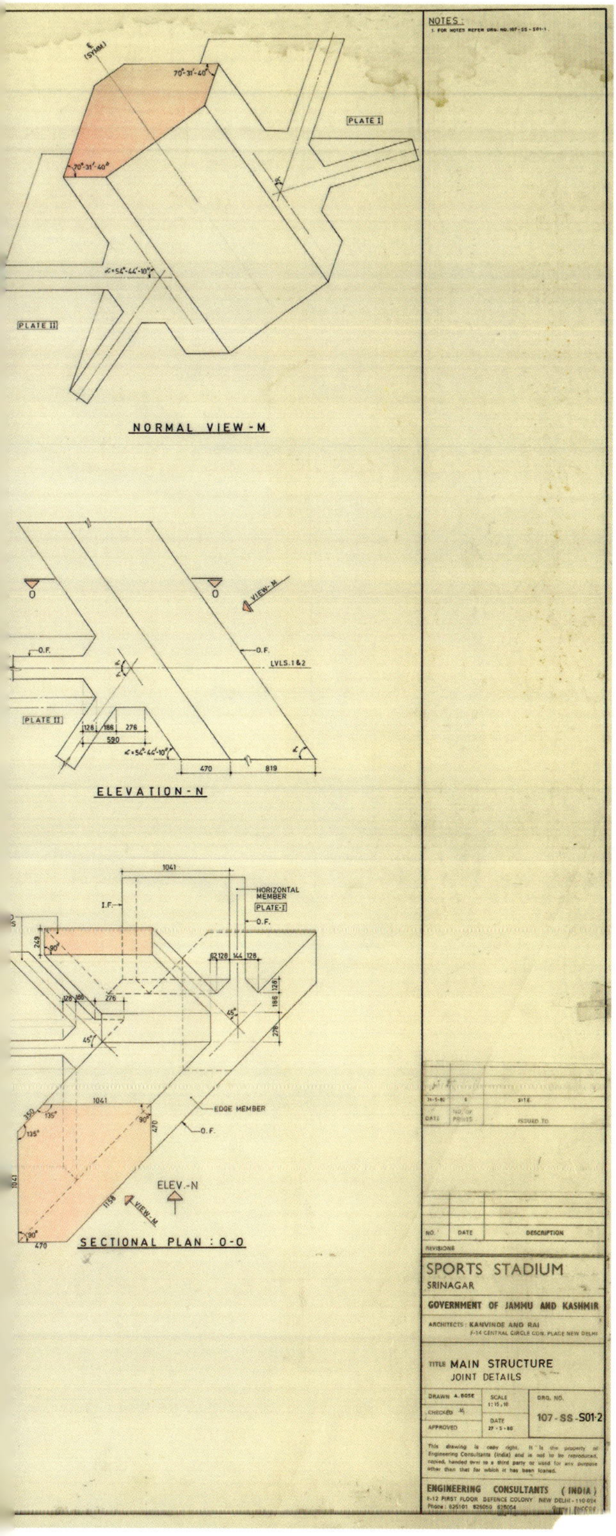

Drawing no. 107-SS-S01-2: Main Structure–Joint Details: 25 May 1980
Drawing meticulously visualising four different joints with their unique junctions, angles and measurements with a combination of section, elevation, sectional plan, sectional elevation and 'Normal View'.

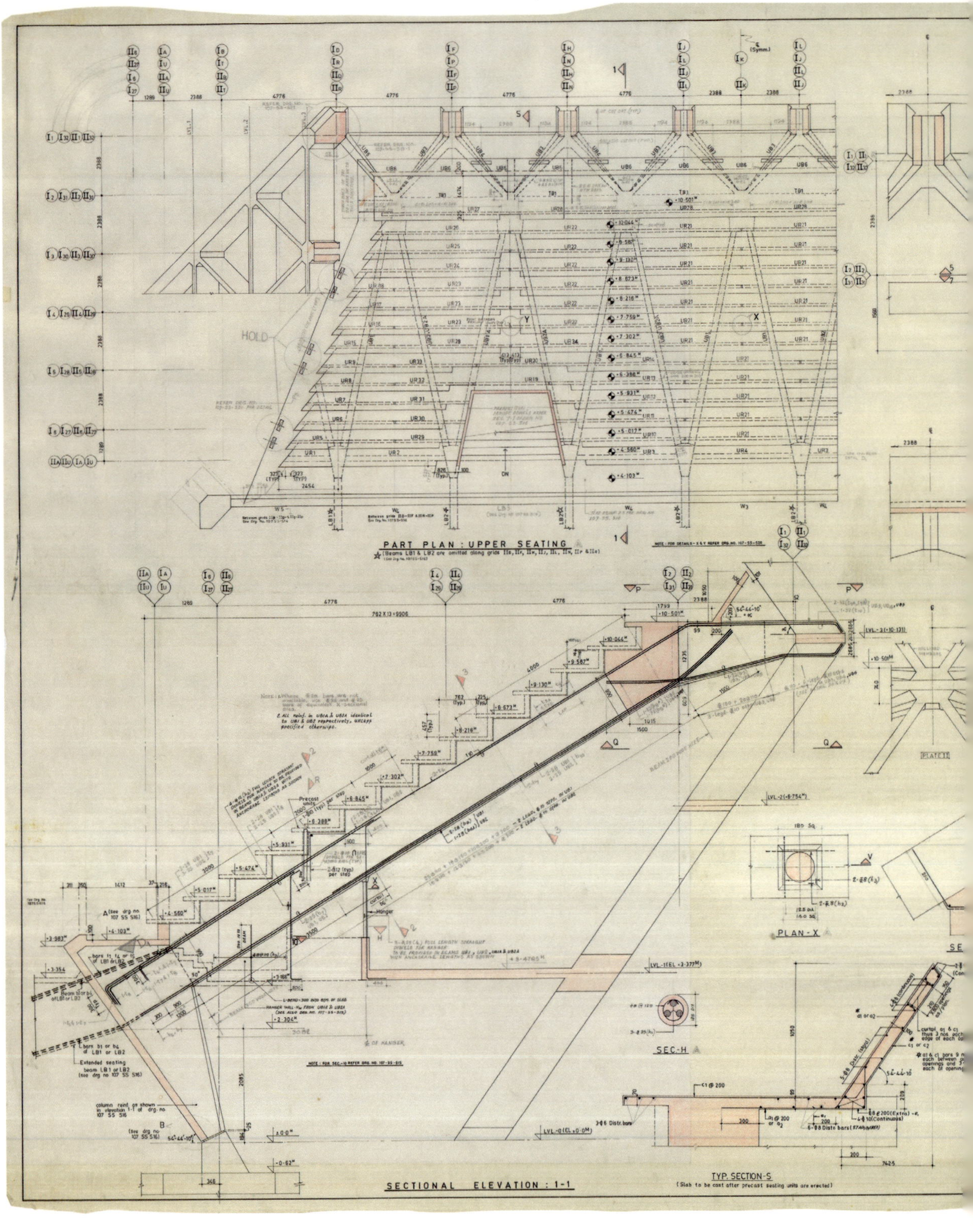
PART PLAN : UPPER SEATING
SECTIONAL ELEVATION : 1-1
TYP. SECTION-S
PLAN-X
SEC-H

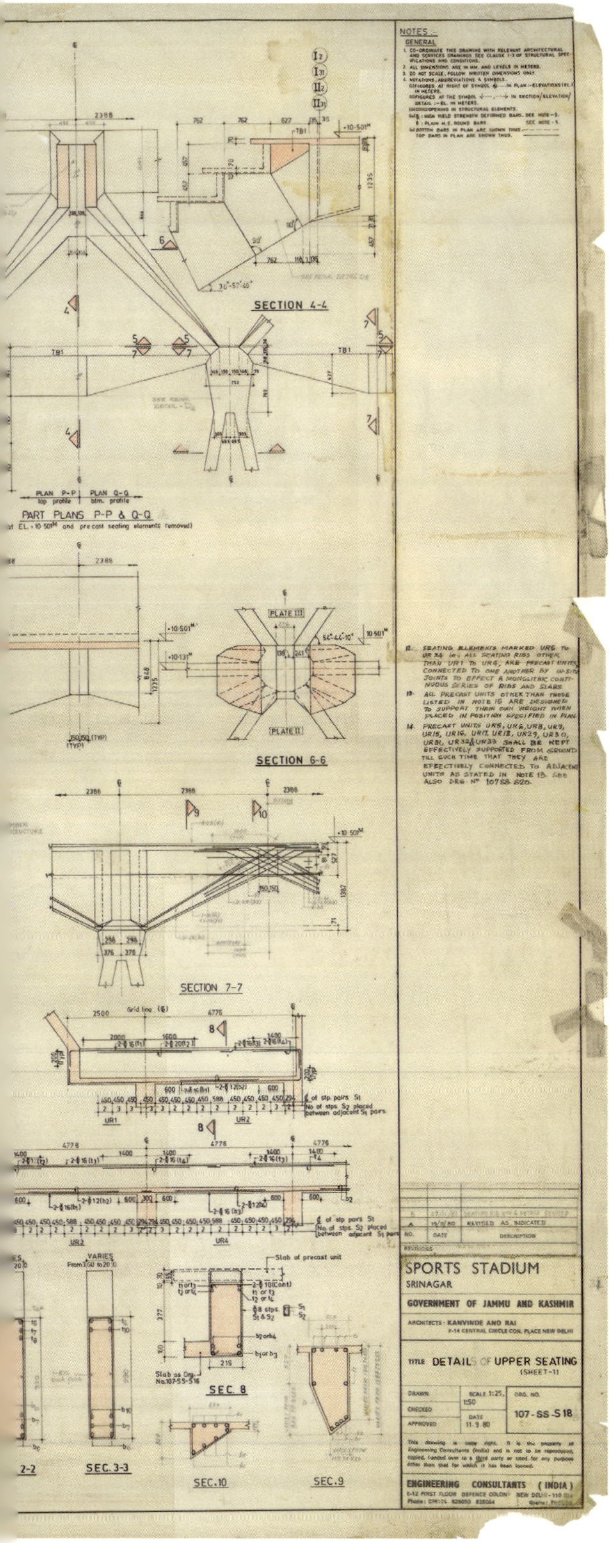

Interior view from the mezzanine floor under the upper seating ribs and their integration with the main structure edge beam at level 3.

Drawing no. 107-SS-S18/B: Details of Upper Seating: 11 September 1980
Details at foundation level, reinforcement of beams carrying the seating and the triangulated configuration that flares out to join the main structure at the nodes on level 3.

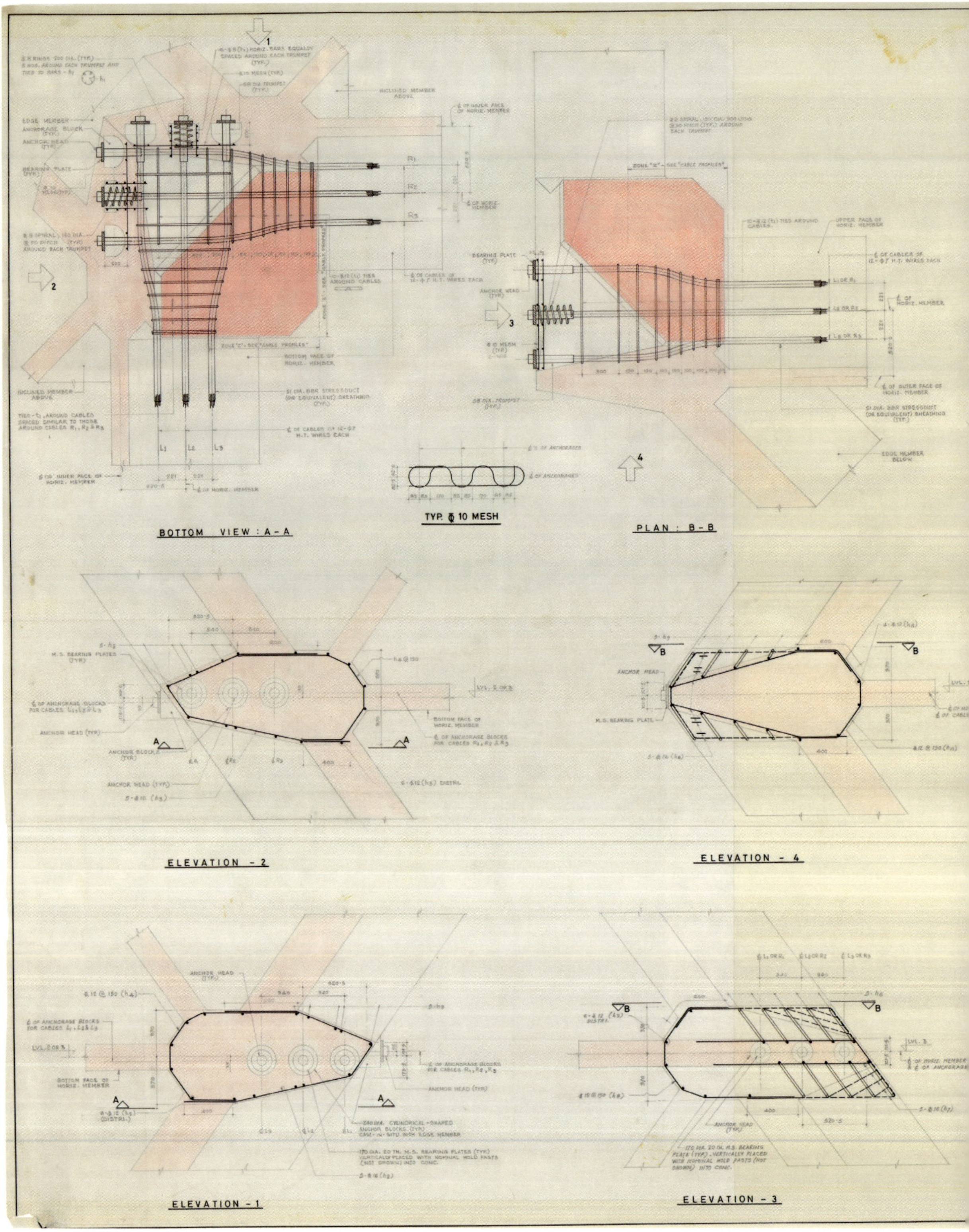
BOTTOM VIEW : A - A
TYP. Φ 10 MESH
PLAN : B - B
ELEVATION - 2
ELEVATION - 4
ELEVATION - 1
ELEVATION - 3

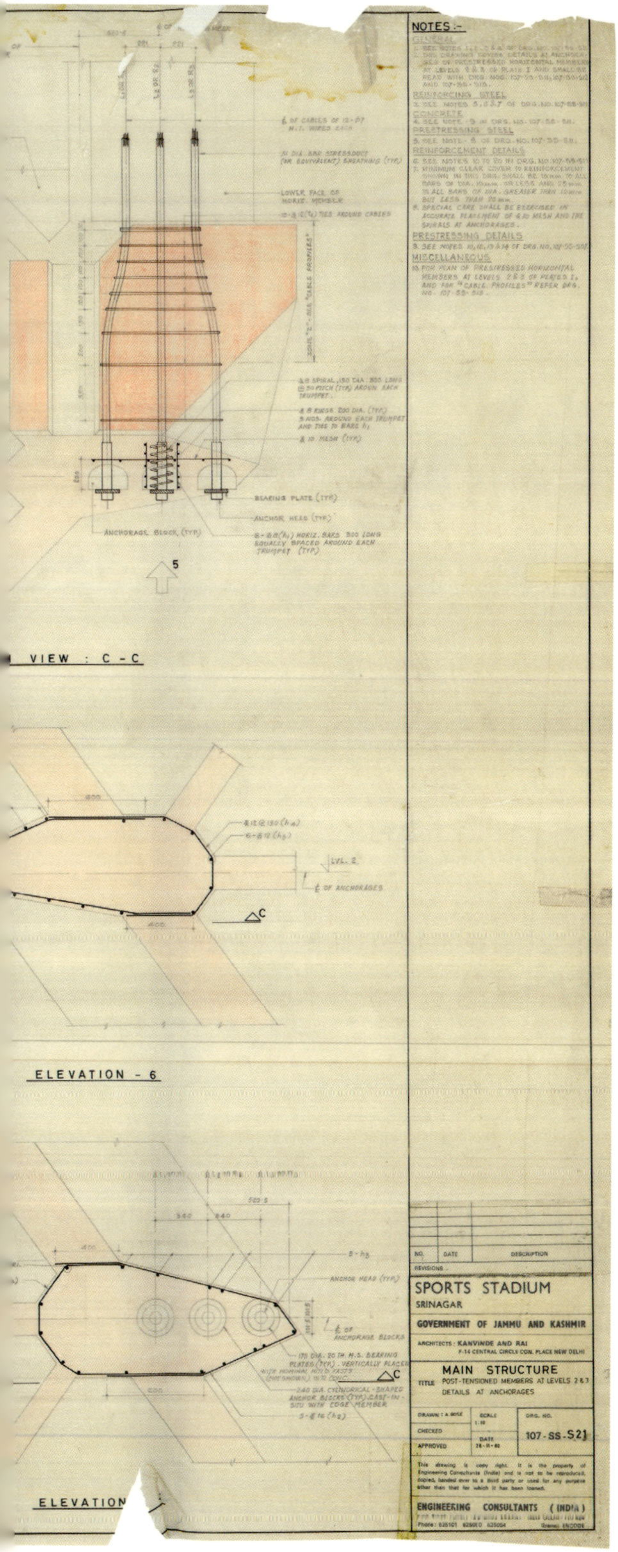

Photo of Plate II during construction where post-tensioned joint at level 2 is completed and post-tensioning is in process at level 3.

Drawing no. 107-SS-S21: Main Structure–Post Tensioned Members at Levels 2, 3, Details at Anchorage: 28 November 1980

Analysis showed many members along the horizontal plane had heavy tensile forces which could not be taken care of by regular reinforcement. Post-tensioning was introduced through the entire structure at three horizontal levels: level 2 (above the openings), level 3 (at the junctions of plates 2 and 3) and at level 5 (at the junction of the roof). To integrate the horizontals of respective plates at various levels, organic shaped joints emerged to manage the geometric resolution and termination of post-tensioned cables.

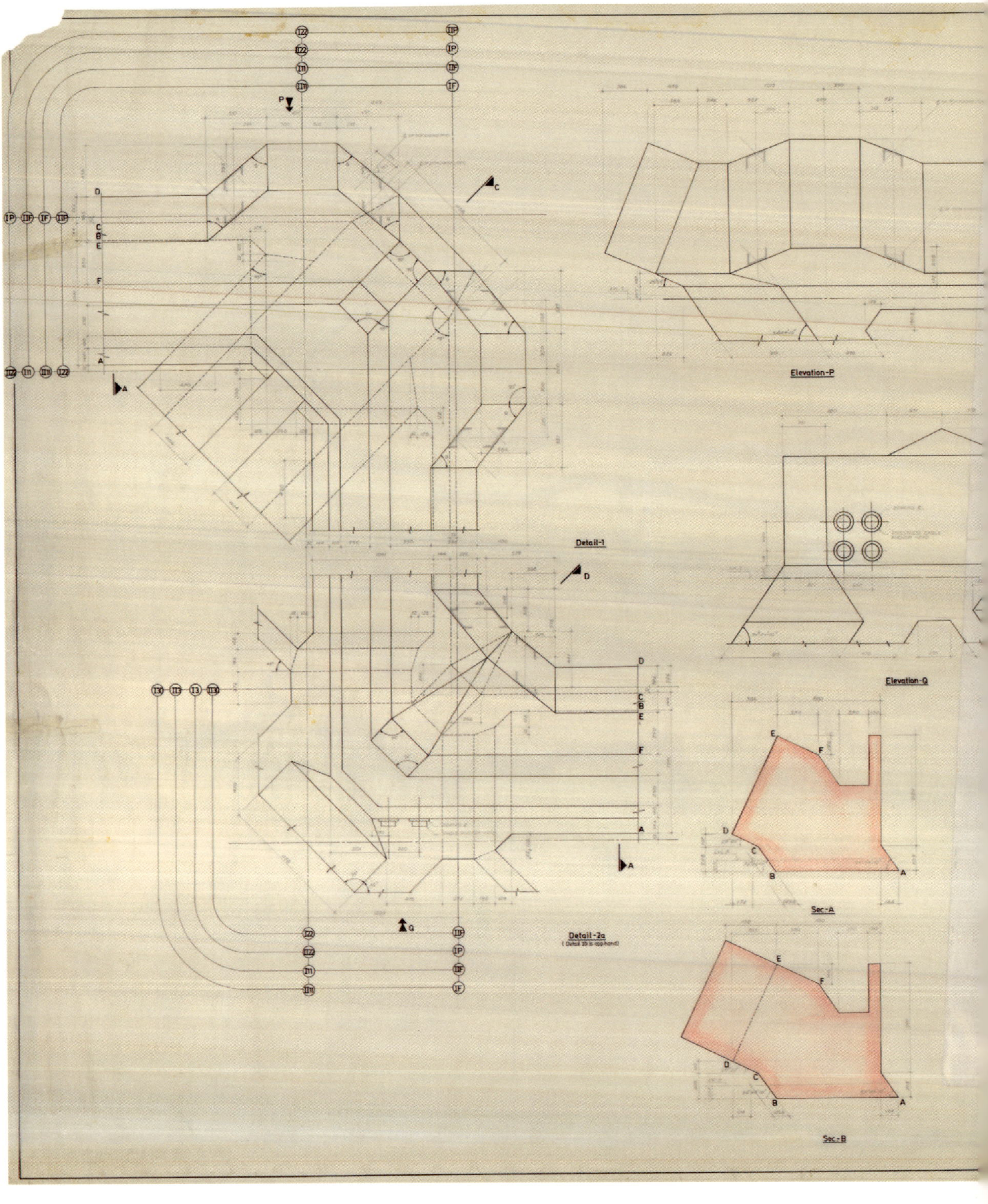
Elevation-P
Elevation-Q
Detail-1
Detail-2a
(Detail 2b is opp hand)
Sec-A
Sec-B

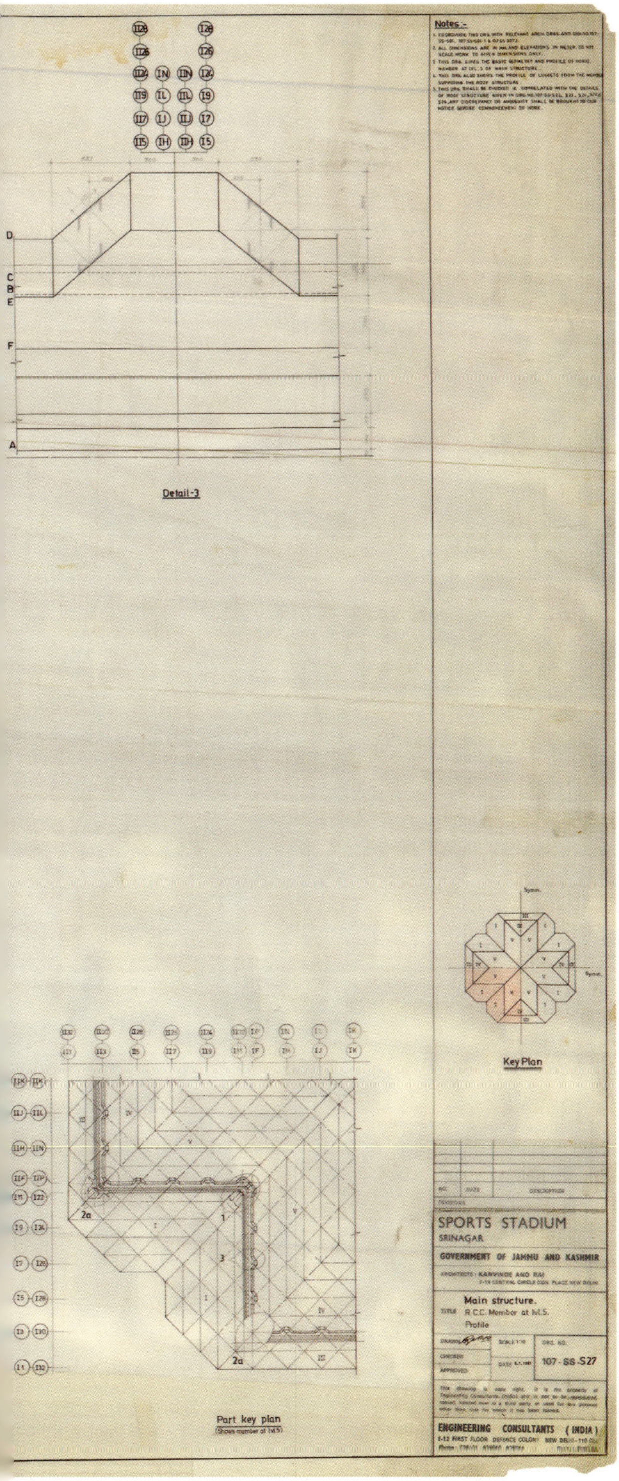

Photos of the edge beam that ties the concrete plates with the steel roof plates.

Drawing no. 107-SS –S27: Main Structure–RCC Member at Level 5–Profile: 8 July 1981

Post-tensioned edge member details at level 5 which connects the roof to the main structure. It also acts as a gutter and contains the outward thrusts of the 16 roof plates as well as the tie member for the main structure

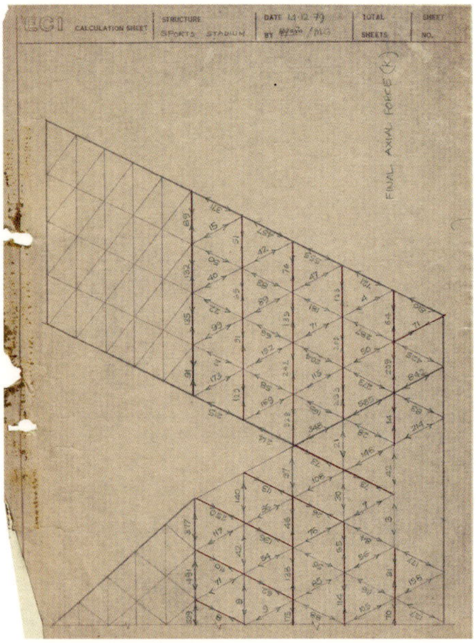

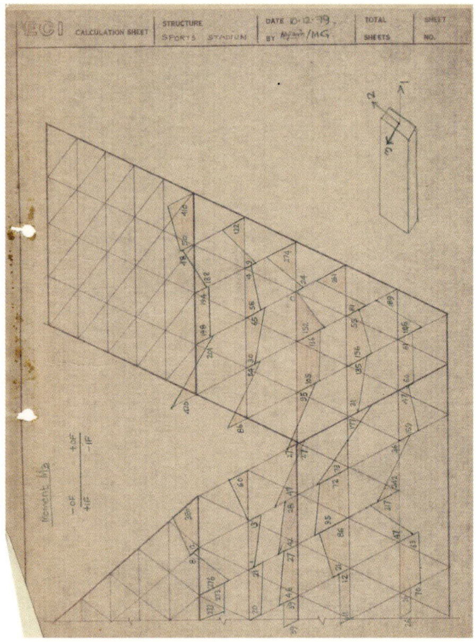

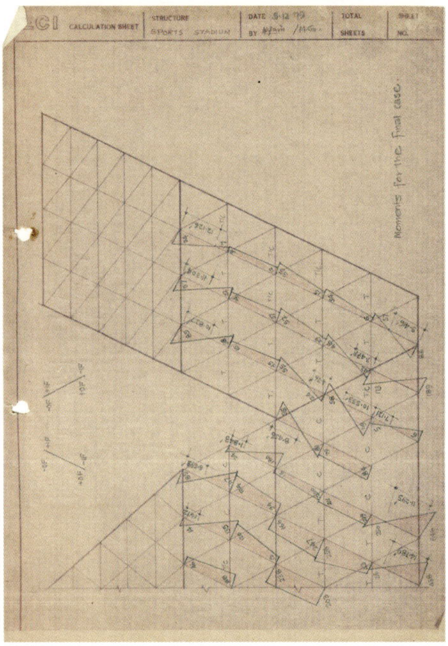

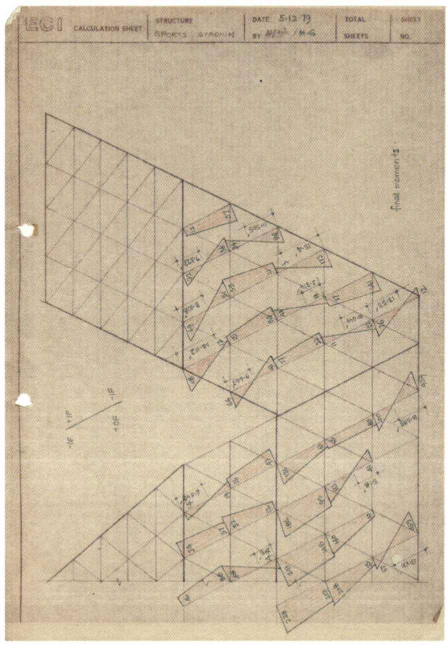

Four representative calculation sheets included here out of the hundreds done for the project. The first is regarding axial forces and the other three are the Final Moment analyses for the different plate inclines. Some of these calculations were confirmed with the help of an in-house computer and indigenous software developed by the firm.

Interior photo of the dramatic inclined member held back by two intersecting plates and post-tensioned ties at different levels. Raj recalls his immense relief when this very challenging element was released off scaffolding.

HALL OF STATES

NEW DELHI | 1982

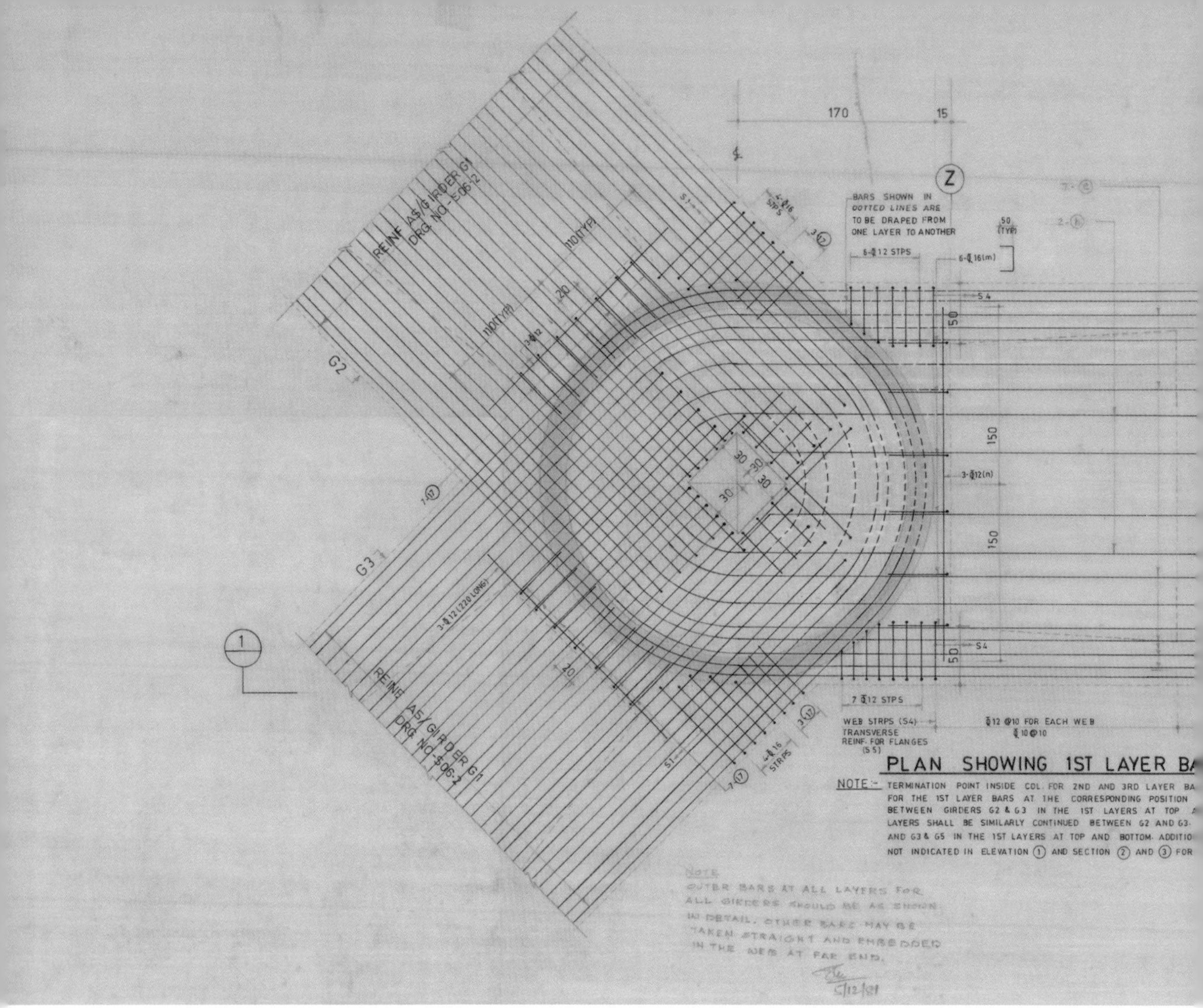

The Hall of States was constructed a decade after the Hall of Nations, in the same complex–Pragati Maidan–for the 1982 Asian Games held in Delhi. Commissioned by the Trade Fair Authority of India, this complex was planned, designed and constructed in 18 months. Inspired by the synergy that produced the Hall of Nations, Mahendra Raj says, 'We wanted to continue the idea of the pyramidal concrete space frame design. However, the CPWD that had to execute it was not willing to build it. So we had to change the (structural) design'.

The building is formed by an unequal but symmetrical octagonal sky-lit atrium space at the centre, which extends on all sides to form four larger exhibition halls and four smaller spaces with services and mezzanines. The central octagonal space spans 44m and has a maximum height of 22m. The surrounding large halls have a span of 22m x 25m. Sixteen large hollow circular columns define the edges of the octagon and larger halls, and also accommodate services, while eight semi-circular columns house ramps and stairwells to the mezzanine floors. All twenty-four have a diameter of 3.4m and together, they support the structure which is spanned by box girders and

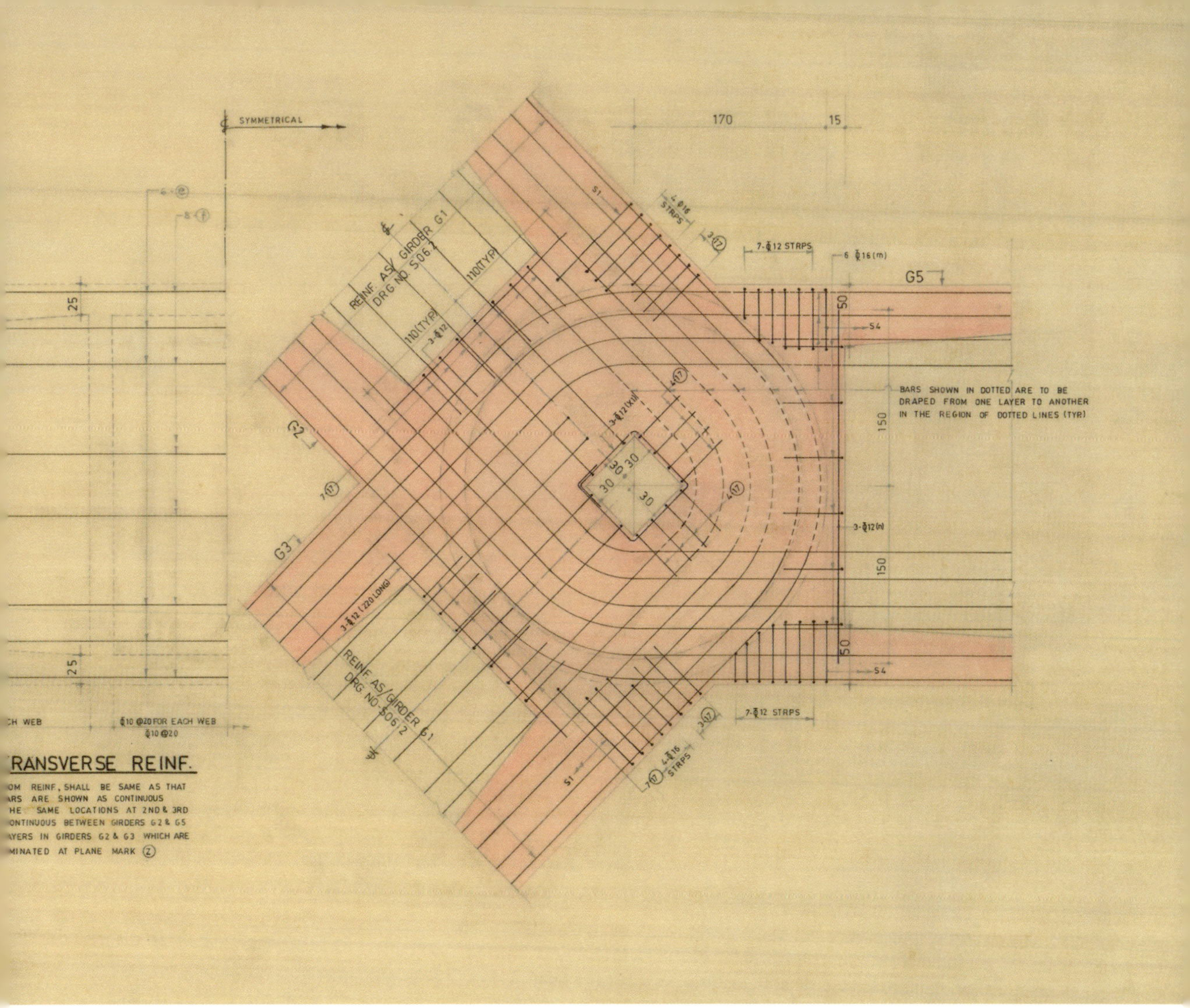

tapered cross-ribs. These unique columns are a forceful structural expression that dominate the articulation of the facade along with the stepped roof structure for the skylight.

The architectural model, highlighting the geometrical form and its clear structural resolution.

(Partial) Drawing no. 100SEC-S07.1: Girder Details (Sheet-2): 30 September 1981, showing different layers of reinforcement bars of the box girders as they terminate into the circular columns.

Previous pages: Interior view highlighting the 22m-high central octagonal atrium with four larger halls at the cross ends of the octagon and the mezzanine floors, services and staircases at the four diagonal ends.

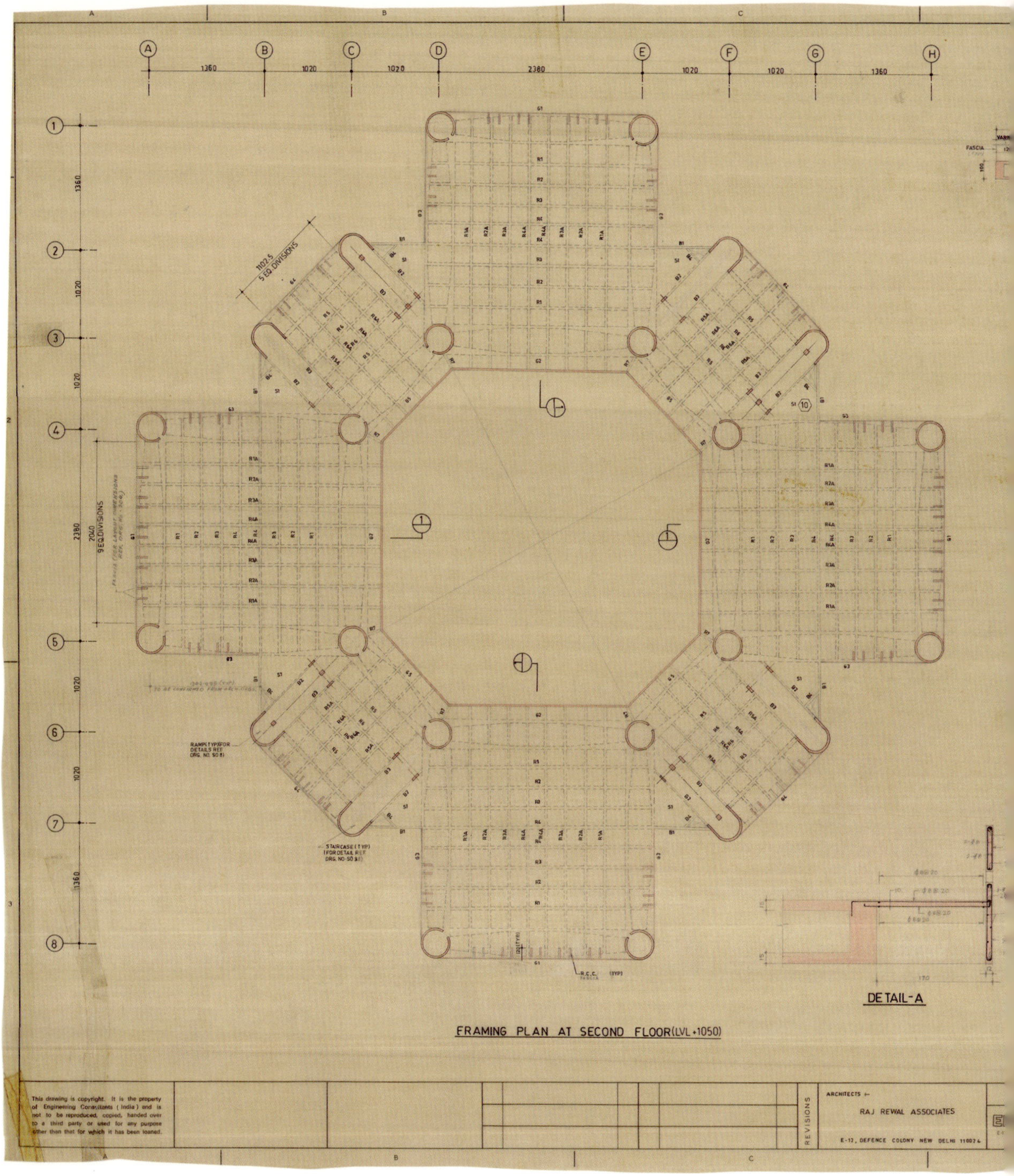

Drawing no. 100SEC-S06: Framing Plan at Second Floor: 4 April 1981

The framing plan of the second floor shows the star-shaped plan with an octagonal-shaped central space formed by the four larger halls at the cross sides and four smaller halls placed diagonally. Despite the strict orthogonal geometry, notice the rib sections taper from 22cm to 16cm in dimension and the box girders attach like muscles to the circular column supports, streamlining as they move away from the supports to soften the edges of the building with the economy and efficiency of expressed structural flows.

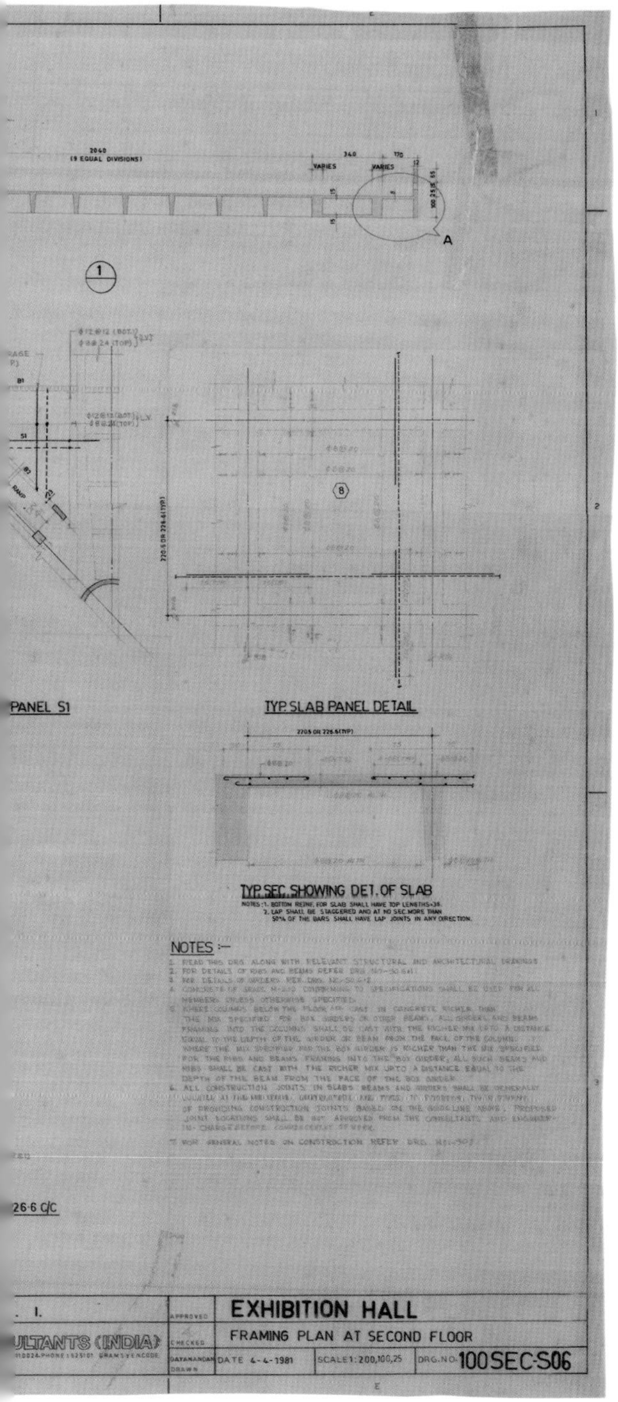

Interior and exterior photos capturing the basic structural components of this large span structure: the coffered slab, the 3.4m-diameter hollow circular columns and box girder beam.

Photo of the exterior view of the overall symmetrical form of the building punctuated by stout circular columns.

Photo of the interior central octagonal space with temporary seating at the larger cross halls for indoor table tennis and boxing matches held for the 9th Asian Games hosted in Delhi.

NDMC CITY CENTRE

NEW DELHI | 1983

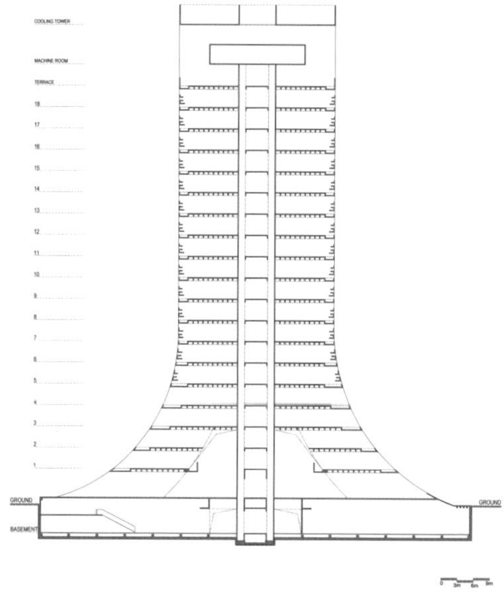

Building section showing the split at the bottom three floors of the inner shear walls.

The landmark eighteen-storey civic block, covering 24,000 square metres, for the New Delhi Municipal Corporation (NDMC) is a unique design distinguished by continuous slender walls that curve up from the ground.

'Raj Rewal and Kuldip Singh won a competition in the 1960s. At that time, they had a tower for the offices and an auditorium. We devised a framing plan for the tower and a cable structure with pre-cast ribs for the auditorium. Then nothing happened for several years,' recalls Mahendra Raj. 'After a few years, when the project was revived, Kuldip and Raj Rewal had split and NDMC came to Kuldip. We designed this eighteen-storey structure, which also faced several delays due to political upheavals in Delhi and problems with contractors.'

The structure is supported on a central core and four parallel shear walls that are 64m wide at the ground and curve up to 28m by the ninth floor. The end shear walls curve along the outside and have narrow vertical slits along their centre. The outer walls are otherwise solid. The two inner walls accommodating a three-storey-high inner atrium are split at the fourth floor and curve out, generating a great degree of horizontal thrust at the base. These forces are stabilised with a storey-high horizontal strut at the fourth floor and a floor-height post-tensioned beam above the foundation level.

Photo of rear and side facade.

Opposite: Photo of the front facade.

Previous pages: Photo of the NDMC civic centre building foregrounding the 18th century astronomical observatory, Jantar Mantar.

The slim, almost paper-thin, appearance of the shear walls is achieved by playing with their thickness which varies from 45 to 70cm. The outer shear walls have been thickened inside in mullion-like sections for strengthening. The elegant ceiling has one-way ribs spanning from wall to wall, tapering from 12cm at the bottom to 18cm at the top. The building is supported on driven cast-in-situ piles and the foundation is connected by horizontal members to cater to earthquake-generated base shears.

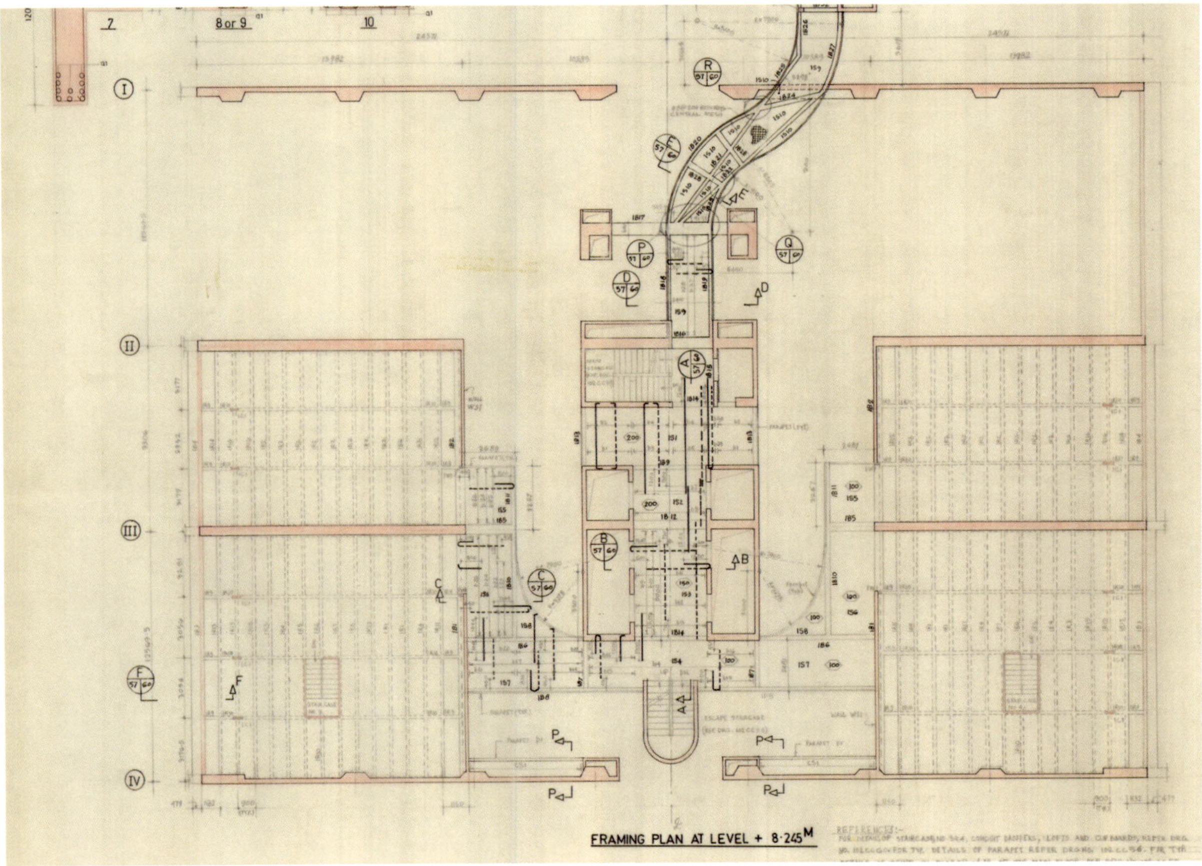

(Partial) Drawing no.102CC57: Framing Plan and Details at Level +8.245m (First floor): 1 December 1978

A typical layout plan of the building.

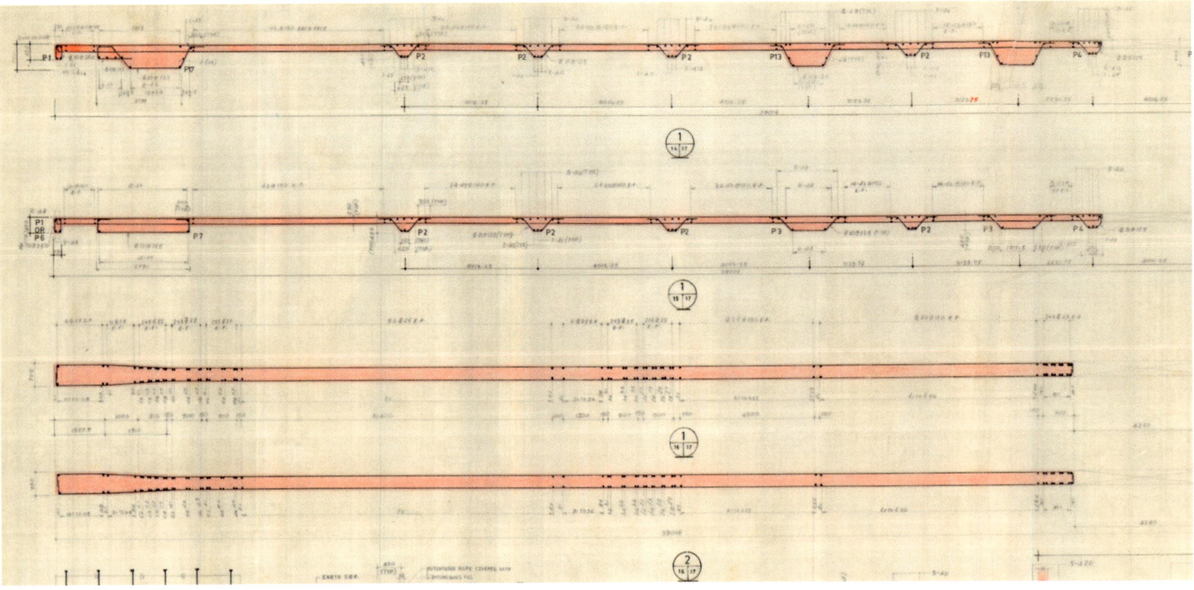

(Partial) Drawing no. 102CC 17 F: Wall Nos. 1, 2, 3 & 4 Details: 19 April 1977

Part plans of the four main shear walls. The thickness of the walls varies from 45cm to 70cm, providing stiffeners to the structure where required while also lending it a slenderness of appearance.

Interior and exterior photos of the three-storey entry foyer.

Photos of the NDMC city centre building at different stages of construction.

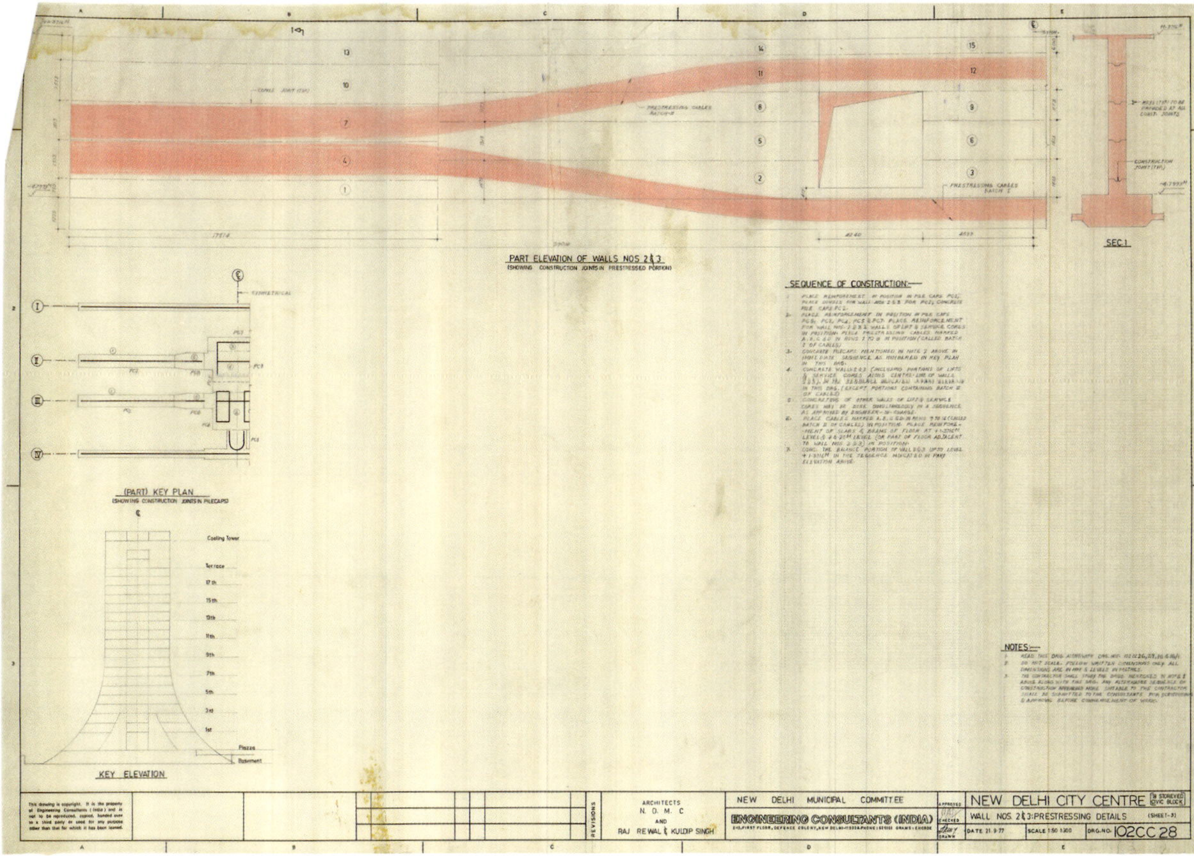

Drawing no. 102CC28: Wall No. 2 & 3 Prestressing Details (Sheet 3): 21 September 1977

Details of the two interior shear walls, No. 2 & 3, showing pre-stressing cables and construction joints, part key plan and elevation. The notes highlight the sequence of construction especially in respect to reinforcement laying at the foundation level, the pile-caps and the connection with the shear walls, construction joints and also the sequence of concreting.

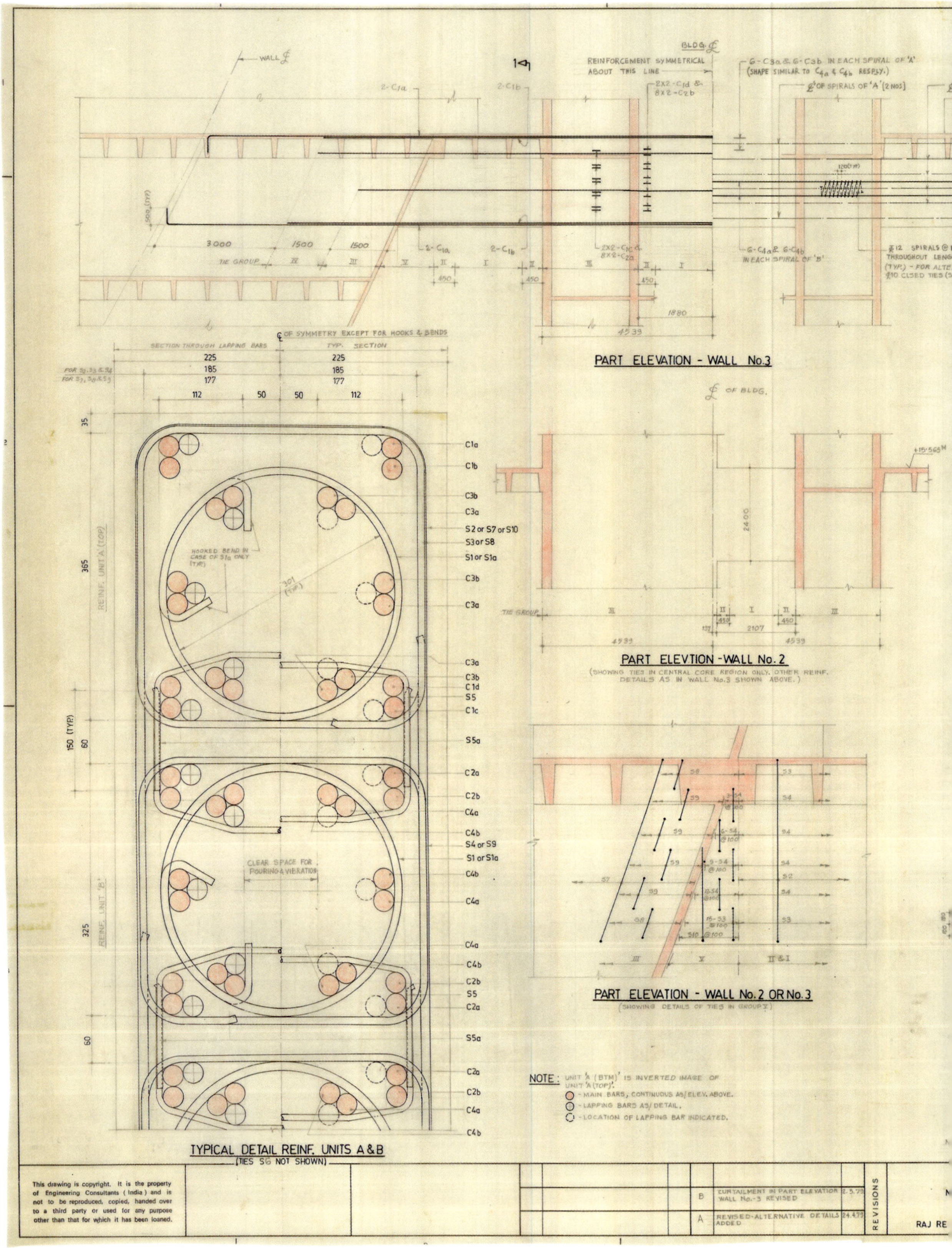

PART ELEVATION - WALL No.3
PART ELEVTION - WALL No. 2
(SHOWING TIES IN CENTRAL CORE REGION ONLY. OTHER REINF. DETAILS AS IN WALL No.3 SHOWN ABOVE.)
PART ELEVATION - WALL No. 2 OR No. 3
(SHOWING DETAILS OF TIES IN GROUP I)
TYPICAL DETAIL REINF. UNITS A & B
(TIES S6 NOT SHOWN)
NOTE: UNIT 'A (BTM)' IS INVERTED IMAGE OF UNIT 'A (TOP)'.
- MAIN BARS, CONTINUOUS AS/ ELEV. ABOVE.
- LAPPING BARS AS/ DETAIL.
- LOCATION OF LAPPING BAR INDICATED.
This drawing is copyright. It is the property of Engineering Consultants (India) and is not to be reproduced, copied, handed over to a third party or used for any purpose other than that for which it has been loaned.
B CURTAILMENT IN PART ELEVATION WALL No.-3 REVISED
A REVISED-ALTERNATIVE DETAILS ADDED 24.4.73
REVISIONS

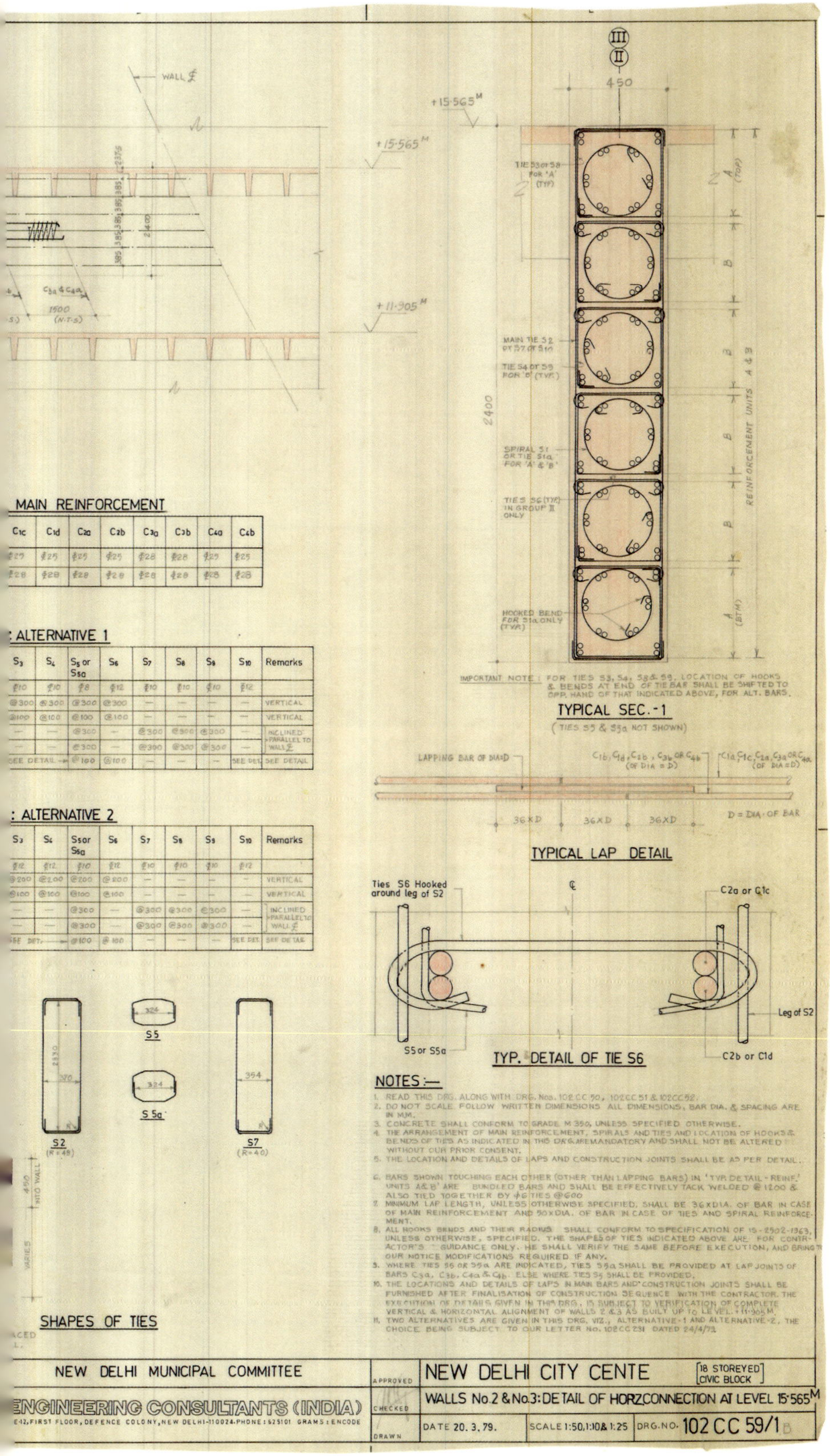

Drawing no. 102CC59/1 B: Wall Nos. 2 & 3, Details of Horizontal Connection Level +15.565m: 20 March 1979

The two interior walls no. 2 and 3 flare open from the ground to the third floor creating forces in the structure that are contained by storey-height horizontal struts at the fourth floor.

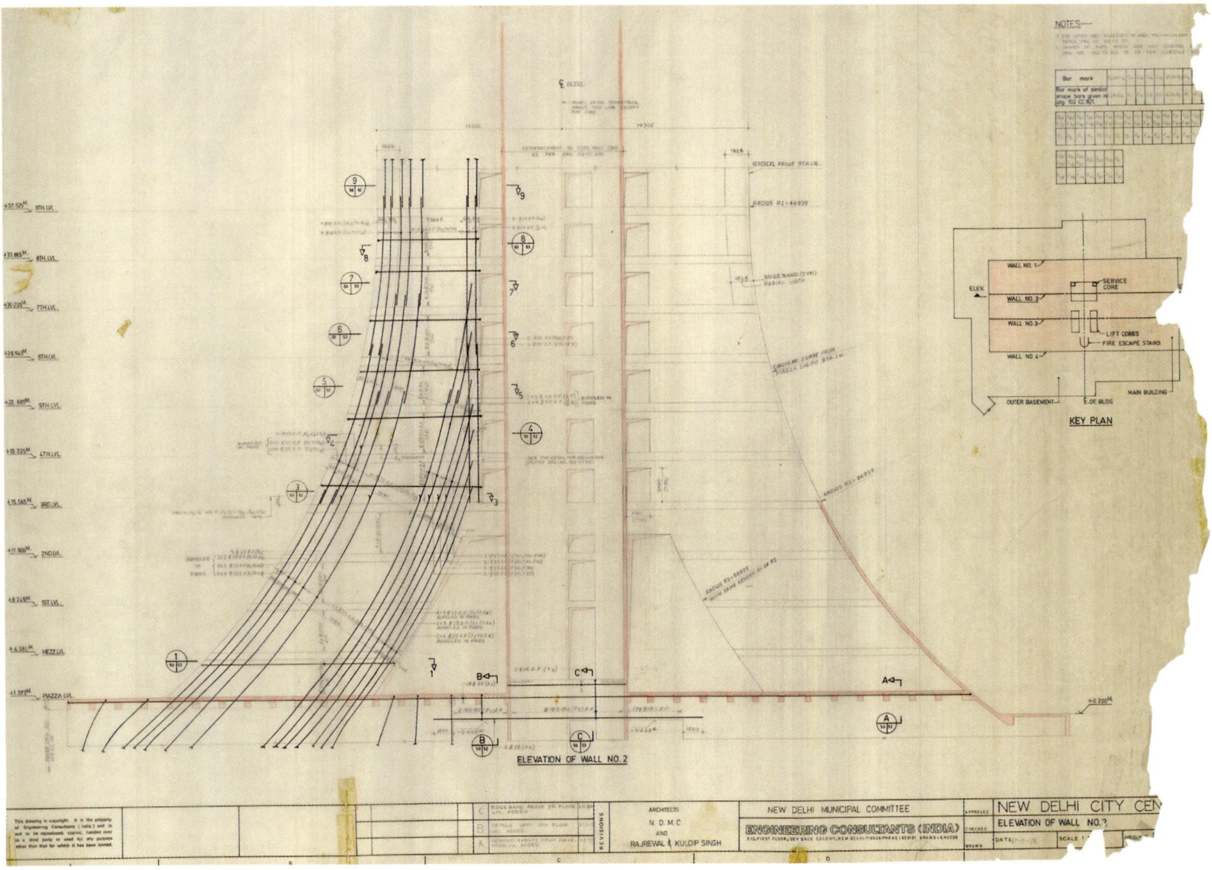
NOTES
KEY PLAN
WALL NO. 1
WALL NO. 2
WALL NO. 3
WALL NO. 4
SERVICE CORE
LIFT CORES
FIRE ESCAPE STAIRS
OUTER BASEMENT
MAIN BUILDING
ELEV.
PIAZZA LVL.
ELEVATION OF WALL NO. 2
REVISIONS
ARCHITECTS
N.D.M.C
AND
RAJREWAL & KULDIP SINGH
NEW DELHI MUNICIPAL COMMITTEE
ENGINEERING CONSULTANTS (INDIA)
ELEVATION OF WALL NO.2

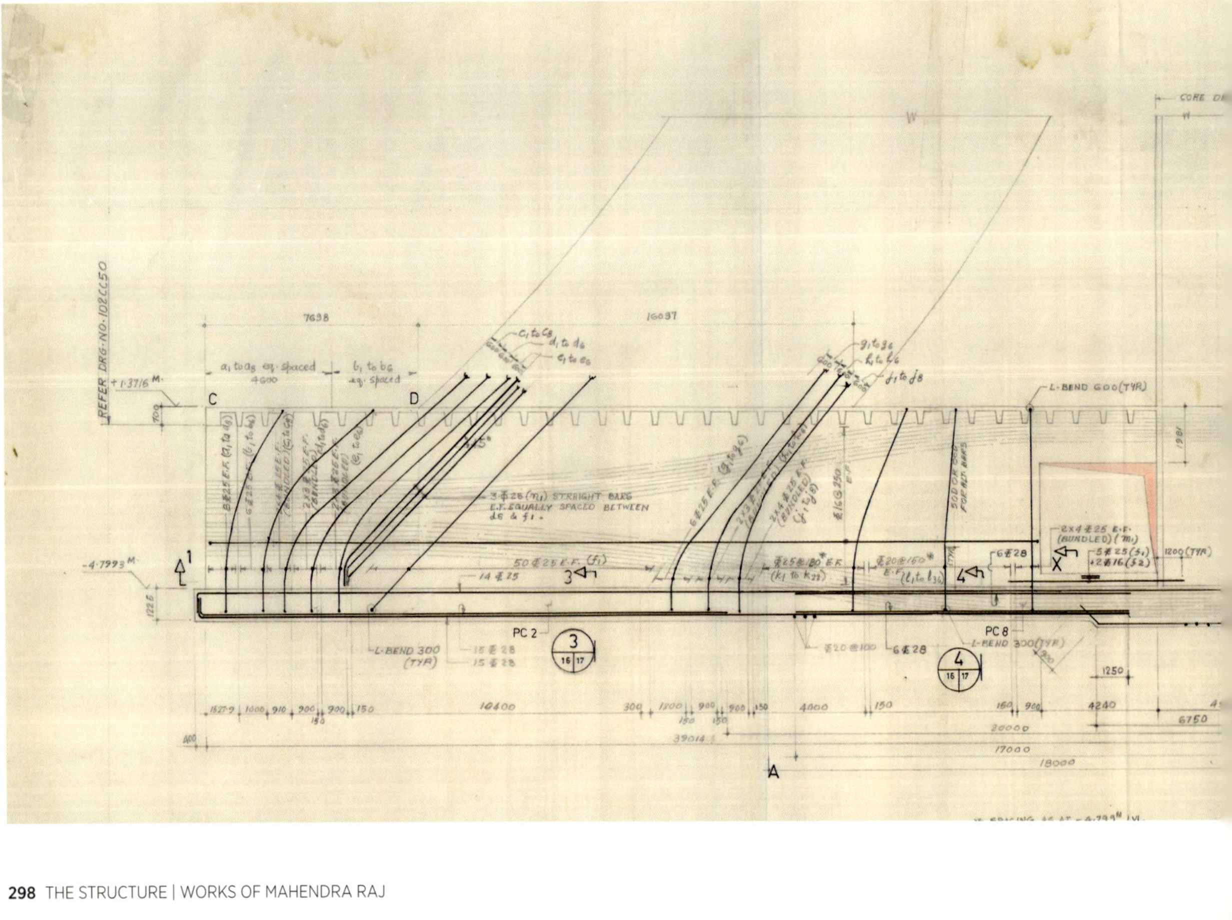
REFER DRG-NO-102CC50
7638
16097
4600
C
D
L-BEND 600 (TYP)
3 # 25 (n1) STRAIGHT BARS E.F. EQUALLY SPACED BETWEEN d6 & f1
50 # 25 E.F. (f1)
14 # 25
-4.7993 M.
1250
L-BEND 300 (TYP)
15 # 28
PC 2
PC 8
10400
4000
4240
6750
39014
20000
17000
18000
A

Drawing: Elevations of Wall No 2: 17 May 1978
(Partial) Drawing no. 102CC16 J: Elevations of Wall 2 & 3: 20 April 1977, showing the elevation and reinforcement of the inner shear wall 2 at the level below ground. The connection of the reinforcement bars and pile caps is shown with post-tensioning cables running along the floor height beam at the basement level.

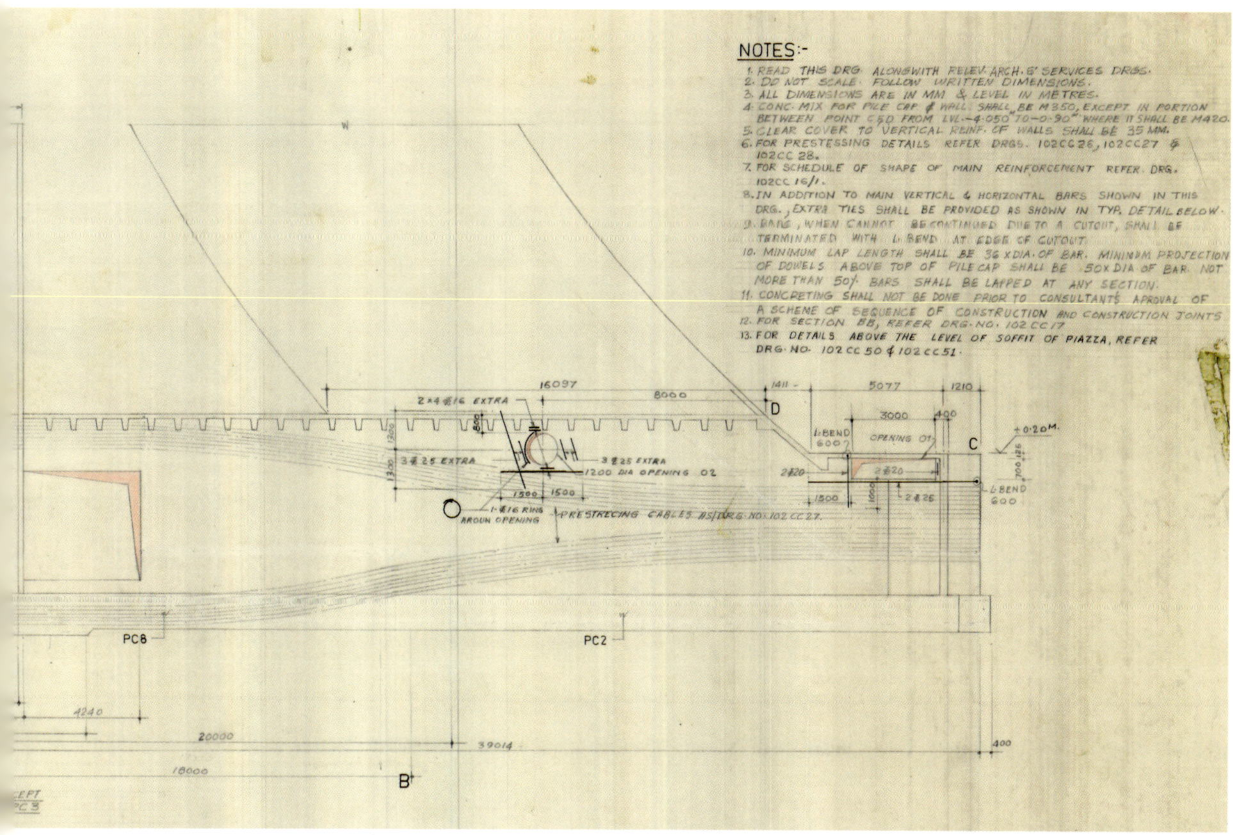

Photo of Mahendra Raj walking through the three-storey foyer 27 years after the project's completion.

Exterior view, including a part of Phase II of the project completed in 2010.

CONVENTION CENTRE
NEW DELHI CITY CENTRE
THE PARK

LIC JEEVAN BHARATI

NEW DELHI | 1986

Located in New Delhi's central business district, the Life Insurance Corporation building is another milestone project by Mahendra Raj, marking one of the first uses of the space frame and curtain wall systems in India.

The 40,000 square metre building is formed of two thirteen-storey towers supported on large-diameter, cast-in-situ piles. The towers, with completely asymmetric cores and shear walls, spring out of a three-level podium with two basements and a terraced plaza. Transfer girders are used to allow entry into the basement levels below the plaza level. An efficient double-dropped flat slab system on an 8m x 8m grid forms the floor plate. Raj, who used the double drop flat slab here for the first time, further developed this system in his subsequent projects as an effective solution for large-span and multi-storey structures.

An open-grid parasol on the roof connects the two towers and gives this building its commanding presence. The space frame system suggested by the contractor replaced the earlier design of a two-way truss with a walkway. The space frame rests on the two main towers and two 45m-high free-standing columns–one circular and the other triangular. The earthquake requirements of the roof parasol were met by allowing movement in two of the four elements supporting it. The erection of the space frame was planned to completely eliminate the need for scaffolding. As the system was being used for the first time in India, a large number of tests of representative elements were carried out at the laboratories of the Indian Institute of Technology, Delhi.

Previous pages: Exterior view of the building.

Opposite: Exterior side view.

Space truss supported on a free-standing circular column.

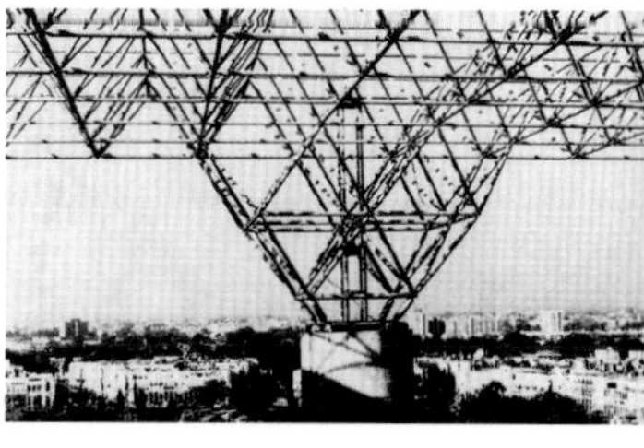

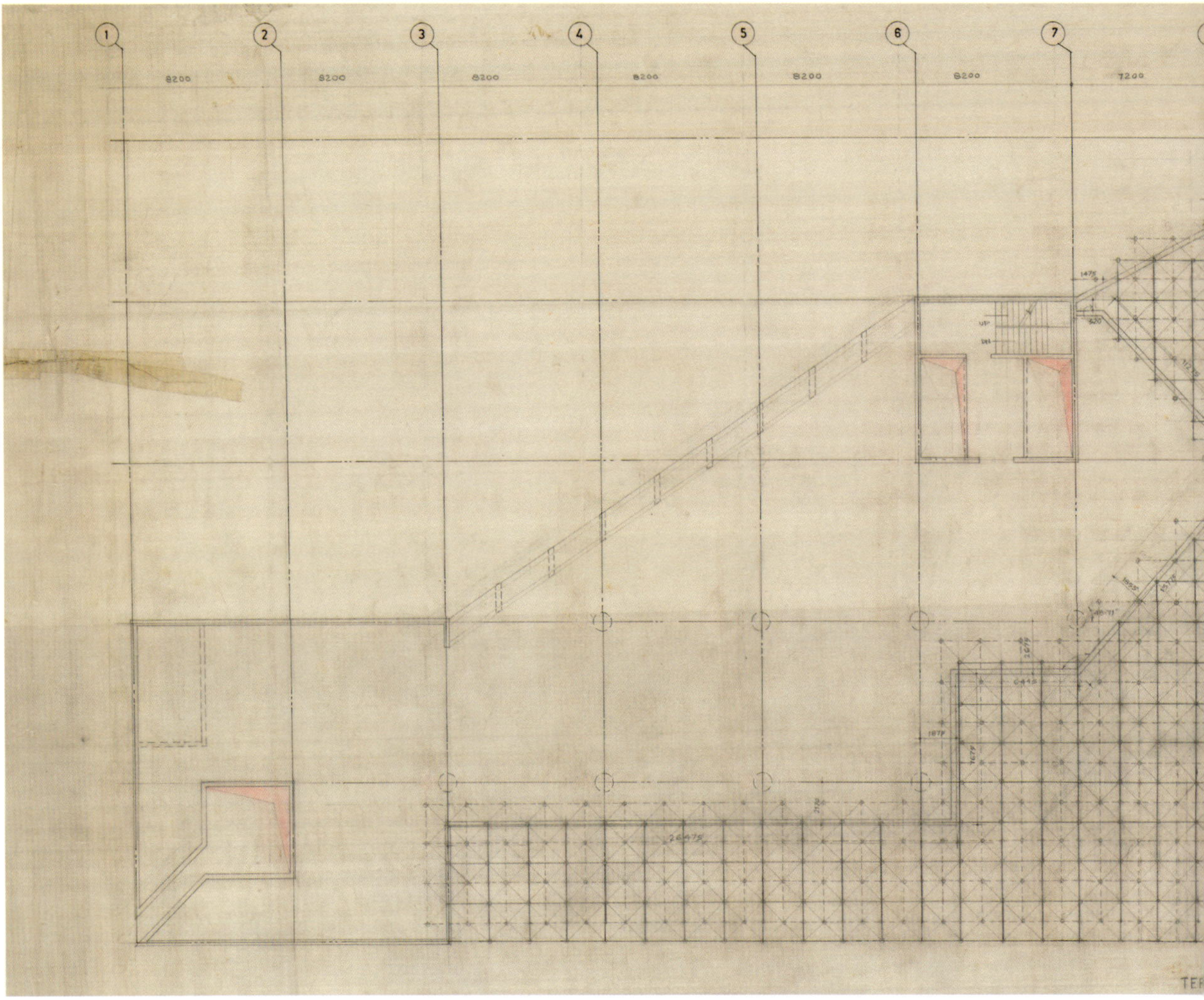

(Partial) Drawing no.121-MC-T/28: Terrace Plan–Space Frame: 16 October 1980
Plan sketch of the space frame in relation to the building profile.

Photos of the on-site space frame fabrication and erection.

Previous pages: Exterior view of the two thirteen-storey towers and the free-standing columns before the installation of the space frame.

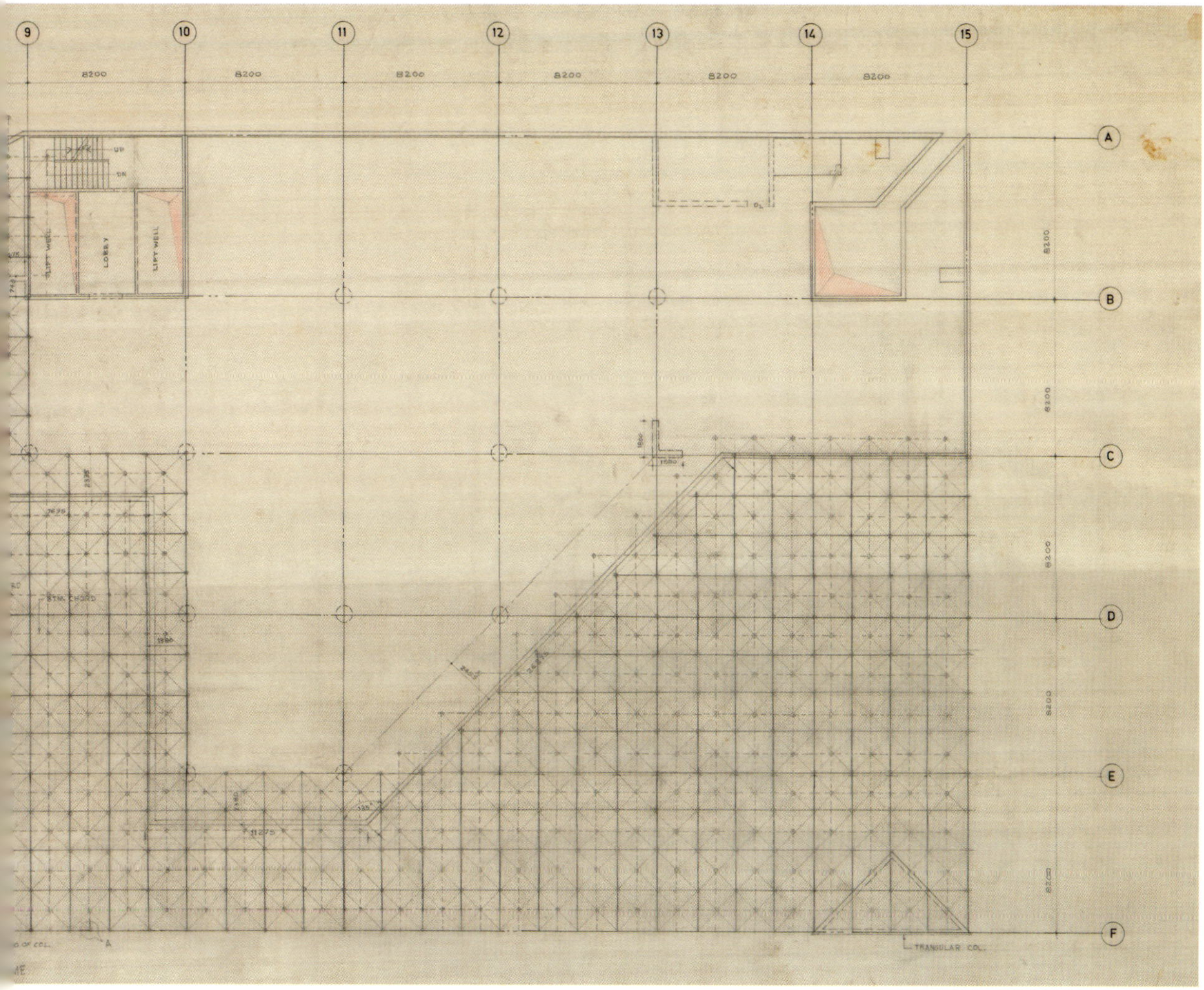
9
10
11
12
13
14
15
8200
8200
8200
8200
8200
8200
UP
DN
LIFT WELL
LOBBY
LIFT WELL
A
B
C
D
E
F
BTM. CHORD
TRIANGULAR COL.

Rear view of the building.

View of the space frame and curtain wall from the plaza level. Also seen here is an innovative system of rails attached on the curtain-wall edge of the space frame to facilitate the cleaning and maintenance of the building.

SCOPE COMPLEX

NEW DELHI | 1987

An eight-storey office complex, including two basements for the Standing Conference of Public Sector Enterprises (SCOPE), provides a total covered area of 100,000 square metres. The main structure is formed by a ribbed system supported on clusters of four columns on a 13m x 13m grid. The four columns, placed 3m apart, form cores that cater to seismic forces and also provide vertical duct space through the building. What emerges is a modular structural system that creates open plan office spaces and accommodates service requirements too. Maintaining the grid, the structural system also allows for longer diagonal connections, cantilevering boxes and large-span facilities such as an auditorium with the use of transfer girders.

Architectural building model.

View of the column clusters that form the geometry and structural system of the building.

Previous pages: Front view of the SCOPE building.

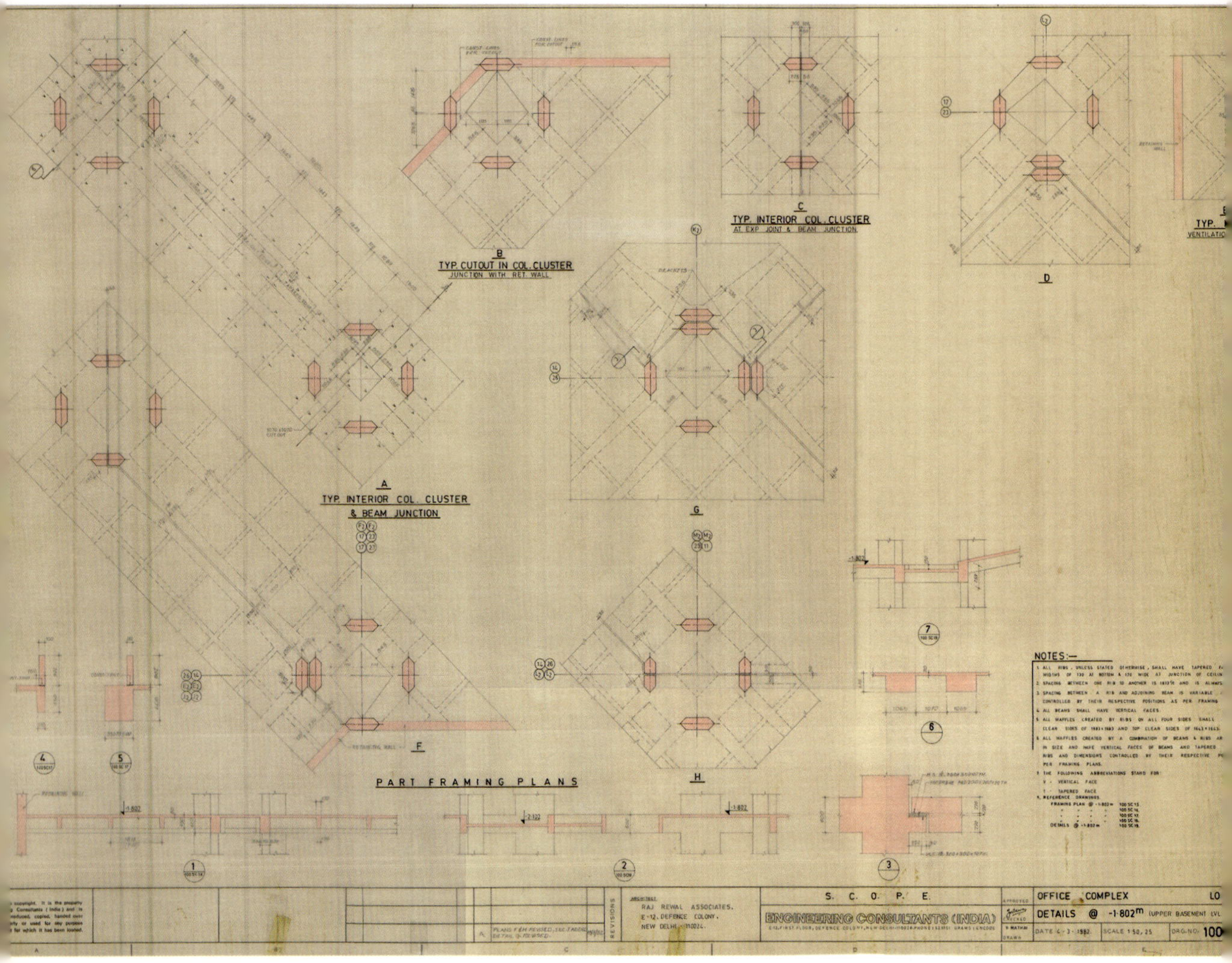

Drawing no. 100SC19 A: Details @-1.802m (Upper Basement Level): 4 March 1982

Typical conditions of the column clusters detailed with their beam junctions in the building.

Opposite: Plans for different levels and corners of the complex.

Drawing no. 100 SC 77: Framing Plan at 25.255m (Sixth Floor): 9 December 1982

Drawing no. 100 SC 67 A: Framing Plan at 18.455m (Fourth Floor): 2 December 1982

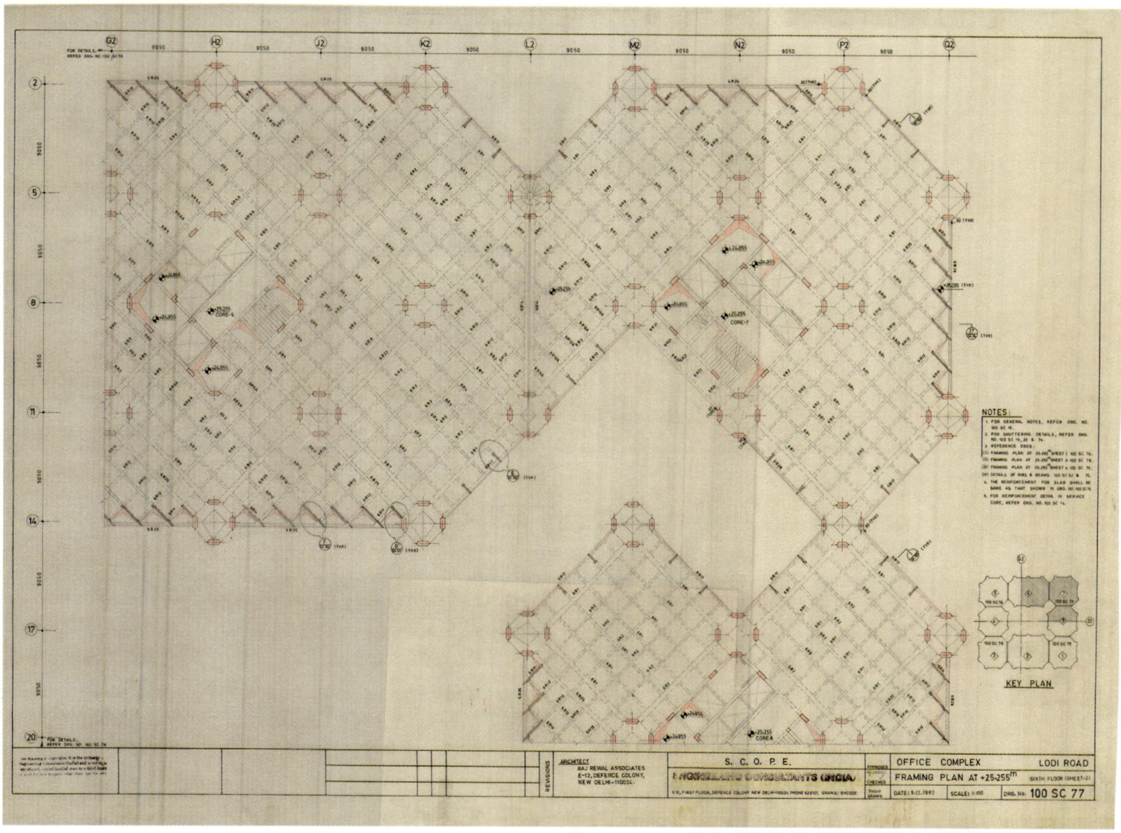
NOTES:
KEY PLAN
ARCHITECT
RAJ REWAL ASSOCIATES
E-12, DEFENCE COLONY,
NEW DELHI-110024.
REVISIONS
S. C. O. P. E.
OFFICE COMPLEX
LODI ROAD
FRAMING PLAN AT +25·255m
SIXTH FLOOR (SHEET-2)
DATE: 9.12.1982
SCALE: 1:100
DRG. No. 100 SC 77

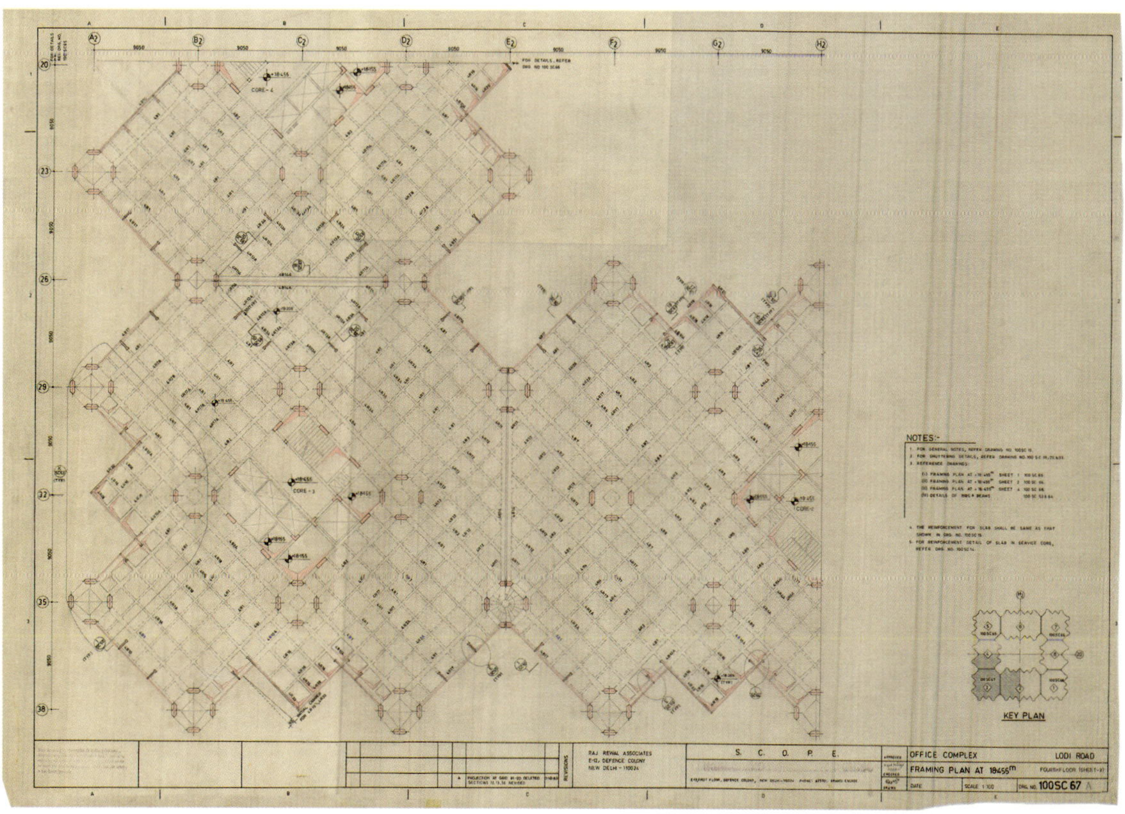
NOTES:-
KEY PLAN
RAJ REWAL ASSOCIATES
E-12, DEFENCE COLONY
NEW DELHI - 110024
REVISIONS
S. C. O. P. E.
OFFICE COMPLEX
LODI ROAD
FRAMING PLAN AT 18·455m
FOURTH FLOOR (SHEET-2)
DATE
SCALE 1:100
DRG. NO. 100SC 67

Building views showing cantilevers and cut-outs created within a system of slender beams and ribs supported on column clusters.

SCOPE Convention Centre
SCOPE

STATE TRADING CORPORATION

NEW DELHI | 1988

The State Trading Corporation's (STC) office building, distinguished by its storey-high Vierendeels that cantilever off vertical service cores, is a defining landmark of central Delhi. The high-rise structure is composed of two office towers of unequal height placed perpendicular to one another. The towers form two edges of the three-storey base structure used for retail activity. Mahendra Raj says, 'The complex posed a number of very challenging and interesting design problems with four service cores carrying the entire building of such large spans. Initially the analysis was all done manually, but as the project took more than a decade to complete we had the chance to verify our results with the help of computers before the construction of the superstructure.'

Each tower is formed by two Vierendeel trusses 14.4 m apart on every alternate floor, resting on 9.6m × 27.2m wide vertical service cores. The taller tower has a 16.1m × 56.1 m floor plate that rises up eighteen storeys above which a floor plate measuring 16.1m x 24.1m rises up to twenty-three storeys. Till the eighteenth floor the tower is supported on two service cores with Vierendeels spanning 22.4m and cantilevering out by 6.4m on both sides. Cantilevers measuring 7.25m x 16.1m

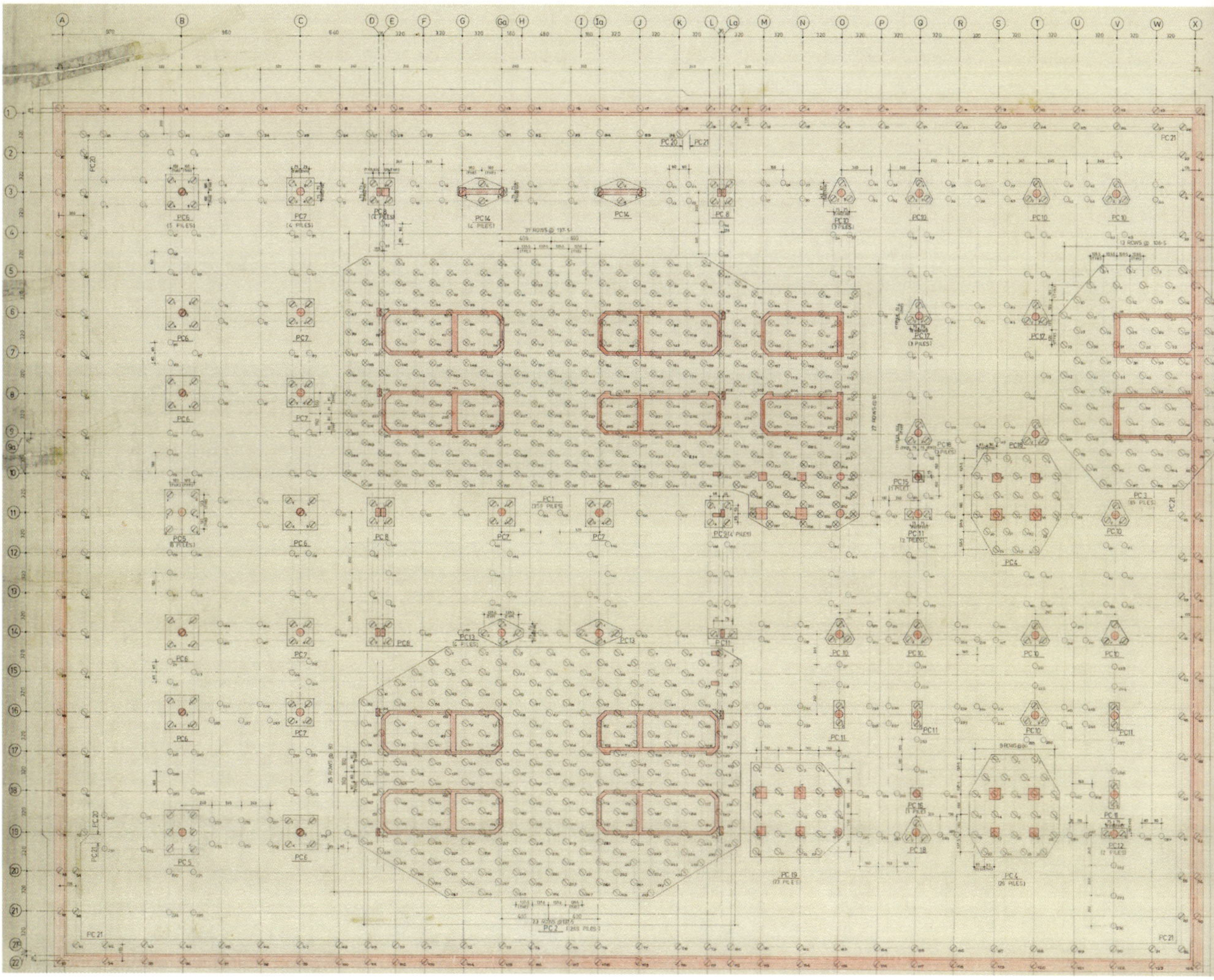

are anchored to a single service shaft forming the top five floors. The second tower is twelve storeys high with a smaller floor plate of 16.1m x 22.4m. This tower is supported on shear walls cantilevering out of two cores spanning 22.4m. The Vierendeel trusses support one-way reinforced concrete ribs placed 1.6m apart, creating open-plan office spaces of over 44,000 square metres.

Bird's-eye view of the building's cantilevered staggered forms rising above central Delhi.

(Partial) Drawing no. 101ST- S01/A: Layout of Piles: 21 December 1977
The high-rise has two basements and is founded on 18–20m deep in-situ reinforced concrete piles.

Previous pages: The cascading 23, 18 and 12-storey towers of the STC building with floor-height Vierendeels and vertical service cores form a forceful structural expression that dominates the building facade.

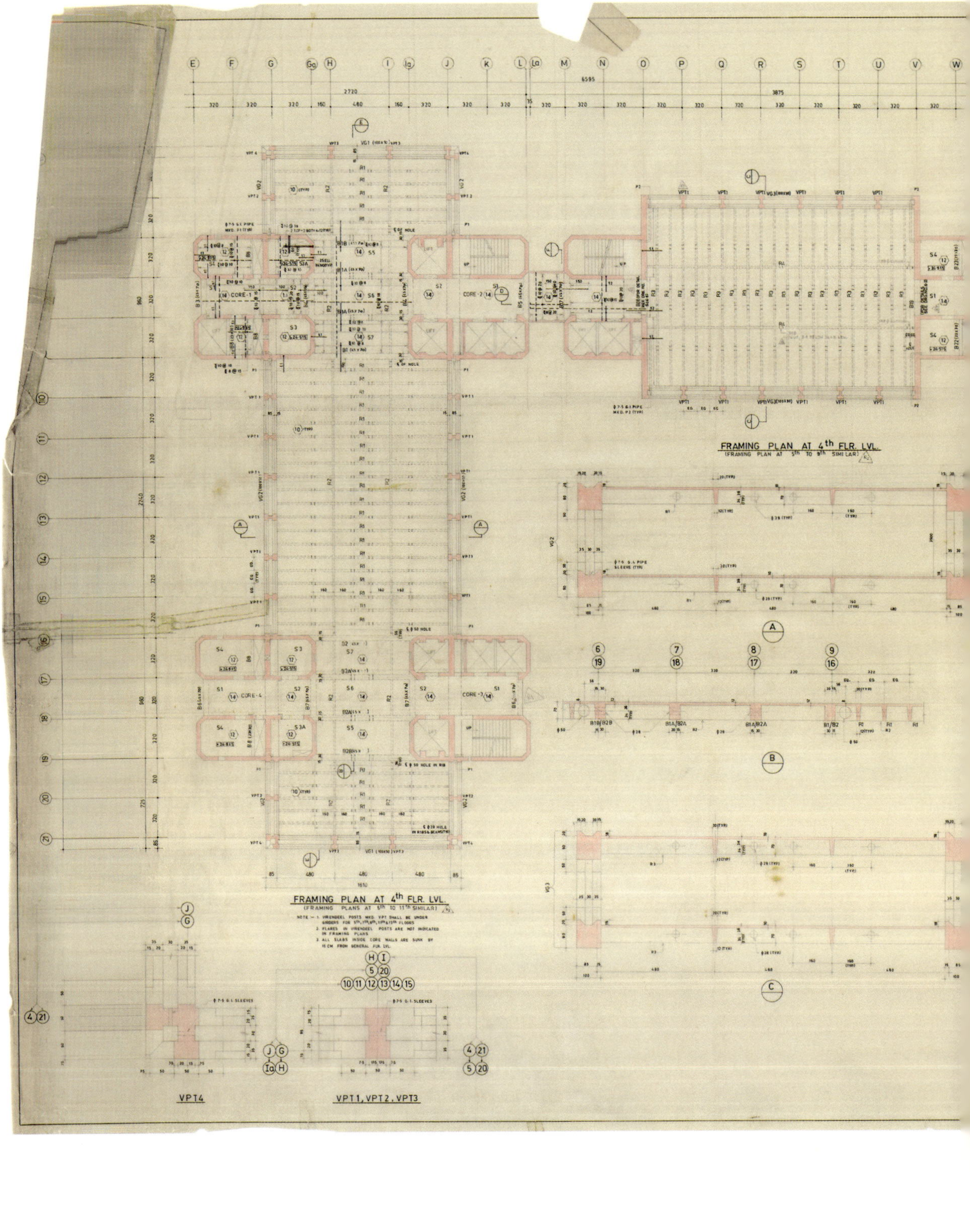
FRAMING PLAN AT 4th FLR. LVL.
(FRAMING PLANS AT 5th TO 11th SIMILAR)
FRAMING PLAN AT 4th FLR. LVL.
(FRAMING PLAN AT 5th TO 9th SIMILAR)
CORE-1
CORE-2
CORE-3
CORE-4
VPT4
VPT1, VPT2, VPT3

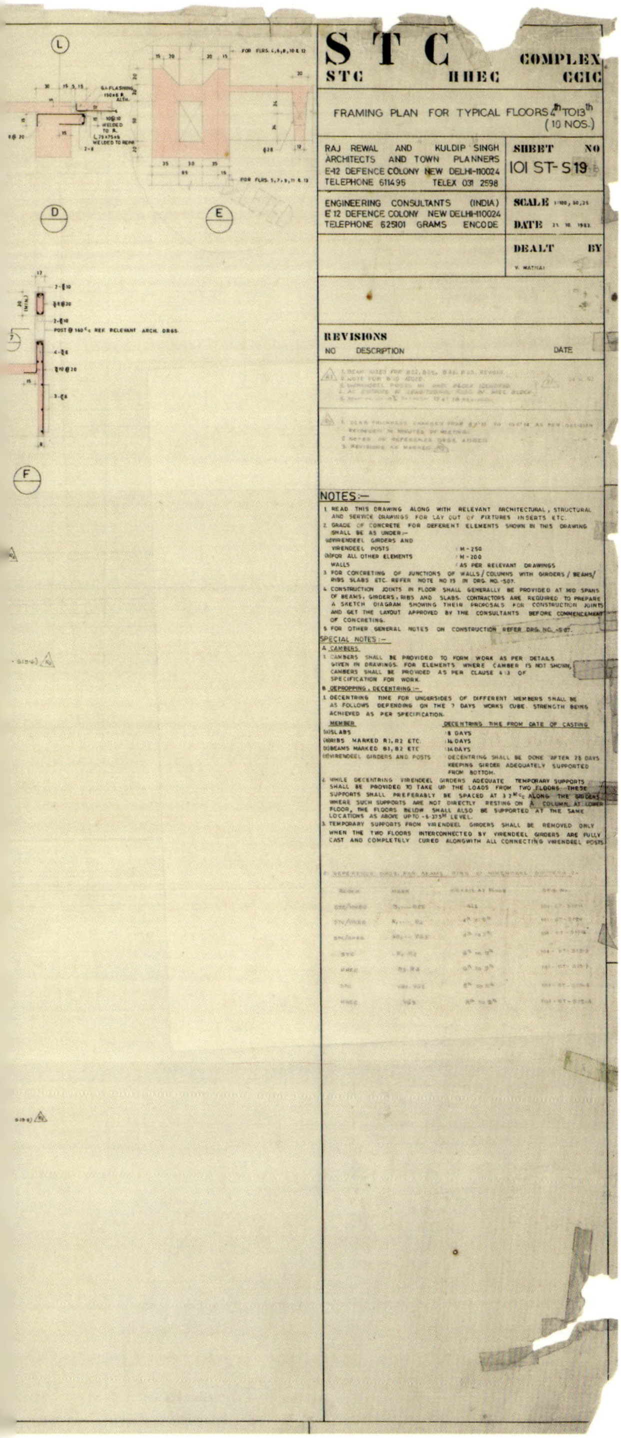

Architectural building model showing the composition of the meandering three-storey base structure and the towers above.

Drawing no. 101ST- S19-B: Framing Plan for Typical Floors 4th-13th: 21 August 1983

A typical framing plan showing the basic structural system of the building. The exterior walls comprising storey-height Vierendeel girders 14.4m apart support one-way reinforced concrete ribs 1.6 m apart that cantilever out or span between the four service cores.

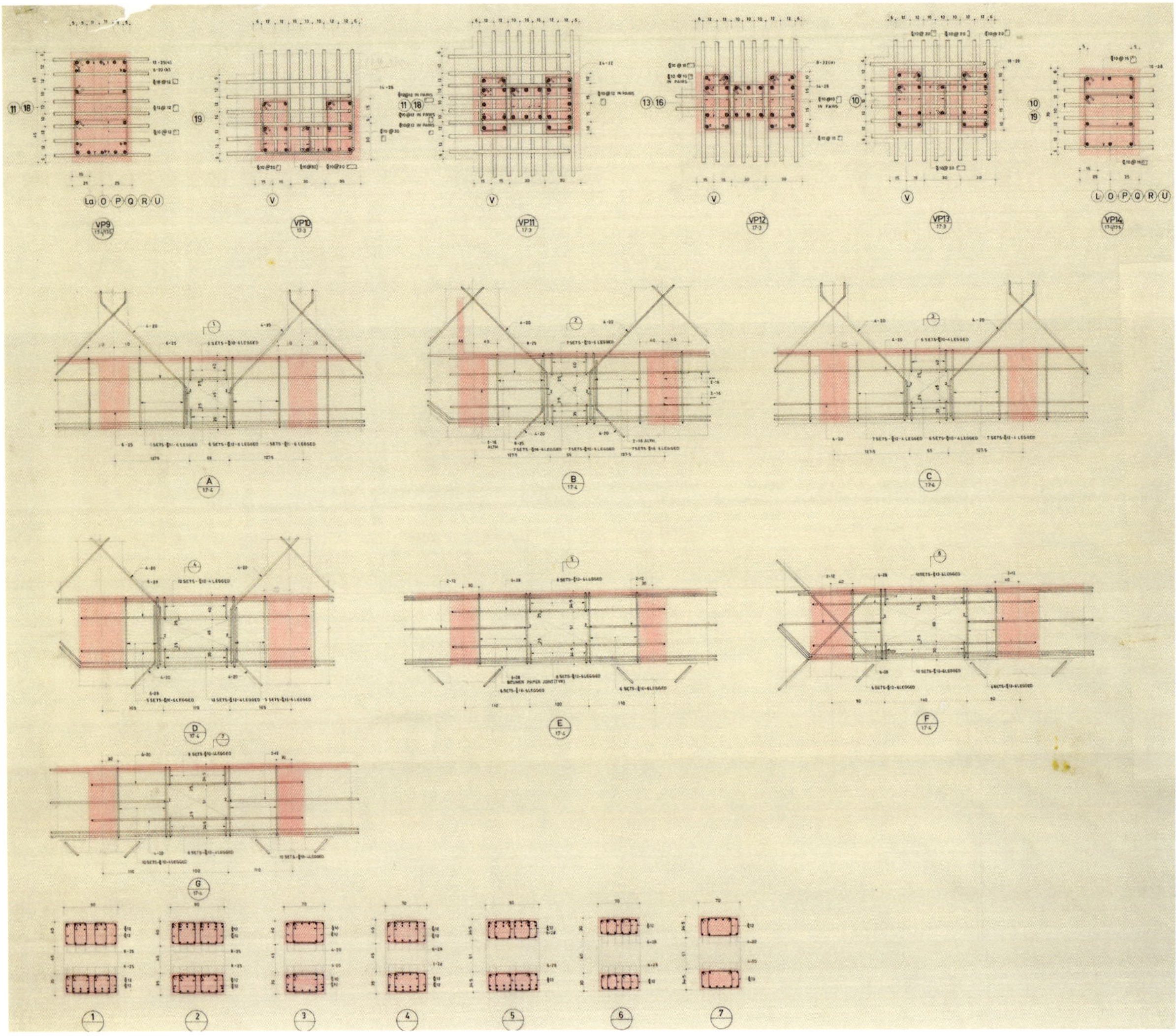

(Partial) Drawing no. 101ST- 17.6A: Details at 2nd and 3rd Floors CCIC Block:
6 February 1985

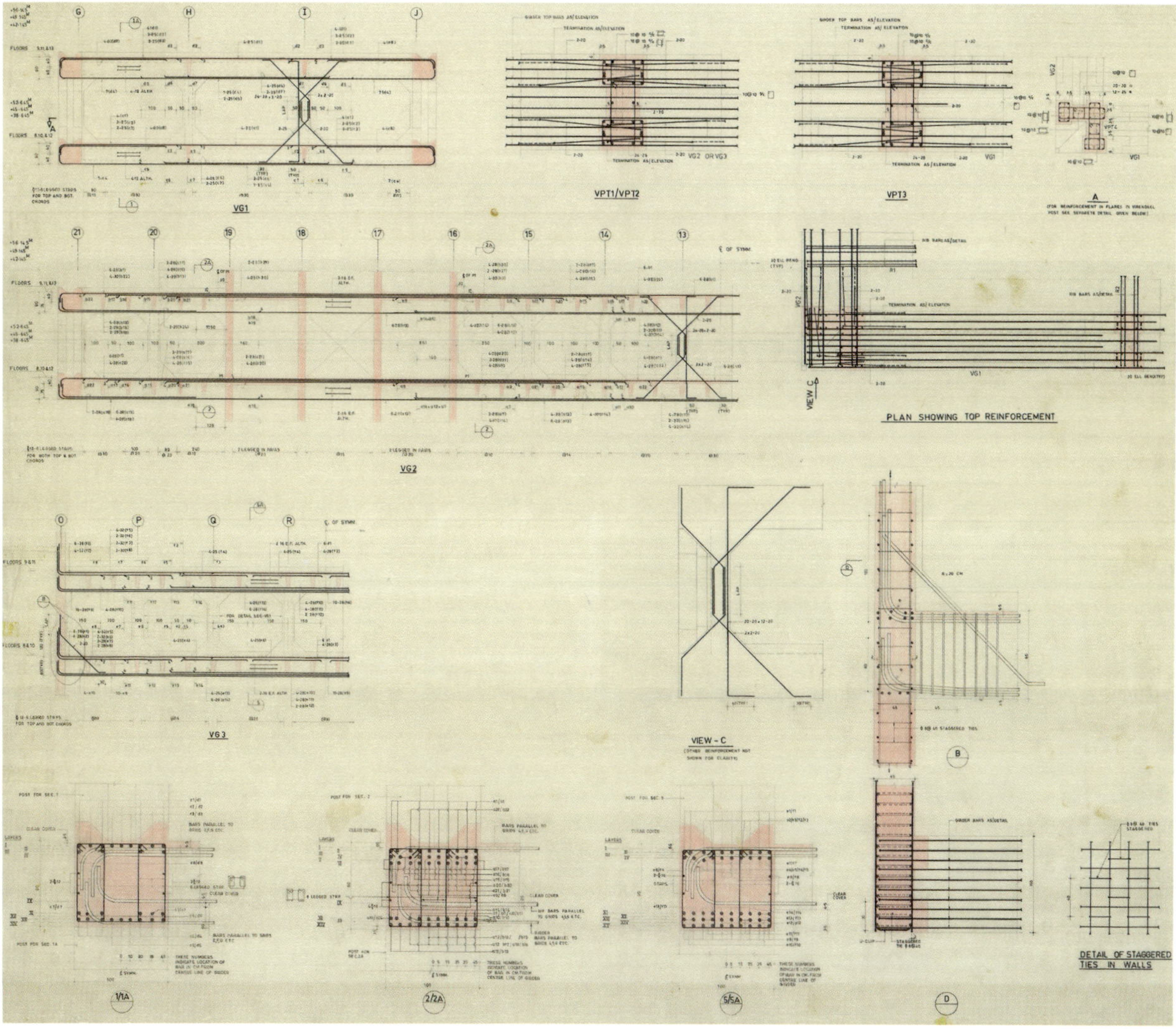

(Partial) Drawing no. 101ST- S19-4: Detail of Girders at 8th to 13th Floor: 8 August 1984

View of the cantilevered Vierendeel trusses emerging from a core with the spanning ribs visible within the building during construction.

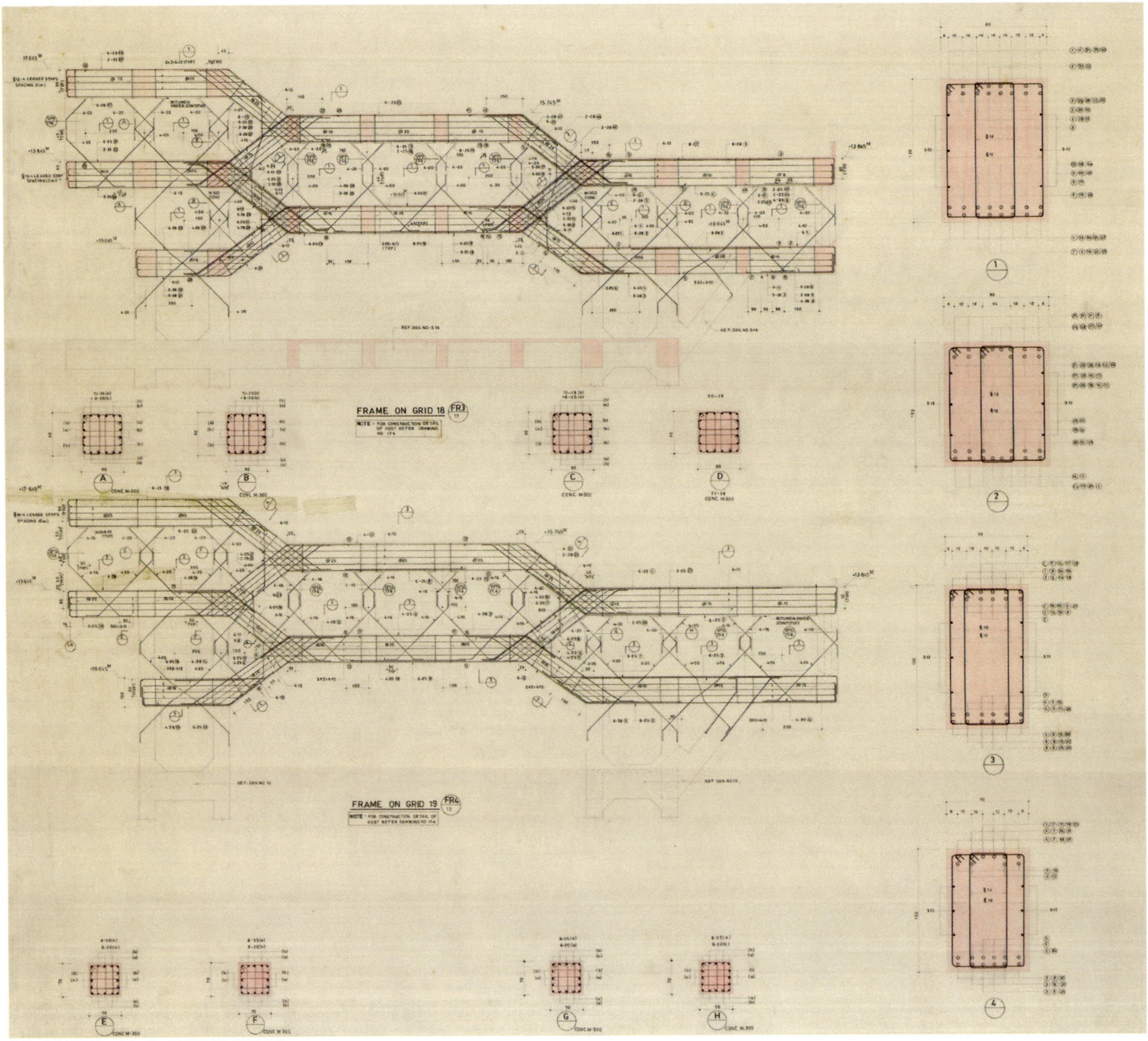

(Partial) Drawing no. 101ST- 17.5 A: Detail of Frames FR3 & FR4 CCIC Block
Drawings of reinforcement detail for the meandering storey-height Vierendeels supporting different levels of the shopping mall on the first three floors.

Photo from across the street showing the three-storey base with meandering Vierendeels and the two towers of staggered heights that rise from it.

FLOWER MARKET

MADRAS | 1989

ஜெயச்சந்திரன் மலர் நிலையம்
JAYACHANDRAN

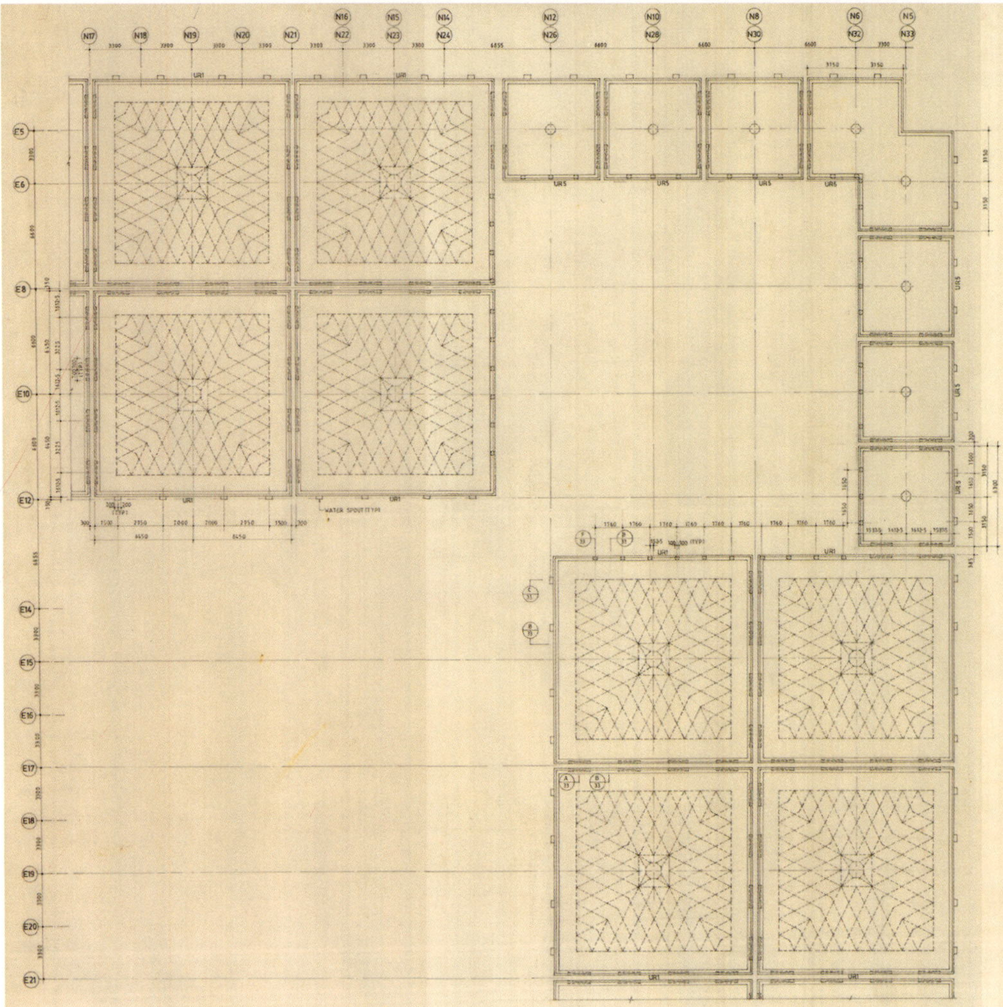

The relief on the umbrellas that form the flower market in Madras is a testament to Mahendra Raj's engineering and aesthetic acumen, which brings immense value to any project. He explains, 'As we have these cantilevers, which require more thickness in the centre and as it is a flower market, and also, as Kuldip Singh wanted to use exposed concrete, we thought of generating this in the image of a flower—a flower over flower that becomes thinner towards the edges.'

What followed were explorations of patterns by Raj to finally arrive at the diamond-shaped petals that stagger and recess back to the edges. An engineering resolution that brought forth an idea of floral relief and the architect's mastery over shuttering systems endows the flower market structure with tremendous elegance and richness. The relief on the cantilevers fulfils structural and aesthetic intentions and the plates supported on a central column create a module that repeats itself in the design of the market. Eight umbrellas, 12.9m x 12.9m in dimension, form the main spaces, and smaller umbrellas of 6.3m x 6.3m line the outer ring of the courtyard.

(Partial) Drawing no. 101FM 34: Plan at +18.59m (showing umbrella roofs) Part C and D : 25 October 1988

Photos of the flower profile on the roofs of the market pavilion.

Previous pages: View from the mezzanine of the central courtyard showing one of the main spaces covered by 12.9m umbrellas.

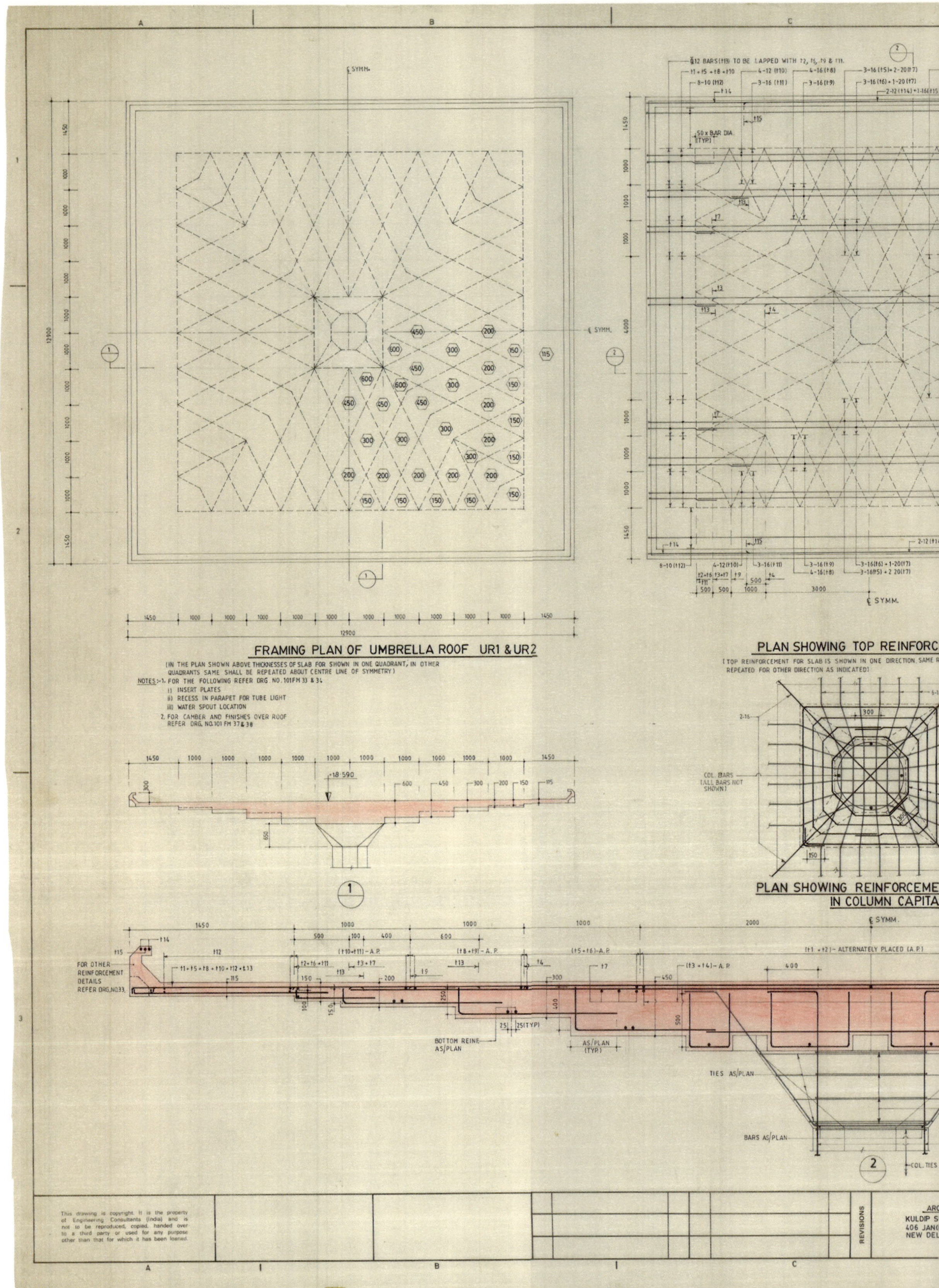
FRAMING PLAN OF UMBRELLA ROOF UR1 & UR2
(IN THE PLAN SHOWN ABOVE THICKNESSES OF SLAB FOR SHOWN IN ONE QUADRANT, IN OTHER QUADRANTS SAME SHALL BE REPEATED ABOUT CENTRE LINE OF SYMMETRY)
NOTES:-1. FOR THE FOLLOWING REFER DRG. NO.101FM 33 & 34
i) INSERT PLATES
ii) RECESS IN PARAPET FOR TUBE LIGHT
iii) WATER SPOUT LOCATION
2. FOR CAMBER AND FINISHES OVER ROOF REFER DRG. NO.101 FM 37 & 38
PLAN SHOWING TOP REINFORC
PLAN SHOWING REINFORCEME
IN COLUMN CAPITA
COL. BARS (ALL BARS NOT SHOWN)
FOR OTHER REINFORCEMENT DETAILS REFER DRG.NO.33.
BOTTOM REINF. AS/PLAN
TIES AS/PLAN
BARS AS/PLAN
COL. TIES
This drawing is copyright. It is the property of Engineering Consultants (India) and is not to be reproduced, copied, handed over to a third party or used for any purpose other than that for which it has been loaned.
REVISIONS
KULDIP S
406 JAN
NEW DEL

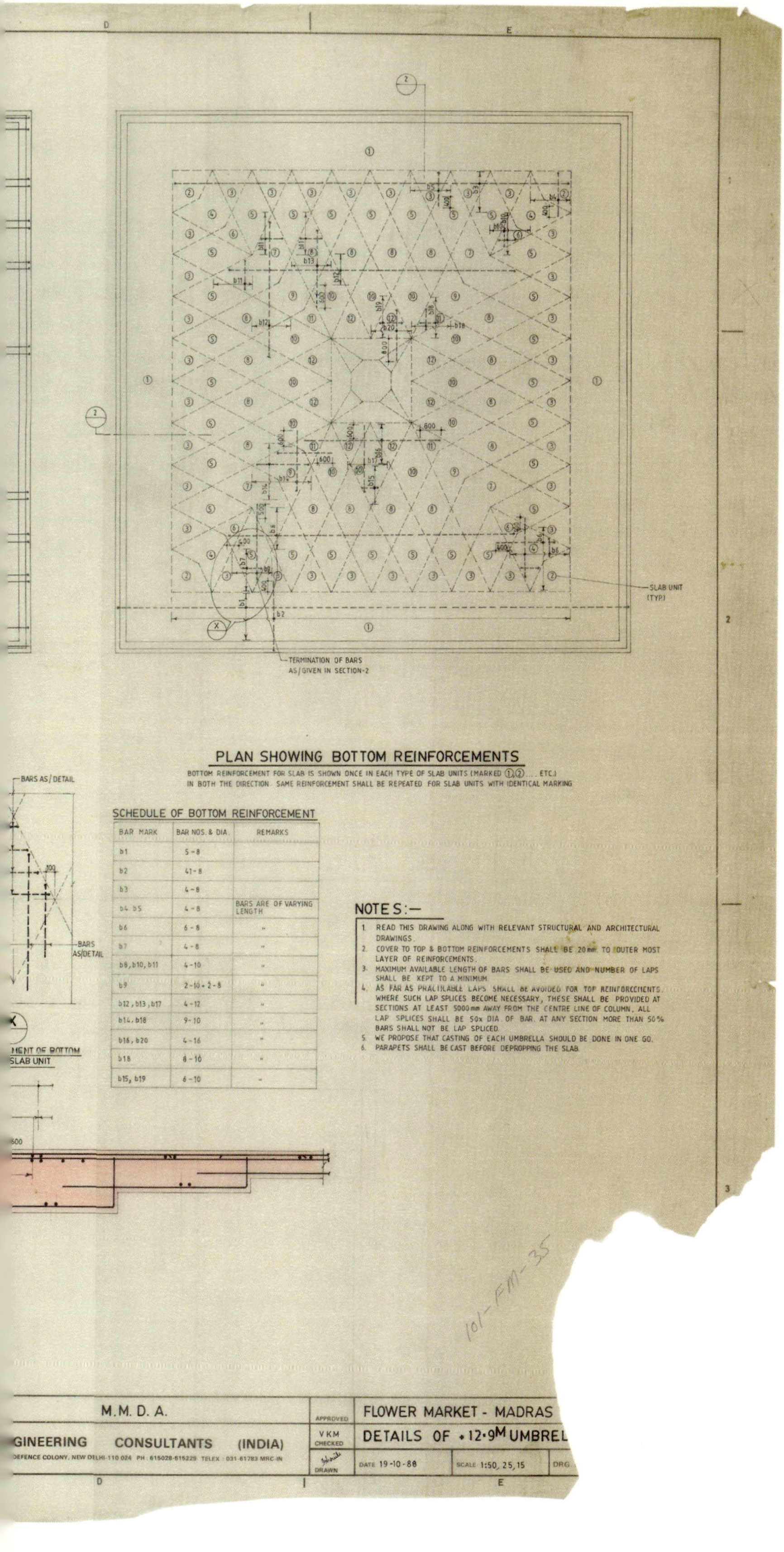

Drawing no. 101FM 35:
Details of 12.9m Umbrella Roofs : 19 October 1988

Drawing of a 12.9m umbrella with a framing plan includes details in plans and sections for the bottom and top reinforcement of the umbrella roof and supporting capital. The floral structure follows the reinforcement stepping up to the edges.

Engineering a Nation

Interview by Hans Ulrich Obrist

When India gained independence in 1947, a generation of architects was tasked with expressing, through their buildings, the identity of a new and modern nation. The most inventive and pioneering among these projects shared a common feature: the engineering genius of Mahendra Raj. Hans Ulrich Obrist interviews one of the founding fathers of Indian modernity.

Hans Ulrich Obrist *Your trajectory has led you to redefine the history of Indian engineering. How did you become an engineer in the first place?*
Mahendra Raj I always wanted be an engineer. Both my father and two older brothers were engineers, and I was really fond of mathematics. I graduated in 1946 and started working in the Punjab Public Works Department, so I began my career with government service. I studied in Lahore when India and Pakistan were still one country, but then the Partition took place and I came to India. In those days, there weren't many avenues for young men. Architecture was not known. Engineering and Medicine were the two most sought-after disciplines, along with the Army, and it was not easy to gain admission into engineering colleges. After the Partition, I soon became involved in the construction of Chandigarh, and the first structural project I worked on was Le Corbusier's High Court–Jane Drew, Maxwell Fry and Pierre Jeanneret were stationed there at the time. Le Corbusier used to come twice a year, and I was lucky enough to be working on his projects–the High Court and the Secretariat building–so I met him twice in the early 1950s. After I had worked in Chandigarh, I thought I should further my studies, and so I went to America in 1955 to pursue a Master's degree in Minnesota, and then Columbia University for a post-Master's. After that, I went to New York and took a job with Ammann and Whitney, who were doing the most forward-looking large-span structures at the time.

Hans *Who were your heroes in those years?*
Raj Foremost was Pier Luigi Nervi. The best-known structural engineers at the time were Eduardo Torroja, Félix Candela and Othmar Ammann. As I said, I worked with Ammann and Whitney, and it was Whitney who devised methods for analysing folded plates. I also had the pleasure of meeting Nervi when I travelled to Italy and introduced myself. I worked with some well-known architects in New York too, such as Minoru Yamasaki, who became a good friend. That's how I went from being a civil engineer to a structural engineer. In 1960, I came back to India, resigned from my previous position and started my own practice.

Hans *1960 is about the same time that Cedric Price opened his office in London. Many things happened in 1960. It was also the beginning for extraordinary architects such as Charles Correa and BV Doshi with whom you have a common history. They would not have been possible without you and vice versa.*
Raj Yes, we were complementary. Both Correa and Doshi became good friends of mine. I knew of Doshi when I was working in Chandigarh on Le Corbusier's building. I used to see these drawings that came from Le Corbusier's office in Paris–they were very stylish, with things that we couldn't decipher. We thought some Frenchman had drawn them, but then we found out it was Doshi making the drawings we were receiving. Doshi didn't stay in France for long because Le Corbusier had offered him a job, working on his buildings in Ahmedabad, so he moved there. When I came back to India from America, I renewed my contact with Doshi. At the time, Charles Correa had just started his practice. He made a point to seek me out. I liked him very much. He lived in Bombay, so I started my practice there too, because at the time, in 1960, there wasn't much going on in Delhi. My very first project was with Charles.

Hans *India in the 1960s felt very modern. Were you aware of belonging to a particular brand of modernity that was distinctly Indian? How did you feel about the idea of the local versus the global? Did you consider yours to be a universal or local language?*
Raj My first exposure to architecture was through Le Corbusier because I worked on his buildings. His language was both local and global. Eero Saarinen also deeply influenced me when I met him in New York. When I came back here, both Charles and Doshi had worked abroad. All of us had had the option to stay overseas, but we had chosen to come back home. Our common objective was to set up practices here, find our own roots and rise to the same stature that other countries had attained. We sought an Indian idiom that expressed our ancient culture, but was in tune with modern times. For us engineers, there was the exposure to the new materials of concrete, steel and pre-cast concrete. We were all idealists when we started out. Not that we aren't anymore!

Hans *As a pioneering master of concrete, you have invented many unexpected and innovative ways of shaping this material. In Europe, they called this use of concrete Brutalism, a style that is currently enjoying renewed interest among young architects. What made you use concrete in India? Was the decision yours, or the architect's? The British architect Peter Smithson spoke of the 'poetry of concrete.' Do you also perceive that poetry?*
Raj Yes, I agree with him. Concrete is a man-made material that you can mould, shape and form. In our projects, you can see that we exploited concrete's mouldability to the extreme. I became very fond of concrete as a consequence of working on Le Corbusier's buildings. Back then, concrete was far more economical than steel, and it still is in India. So as far as India is concerned, that was an issue. When you want to create large spans with different shapes and forms, concrete is more amenable than steel. I worked on Louis Kahn's Indian Institute of Management in Ahmedabad, for which Doshi was a consultant and I was a structural engineering consultant. It was around 1962. When Kahn came to India, he wanted to work with brick. We were contrary to the idea, but he felt that we didn't know the art of working with concrete. He underestimated us. He'd been working in Islamabad (Pakistan) and Dhaka (Bangladesh), and there he had chosen brick as his construction medium. He was still in that mental make-up. When he came here and looked around, of course, the old houses were made of brick and stone. But concrete was there too, so we had long arguments with him because we knew that the Indian Institute would require large spans, big buildings, and you can't make them with brick. He was insistent, and as you can see, he won in the end. But brick buildings are hard to maintain. His dream was to create brick buildings on stilts, but in an earthquake area that's very difficult. Eventually, he employed the concepts we offered him for achieving wide arches. It was very educative to work with Kahn.

Hans *Once the initial problem of concrete versus brick had been resolved, what was it like to collaborate with him?*
Raj We had to devise the means to make it safe against earthquakes. Horizontal elements of reinforcement were all made in concrete. Vertical elements were brick. To me, for any building or creative urban form to be successful, there has to be intense collaboration between the architect and the structural engineer. Some young architects feel that the structural engineer's only task is to devise the structure for their creations. My stand is that in order to create great works, an architect has to honour the requirements of engineering. Le Corbusier was an *égoïste absolu*: once the line had been drawn, you couldn't tell him to change it. If you did, he'd tell you to get lost.

But there were cases where we, as engineers, had to argue with him, and he would agree to make changes.

Hans *You are the visionary structural engineer of your generation. In Europe, there is Cecil Balmond (1943). He is Sri Lankan and you are Indian. Is it a coincidence or is there some connection between structural engineering, India and Sri Lanka?*
Raj Now that you mention it, there was also Fazlur Khan, who was Bangladeshi. I don't know what the relationship is. I actually didn't know that Cecil was Sri Lankan.

Hans *Let's talk about the Hall of Nations.*
Raj The Hall of Nations (New Delhi, 1971) was a competition I did with Raj Rewal for a permanent exhibition space. It was very challenging. It was designed as a steel space frame, but at that time there was a scarcity of steel. Many sections were unavailable, so we came up with the idea of using pre-cast concrete elements. The joints would also be pre-cast and bolted together. That was our competition entry, but then no contractor was willing to construct it, because no one had experience in pre-cast concrete. They felt they couldn't achieve the high level of accuracy that was required. So then we had to build the space frame from in-situ concrete and devise a way to get a sufficient number of bars crossing through it.

Hans *It's gorgeous. There's definitely a link between this structure and Buckminster Fuller's work.*
Raj I knew him–he came to visit us regularly here at our house and he was very fond of us, although we never collaborated directly. The Hall of Nations was under construction during one of his visits, so I proudly took him to visit the space frame structure. But all he saw was scaffolding, labour, women and men hauling buckets of concrete on their heads. The place was swarming. He was furious. He said, 'What are you doing? If you want to make modern buildings, you have to devise modern construction techniques! I saw the model, this is an ultra-modern structure, but you are following the same old techniques to build it!' He was very upset.

Hans *Tell us about the 1961 Hindustan Lever Pavilion for the India International Trade Fair. It's another truly visionary design.*
Raj That's a project by Correa. It's like a crumpled piece of paper, and I had to engineer it. We used the folded-plate concept in this project, but in quite an ingenious manner. Plates were made into valleys and ridges. After that exhibition, we didn't do many pavilions. Another interesting project was the NCDC (National Cooperative Development Corporation) office building in 1978, with Kuldip Singh. It was a competition, and he showed me the design and asked me if it was possible. I looked at it, figured out how it could be made, and told him yes. It would require some modifications, but it was feasible. Kuldip Singh submitted his entry and the jury thought it was very impressive, but they wanted to know if it could be built. So I went with him to the jury members and explained how it would be possible. Then he won the competition. We used post-tensioned concrete with cables. It's quite an interesting structure because it leans on both sides and the columns also twist and turn.

Hans *Has an architect ever come to you asking if a design were possible and you've had to say no?*
Raj I think that whatever it was, it could somehow or the other be done! In that sense, the only one with whom I had many an argument was Louis Kahn because of his preference for bricks for construction material.

Hans *It's fascinating that the building's aesthetic is an integral part of its engineering. That's a specific element of your buildings: one cannot separate the engineering from the aesthetic appearance. What is the building you're most proud of?*
Raj That's a difficult question. The greater the challenge, the more I enjoy working on a building. At one time, all the projects were challenging, but not anymore. Now things have changed completely. These buildings were done before the age of computers. Life for a structural engineer has become much easier now. Everybody has sophisticated computer programmes. When I started working on projects in Bombay, there was one single computer in the university, at the Tata Institute of

Fundamental Research, and it was only available to private people during the night. When I moved to Delhi, there was no computer available at all. There was one in Kanpur, so I went to the IIT (Indian Institute of Technology) in Kanpur and made arrangements with a professor there. A professor and his wife would take on private work, so we gave them assignments.

Hans *What is your advice today to a young structural engineer?*
Raj That's a good question, but I can't give you a simple answer. The problem with structural engineers today is that they have been educated in the computer era. Computers have their good and bad points. Engineers graduating now have not learned the art of working with their own hands and mind. They're not used to understanding the behaviour of structures on their own. They are so dependent on computers that they just do their work mechanically. When these engineers come to our office, it takes quite a lot of effort to break them in. Don't just go to the computer to find the answers–first try to realise things and work with your own hands, and then go to the computer.

Originally published in Domus, *November, 2011*

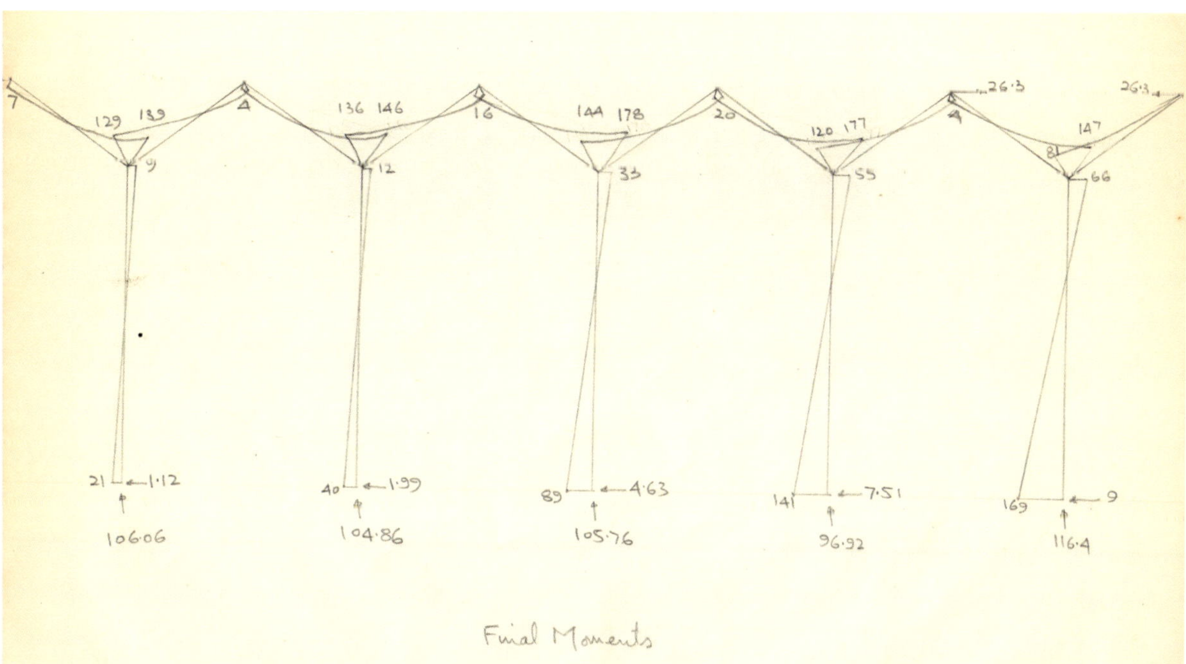

A sample calculation sheet showing moment diagrams for the Municipal Stadium, Ahmedabad.

1990s
onwards

VIDHAN BHAVAN

BHOPAL | 1994

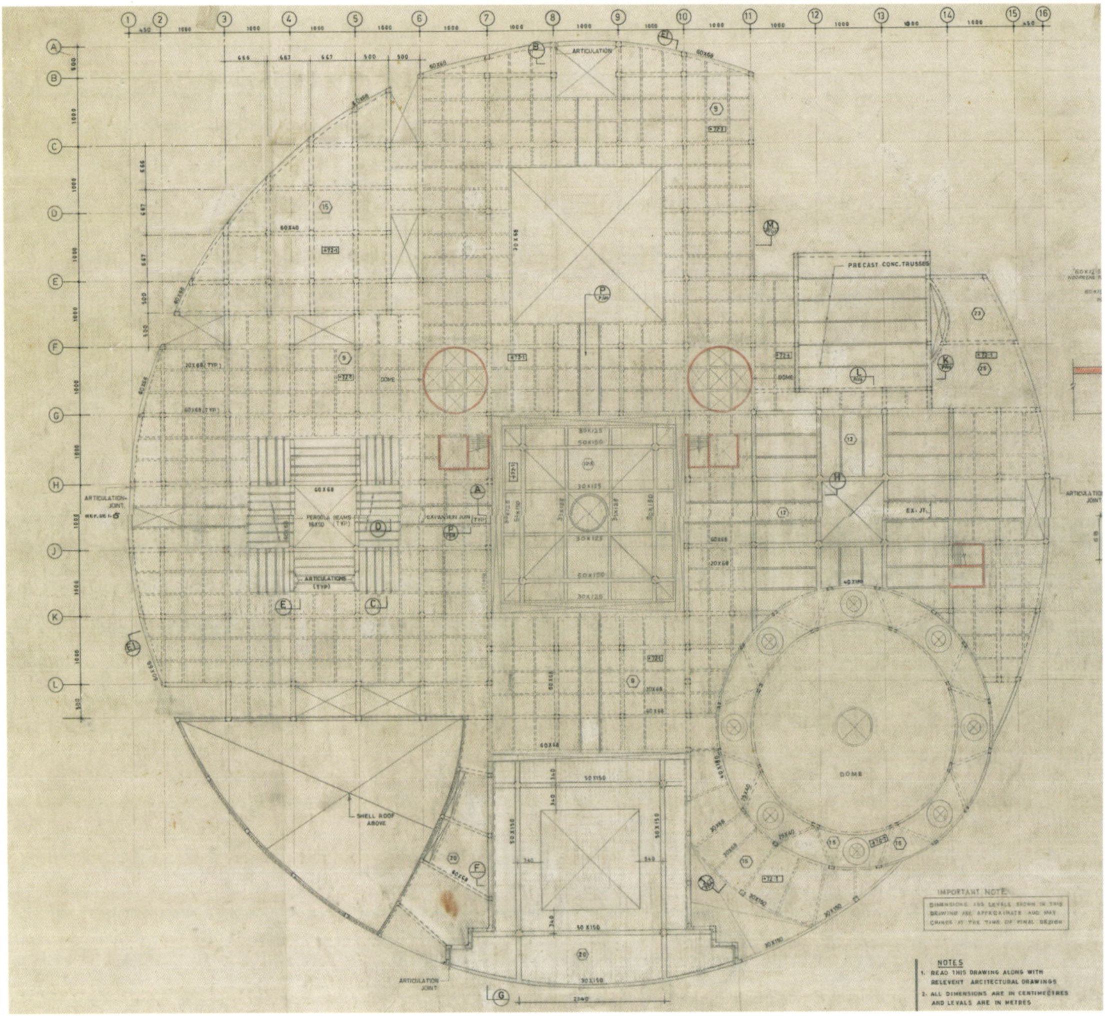

(Partial) Drawing no.121/VB/(P) 03: Framing Plan and Sections at Roof Level: 22 July 1982
Overall, a nine-sector plan with a two-way ribbed grid system of 10m x 10m, punctuated by different roof forms.

Previous pages: The doubly curved shell roof of the multipurpose hall during construction.

Built on the undulating terrain of Central India, the Vidhan Bhavan of Bhopal, with its circular form, represents architect Charles Correa's move toward prioritising materials and narrative over structural expression. Still, this building's main programmatic spaces, with bold structural forms that create a roof-scape of domes, shells and a tower, bear Mahendra Raj's distinctive stamp.

A doubly curved shell roof for the Multipurpose Hall is set on a triangular base of 43m x 43m x 36m with a thickness of 15–20cm; the main Vidhan Sabha hall is a 31m-diameter hemispherical dome in concrete with thickness varying from 10 to 20cm; and the Vidhan Parishad has a composite space structure in concrete and steel covered by a 14m x 14m roof. Two main lobby areas have impressive large-span spaces of 30m x 30m, one of which is supported on only four columns.

These large-span structurally challenging forms punctuate a circular plan of 140m diameter circumscribing an area of about 29,000 square metres, which is divided into nine sectors. A grid, of 10m x 10m with two-way beams and secondary ribs at

one-third points in both directions, forms the basic structural system of the rest of the building. It is worth noting that the construction of the complex took more than a decade to complete because of various administrative and other challenges.

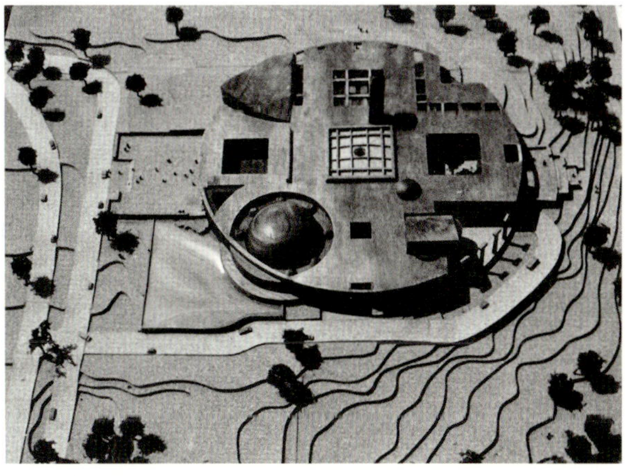

Exterior views from the north-east and the south-west. Within the extent of the building, the existing ground level falls by approximately 8m, which was integrated in the design by creating two partial basements under the complex.

Architectural building model.

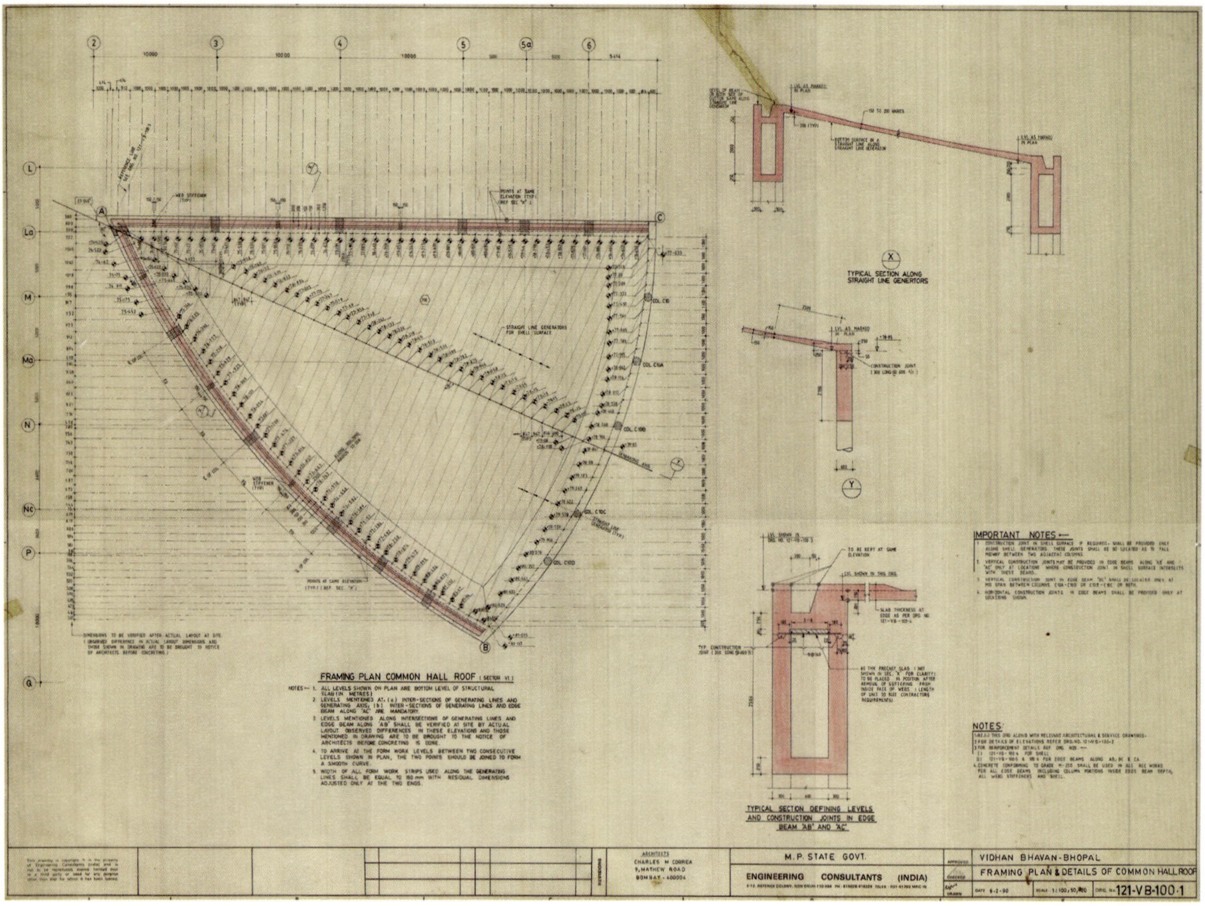

Drawing no.121-VB-100-1: Framing Plan & Details of Common Hall Roof: 6 February 1990

Opposite: Drawing no.121-VB-100 A: Framing Plan & Details of Common Hall Roof: 27 May 1986

The roof of the Multipurpose Hall was first resolved as a rotational shell and then later developed as a series of conoids. Their curved edges spring from the two adjacent 43m-long sides of the triangle and the straight edges meet in the centre. The curved ends on each side of the triangle form a serpentine shape. The ridge part of this curve on one side faces the valley part on the other side of the triangle. The resultant configuration is a dramatic doubly curved form with a straight surface in the centre of the shell.

The shell, analysed using a finite element technique, indicated generation of very heavy tensile forces along the edges. To verify the analytical results, a model was tested at the Indian Institute of Technology, Kanpur.

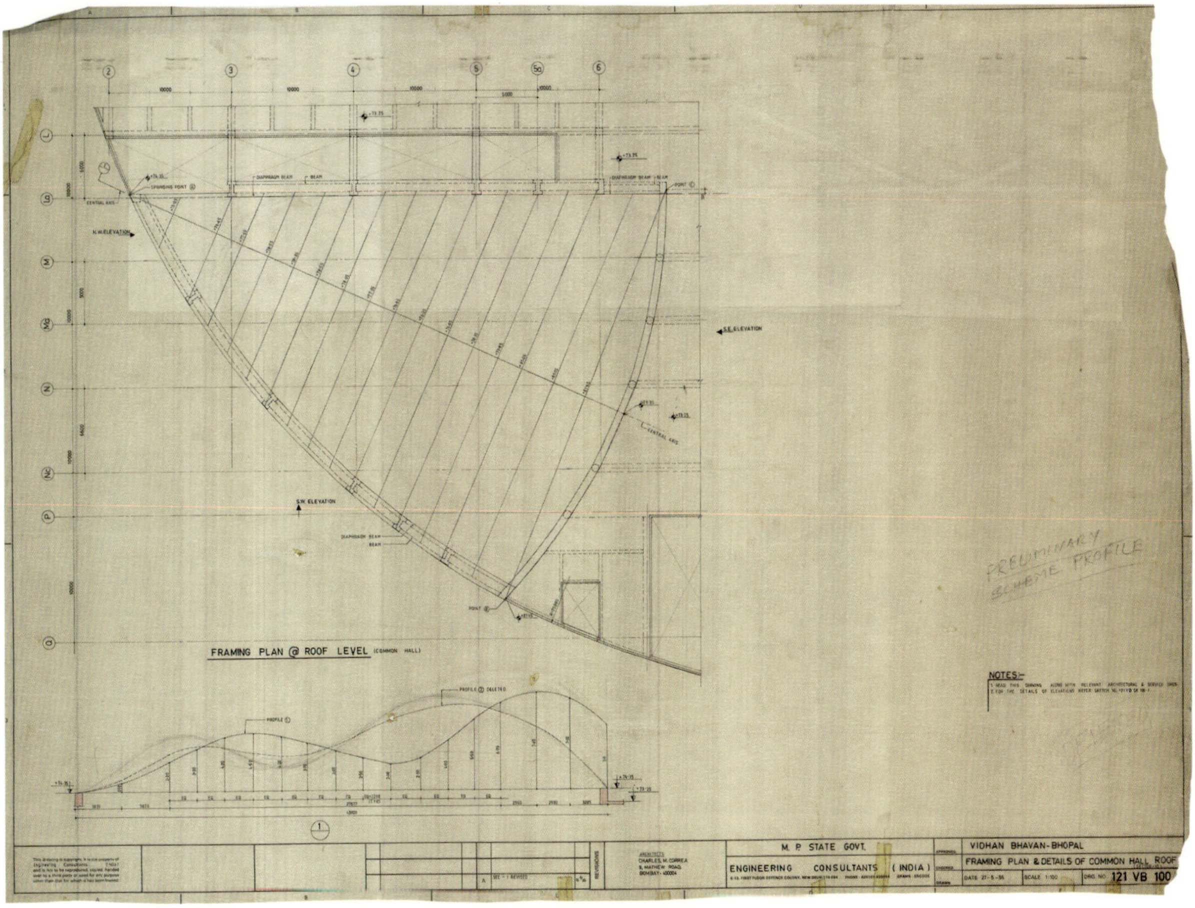

Photo taken during construction of the shell.

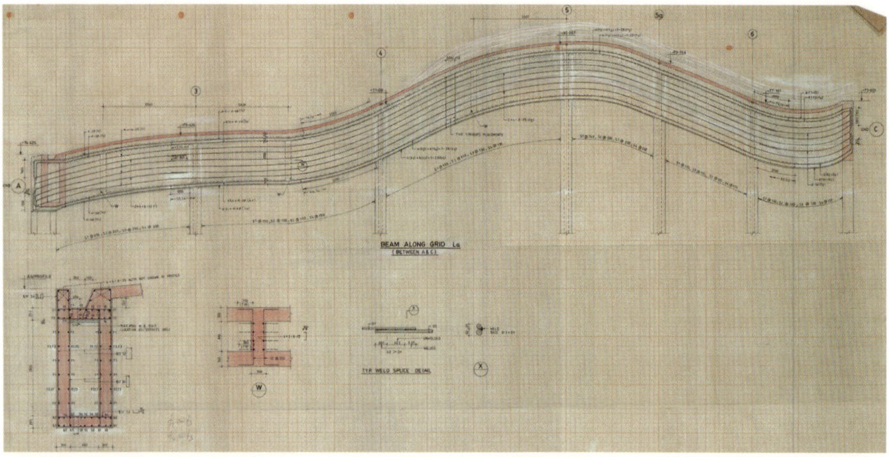

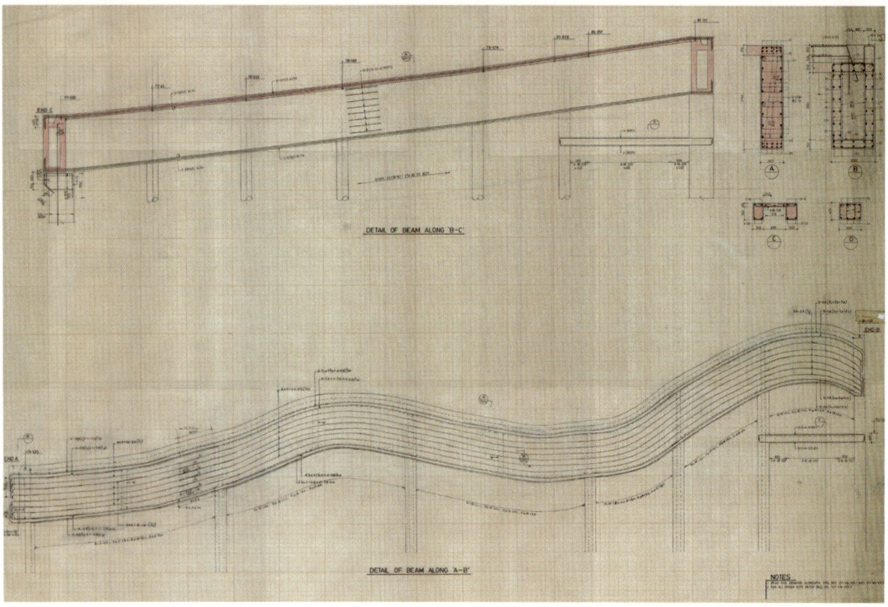

(Partial) Drawing no. 121-VB-100-5: Shell Edge Beam Details (Sheet-1): 13 August 1990

(Partial) Drawing no. 121-VB-100-6: Shell Edge Beam Details (Sheet-2): 20 August 1990

Drawings of the serpentine-shaped beams that form the edges of the triangular base of the shell.

Photo taken during construction of the shell with reinforcement layout in progress.

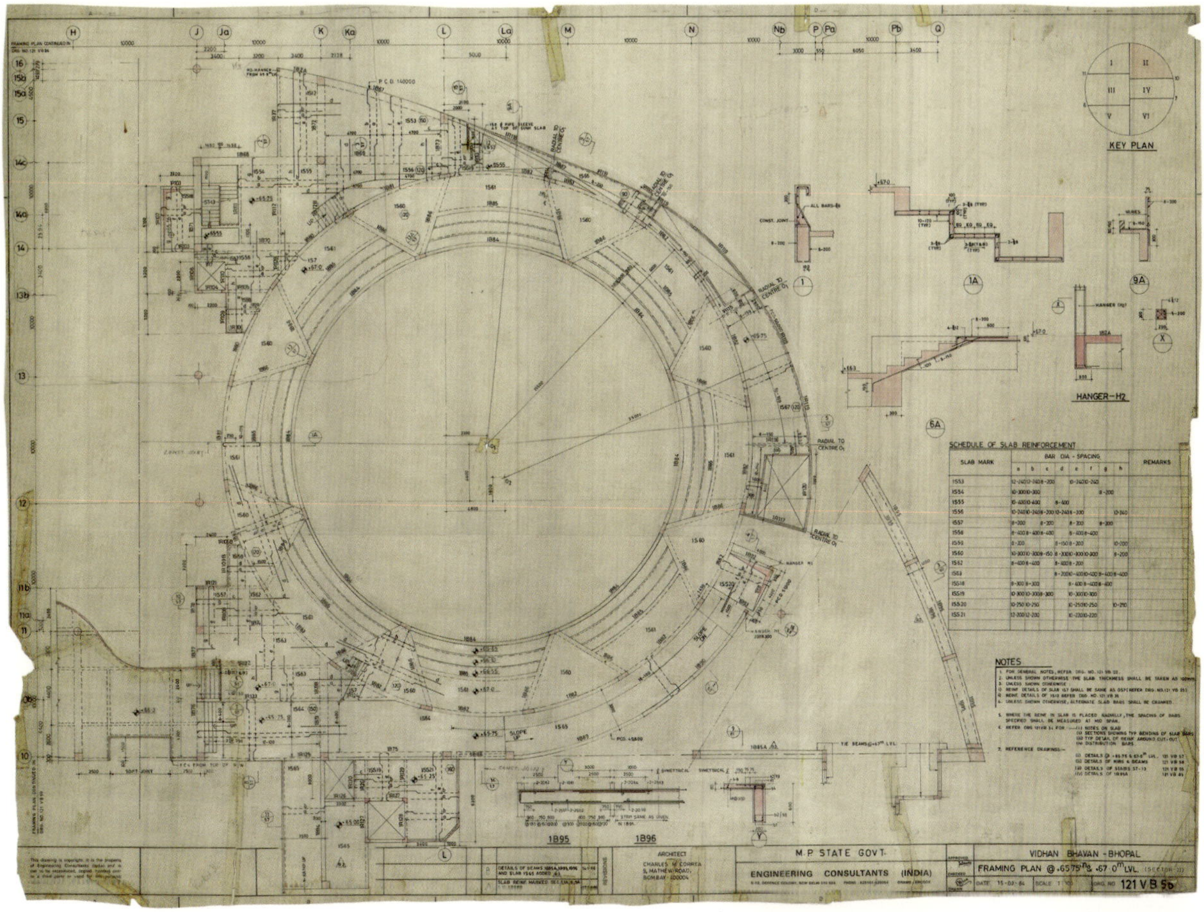

Drawing no.121VB 56B: Framing Plan @+65.75m &+67.0m Level: 15 March 1984

Plan section and construction sequence drawings of Vidhan Sabha's hemispherical dome roof with a circular skylight opening at the top.

Interior view of the dome and skylight.

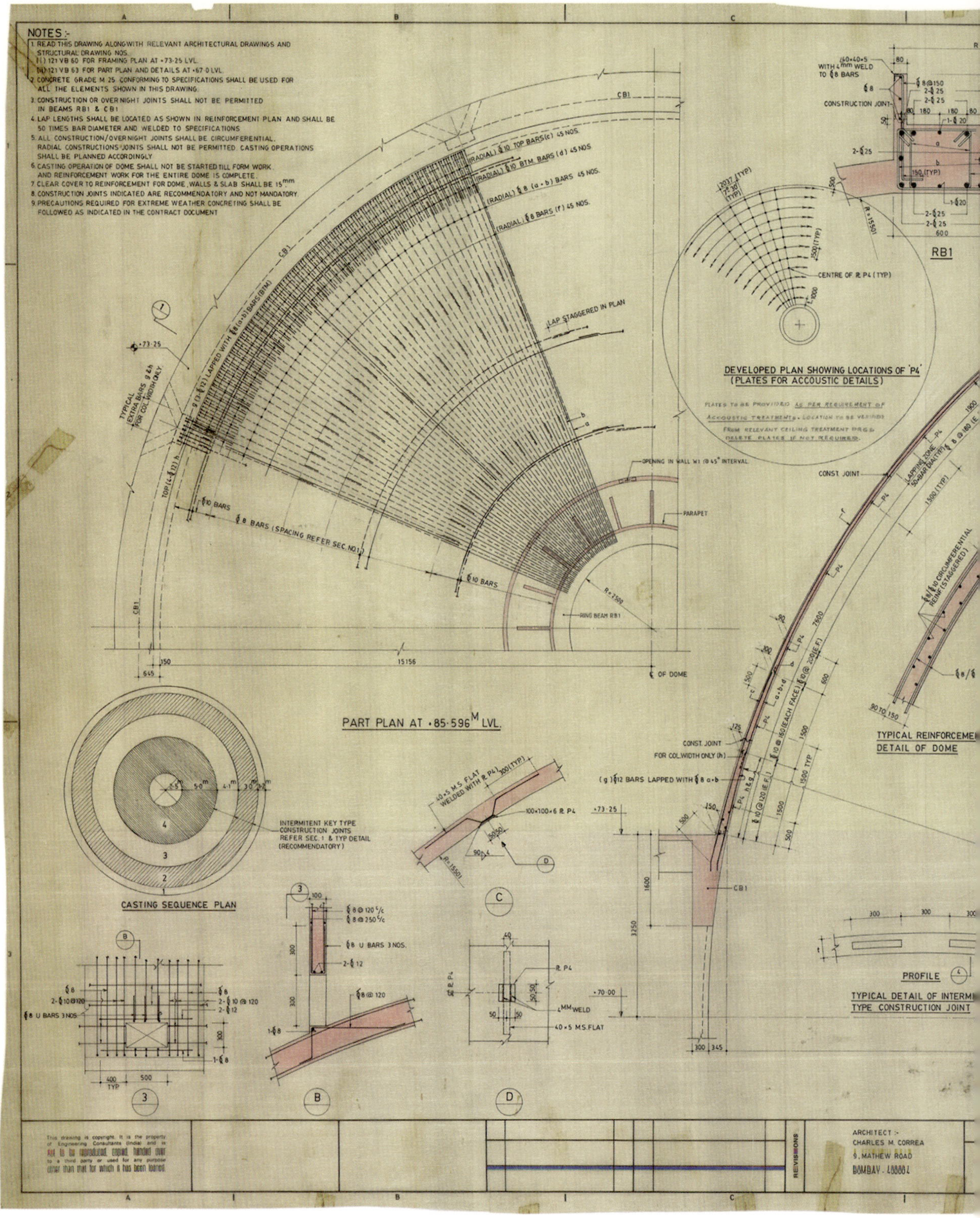
NOTES :-
1. READ THIS DRAWING ALONGWITH RELEVANT ARCHITECTURAL DRAWINGS AND STRUCTURAL DRAWING NOS.
(i) 121 VB 60 FOR FRAMING PLAN AT +73·25 LVL.
(ii) 121 VB 63 FOR PART PLAN AND DETAILS AT +67·0 LVL.
2. CONCRETE GRADE M 25 CONFORMING TO SPECIFICATIONS SHALL BE USED FOR ALL THE ELEMENTS SHOWN IN THIS DRAWING.
3. CONSTRUCTION OR OVERNIGHT JOINTS SHALL NOT BE PERMITTED IN BEAMS RB1 & CB1
4. LAP LENGTHS SHALL BE LOCATED AS SHOWN IN REINFORCEMENT PLAN AND SHALL BE 50 TIMES BAR DIAMETER AND WELDED TO SPECIFICATIONS
5. ALL CONSTRUCTION/OVERNIGHT JOINTS SHALL BE CIRCUMFERENTIAL. RADIAL CONSTRUCTIONS JOINTS SHALL NOT BE PERMITTED. CASTING OPERATIONS SHALL BE PLANNED ACCORDINGLY
6. CASTING OPERATION OF DOME SHALL NOT BE STARTED TILL FORM WORK AND REINFORCEMENT WORK FOR THE ENTIRE DOME IS COMPLETE.
7. CLEAR COVER TO REINFORCEMENT FOR DOME, WALLS & SLAB SHALL BE 15mm
8. CONSTRUCTION JOINTS INDICATED ARE RECOMMENDATORY AND NOT MANDATORY.
9. PRECAUTIONS REQUIRED FOR EXTREME WEATHER CONCRETING SHALL BE FOLLOWED AS INDICATED IN THE CONTRACT DOCUMENT
LAP STAGGERED IN PLAN
OPENING IN WALL W1 @ 45° INTERVAL
PARAPET
RING BEAM RB1
℄ OF DOME
PART PLAN AT +85·596M LVL.
DEVELOPED PLAN SHOWING LOCATIONS OF 'P4' (PLATES FOR ACCOUSTIC DETAILS)
CENTRE OF ℞ P4 (TYP)
RB1
CONST. JOINT
TYPICAL REINFORCEMENT DETAIL OF DOME
CONST. JOINT FOR COL.WIDTH ONLY (h)
CASTING SEQUENCE PLAN
INTERMITENT KEY TYPE CONSTRUCTION JOINTS REFER SEC.1 & TYP DETAIL (RECOMMENDATORY)
PROFILE
TYPICAL DETAIL OF INTERMI[TENT KEY] TYPE CONSTRUCTION JOINT
40×5 M.S.FLAT
ARCHITECT :-
CHARLES M. CORREA
9, MATHEW ROAD
BOMBAY - 400004
REVISIONS

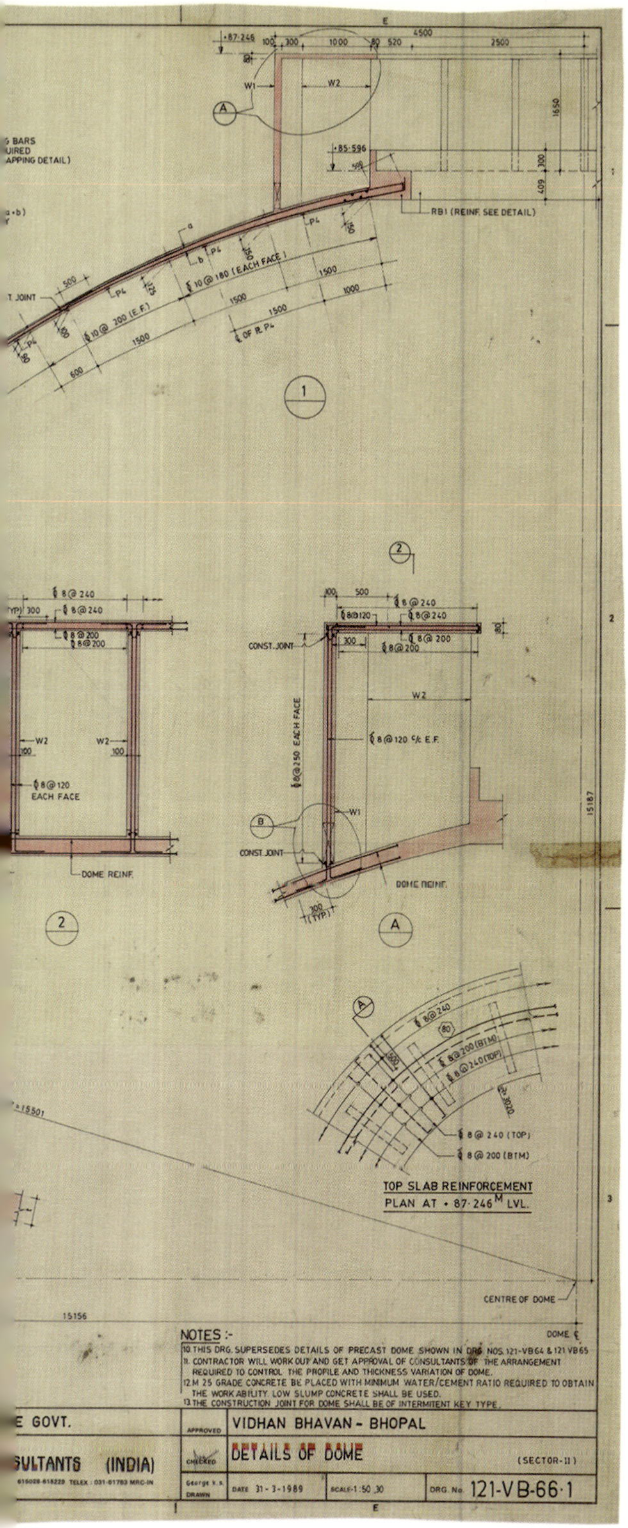

The dome springs out of a ring beam supported on eight columns. Its thickness at the springing point is 20cm, tapering to 10cm and increasing to 15cm around the circular skylight opening, 5m in diameter.

Drawing no.121VB-66-1: Details of Dome: 31 March 1989

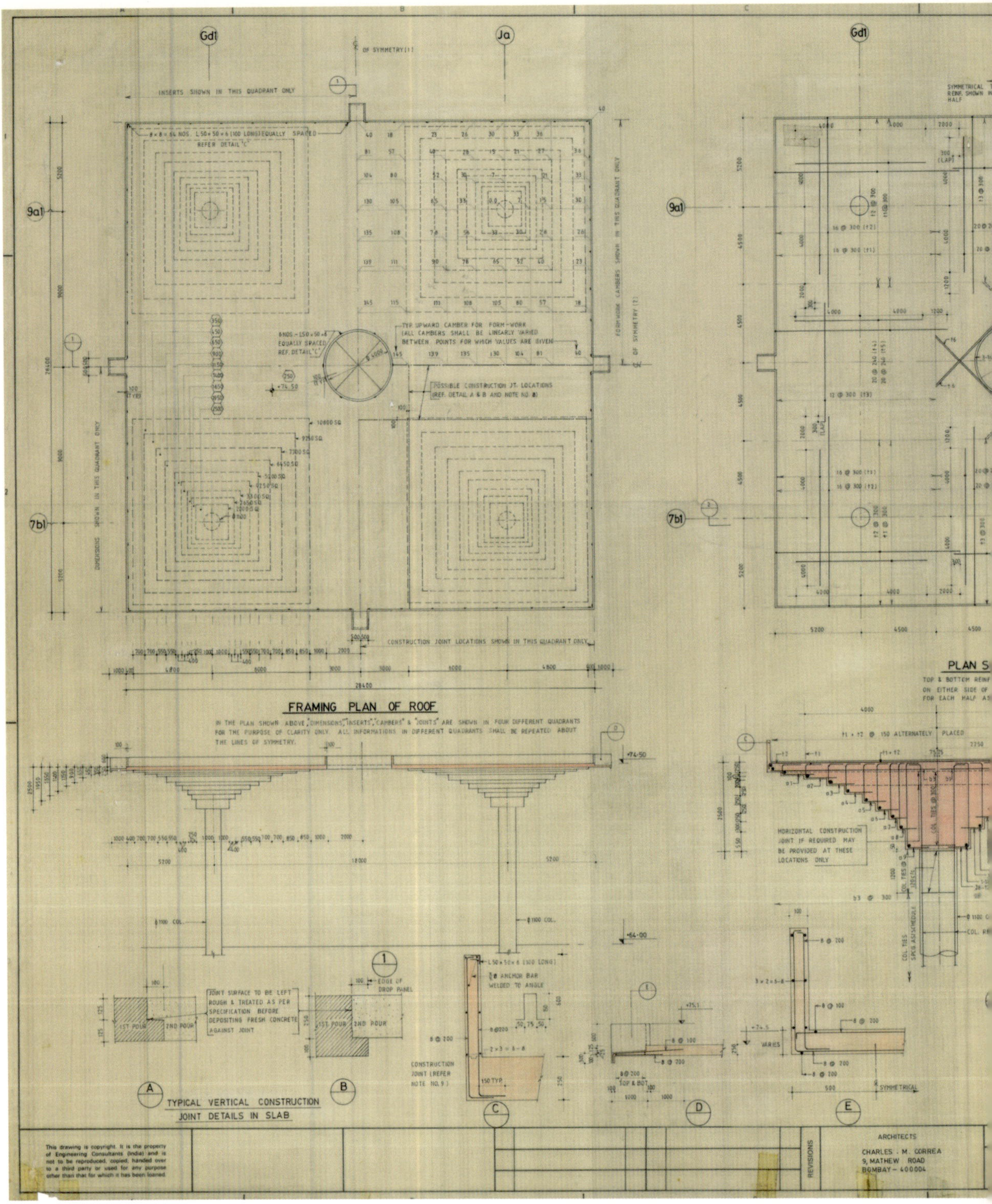

Gd1
Ja
9a1
7b1
INSERTS SHOWN IN THIS QUADRANT ONLY
TYP. UPWARD CAMBER FOR FORM-WORK (ALL CAMBERS SHALL BE LINEARLY VARIED BETWEEN POINTS FOR WHICH VALUES ARE GIVEN)
POSSIBLE CONSTRUCTION JT. LOCATIONS (REF. DETAIL A & B AND NOTE NO. 8)
DIMENSIONS SHOWN IN THIS QUADRANT ONLY
FORMWORK CAMBERS SHOWN IN THIS QUADRANT ONLY
CONSTRUCTION JOINT LOCATIONS SHOWN IN THIS QUADRANT ONLY
FRAMING PLAN OF ROOF
IN THE PLAN SHOWN ABOVE, DIMENSIONS, INSERTS, CAMBERS & JOINTS ARE SHOWN IN FOUR DIFFERENT QUADRANTS FOR THE PURPOSE OF CLARITY ONLY. ALL INFORMATIONS IN DIFFERENT QUADRANTS SHALL BE REPEATED ABOUT THE LINES OF SYMMETRY.
JOINT SURFACE TO BE LEFT ROUGH & TREATED AS PER SPECIFICATION BEFORE DEPOSITING FRESH CONCRETE AGAINST JOINT
1ST POUR
2ND POUR
EDGE OF DROP PANEL
TYPICAL VERTICAL CONSTRUCTION JOINT DETAILS IN SLAB
CONSTRUCTION JOINT (REFER NOTE NO. 9)
HORIZONTAL CONSTRUCTION JOINT IF REQUIRED MAY BE PROVIDED AT THESE LOCATIONS ONLY
PLAN S
This drawing is copyright. It is the property of Engineering Consultants (India) and is not to be reproduced, copied, handed over to a third party or used for any purpose other than that for which it has been loaned.
REVISIONS
ARCHITECTS
CHARLES M. CORREA
9, MATHEW ROAD
BOMBAY - 400004

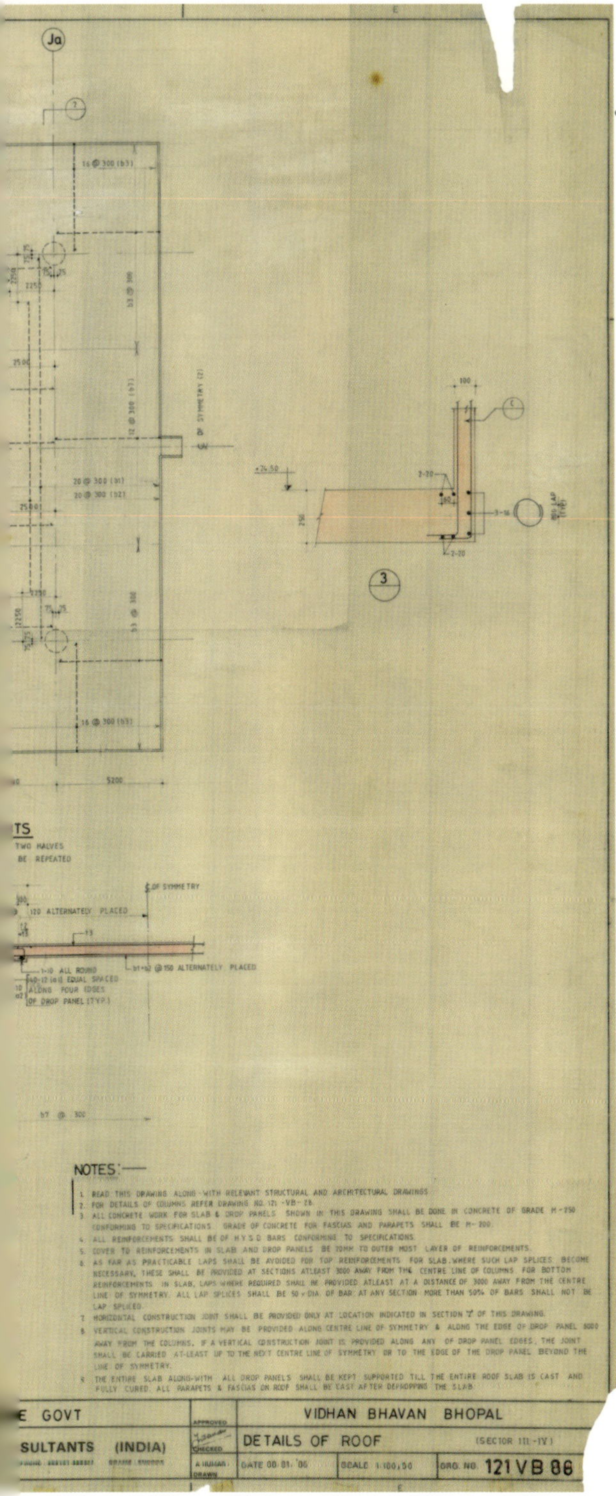

Interior view of the lobby showing the corbelled column capitals.

Drawing no. 121-VB-86: Details of Roof: 8 January 1986

Roof details of the lobby area in Sectors III and IV. The roof is a 30m x 30m flat plate slab supported by four columns with corbelled capitals.

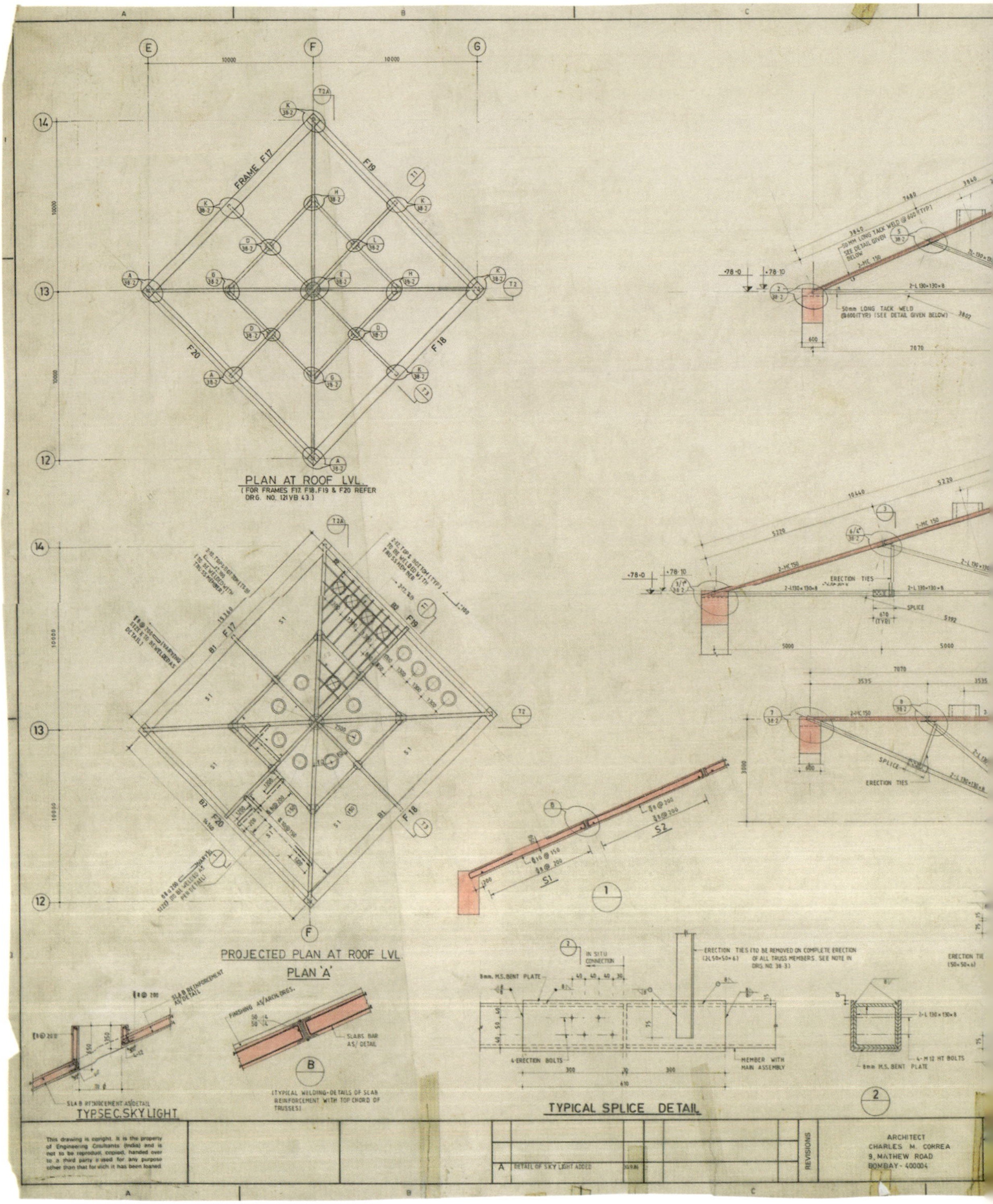
PLAN AT ROOF LVL.
(FOR FRAMES F17, F18, F19 & F20 REFER DRG. NO. 121VB 43.)
FRAME F17
F19
F18
F20
PROJECTED PLAN AT ROOF LVL.
PLAN 'A'
TYP.SEC.SKY LIGHT
SLAB REINFORCEMENT AS DETAIL
FINISHING AS/ARCH. DRGS.
SLABS BAR AS/ DETAIL
(TYPICAL WELDING-DETAILS OF SLAB REINFORCEMENT WITH TOP CHORD OF TRUSSES)
TYPICAL SPLICE DETAIL
IN SITU CONNECTION
8mm. M.S. BENT PLATE
4-ERECTION BOLTS
MEMBER WITH MAIN ASSEMBLY
ERECTION TIES
SPLICE
50mm LONG TACK WELD
This drawing is copyright. It is the property of Engineering Consultants (India) and is not to be reproduced, copied, handed over to a third party or used for any purpose other than that for which it has been loaned.
DETAIL OF SKY LIGHT ADDED
REVISIONS
ARCHITECT
CHARLES M. CORREA
9, MATHEW ROAD
BOMBAY - 400004

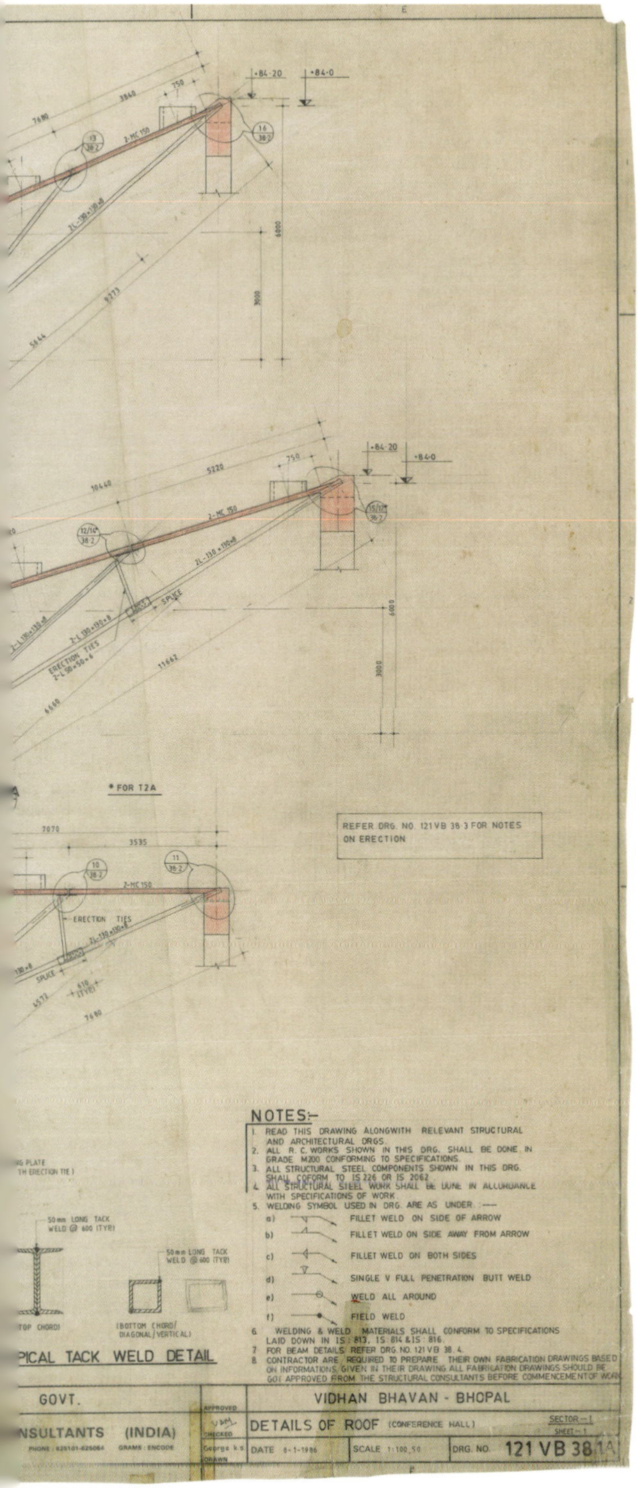

Construction photos of the Vidhan Parishad roof.

Drawing no.121-VB-38.1A: Details of Roof (Conference Hall): 8 January 1986
Roof plan, section and details of the 14m x 14m conference hall for the Vidhan Parishad in Sector I, with a composite structure comprising a concrete frame, three intersecting steel trusses and a concrete slab.

Exterior photos of the Vidhan Bhavan.

ALLIANCE FRANÇAISE

NEW DELHI | 2004

ance
aise

The Indo-French cultural centre, approximately 2900 square metres in area, is an elegant three-storey terraced building with a large dramatic roof parasol floating above it.

Mahendra Raj's penchant for structural clarity unified all the floors of the building, which the architects had initially designed with multiple structural systems due to the varied span requirements of the administrative areas and classrooms, halls, café and even an auditorium. Reworked elegantly, the structural system uses 30cm-deep pre-cast waffles to form 8m x 10m slab panels. Beam depths were altered in larger rooms and a transfer girder was used in the auditorium while maintaining the waffle slab on all floors. By separating the roof structure from the rest of the building, Raj also reduced the number of columns supporting the roof pergola to four. Finally, the roof is a steel structure of 60m x 17.5m, supported and cantilevered off four cross-shaped concrete columns set 30m x 9m apart, hovering elegantly above the entire building.

View from the roof terrace.

Previous pages: Front view of the building with pergola above and a 13m-high inverted L-shaped display wall on the right of the entrance.

Interior view of the cafeteria's waffle slab ceiling.

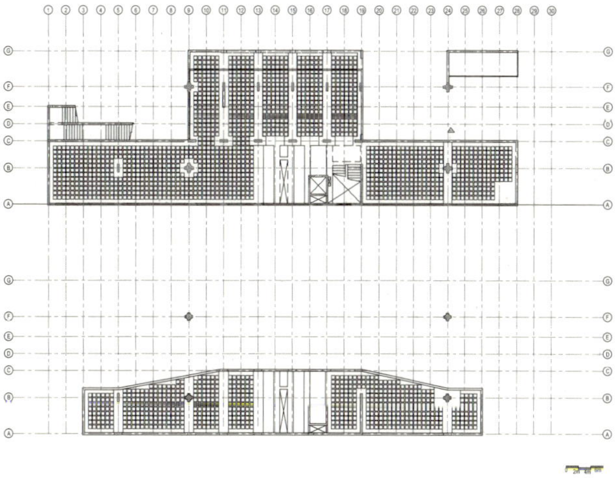

(Partial) Drawing no.ST-2.4: Second Floor Framing Plan: 15 November 2002

(Partial) Drawing no.ST-2.6: Framing Plan at Roof Level: 15 November 2002

BEARYS GLOBAL
RESEARCH TRIANGLE
BANGALORE | 2008

A LEED Platinum-certified Green building, the Bearys Global Research Triangle (BGRT) office complex covers a total area of 112,000 square metres in three, nine or ten-storey towers and two basements. Large floor spans have been achieved with a two-way 16m x 16m grid on the upper floors and an 8m x 8m grid at the basement level. An innovative flat slab of non-pre-stressed concrete, combining the waffle slab in span and double-drop solid panels in the columns, achieves structural efficiency and economy unusual for such office buildings.

Exterior view of the building.

Photo of a typical office floor.

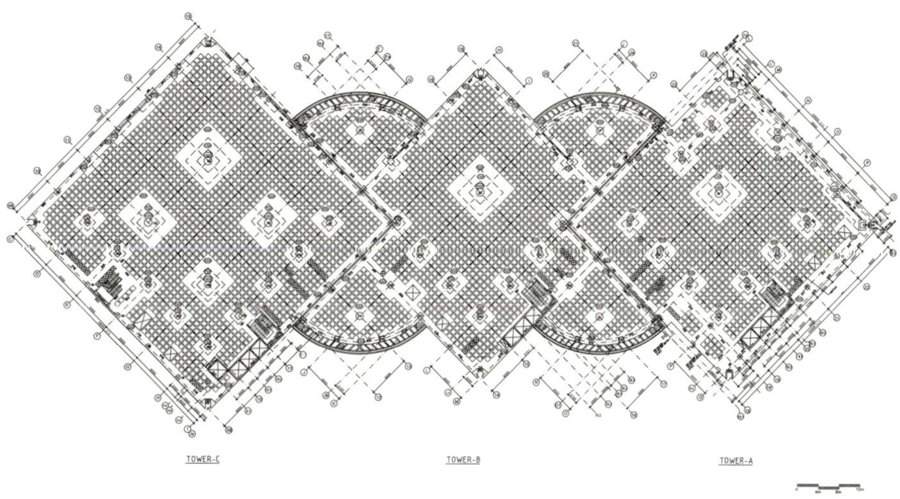

Drawing no.S.07: Framing Plan at Second Floor Level (+111.20): 12 December 2007

Opposite: Exterior side view of the BGRT building under construction.

Previous pages: View of the double-height lobby area with combined waffle slab and double-drop flat slab at columns allowing for a large span area.

SOLAR PAVILION

YARRA VILLAGE, MADHYA PRADESH | 2011

Many materials were considered for the Solar Pavilion, a prototype for introducing flexible community spaces within solar energy structures in remote rural areas.

'The architects had options in steel, concrete and brick. Brick was the least favoured material of the client but made most sense in the context,' says Mahendra Raj. 'I pushed them to work with brick.'

The realised pavilion is a 33m x 4.8m long rectangular building formed through a system of arched modules that repeat themselves for the length required for the solar panels. Due to various structural considerations such as unsymmetrical loading from the roof, all the arches have buttresses. The slope of the roof required in relation to the position of the sun is achieved by the different heights of the north and south facing walls. Spanning between the north and south walls are elliptical arches on which solar panels set on aluminium purlins are fixed.

View of the southern wall, articulated by a set of semi-circular arches and buttresses.

(Partial) Drawing no. A-T-101: Plans–Foundation Plan, Ground Floor Plan and Roof Plan: 1 September 2010

Previous pages: Community pavilion comprising elliptical brick arches with buttresses, supporting the solar panel roof system.

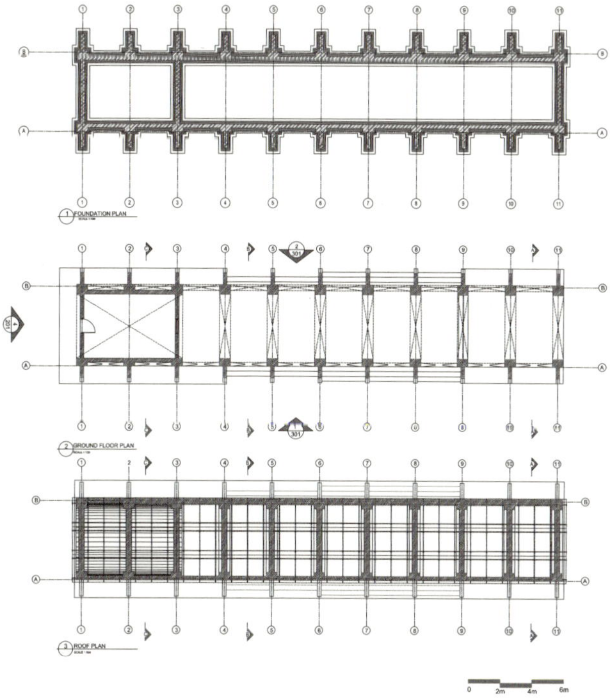

The northern wall extends up much higher to the pitch of the solar roof and is further articulated by circular openings above that connect the space inside to the sky above.

A dynamic yet simple semi-open community space created below the solar panel roof.

JAISALMER AIRPORT

JAISALMER | 2013

Fourteen kilometres from the historic city of Jaisalmer, the new, 10,000 square metre airport clad in the yellow sandstone typical of the region, resonates with local traditions and geography. Situated as it is in the Rajasthan desert, Jaisalmer is subject to frequent dust storms of high velocity, large fluctuations in daily temperatures as well as earthquakes. The structure was designed to cater to the stresses generated by these forces. According to the architect, 'The original idea was a sort of flattened arch, which my father [Raj] said violated structural clarity. But by introducing a continuous curvature into the arch, he made it possible to preserve and even enhance the key element in the design.'

Ten asymmetrical trapezoidal steel arches of 66m span form the roof of the airport, echoing the rise and fall of the surrounding dune-scape. Each arch springs from a 500mm x 500mm square hinge and rises to a height of 23m over a distance of 53m before descending steeply to a second hinge on the other side. The final curve was the result of a collaborative effort between Mahendra Raj and Studio VanRO, bringing together structural efficiency and aesthetic vision.

Each arch has a triangulated warren truss face and independent concrete footings that are tied together with reinforced tie beams. The ten arches are interconnected with four triangular cross-trusses to contain lateral stresses. The external faces of the arches on the gable ends of the building have vertical wind trusses with sliding connections which support various architectural details such as cladding and stone *jaali* work.

The undulating roof form of the Jaisalmer Airport with an intricate *jaali* facade at the entrance foyer.

(Partial) Drawing no. S-G-JTB-06.4A: Truss Elevations with Dimensions: 26 April 2010

Previous pages (356-357; 358-359): Photos taken during the construction process and erection of the asymmetrical arches; Details of the *jaali* work at the entrance portico that gives the airport its distinctive appearance with mushroom shaped columns and dropped-slab structure.

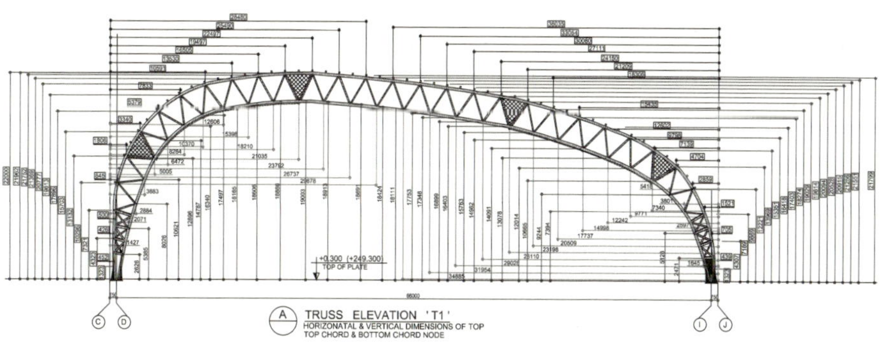

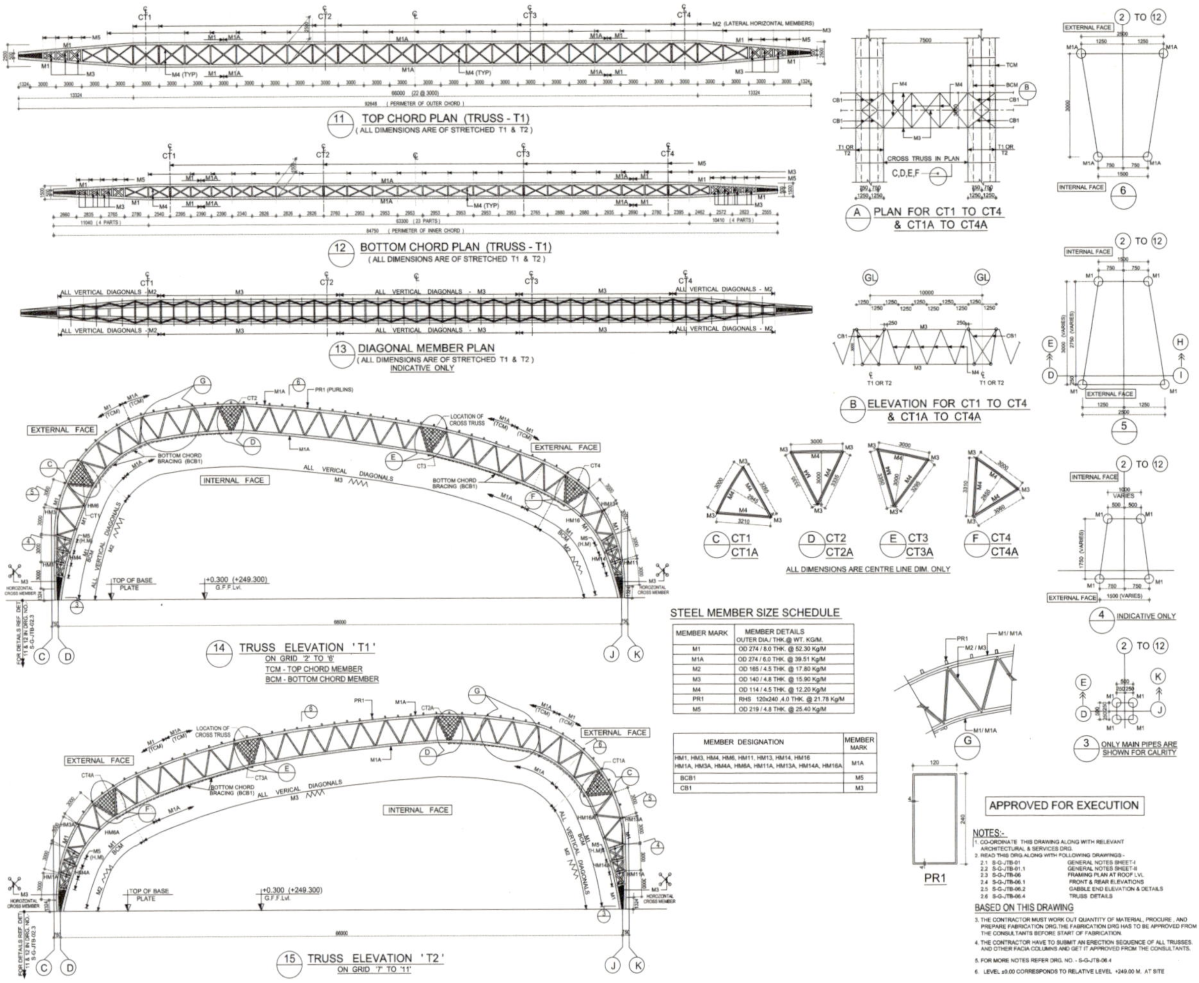

Drawing no. S-G-JTB-06.3A: Truss Elevations & Details: 5 May 2010, showing dimensions, plans and elevations of the asymmetrical arches. Each arch starts as a 50cm x 50cm square at the hinges and gradually flares into a trapezoidal form (with a 2.5m top chord, 1.5m bottom chord and maximum depth of 3m). The ten arches are interconnected with four triangle-shaped cross trusses to contain all horizontal forces generated by earthquakes, and wind and temperature variations.

Below and opposite: Construction photos of the Jaisalmer Airport.

View of the arrival hall.

Night photo from the front of the building with the stone *jaali* and the split roof above.

The asymmetrical arches as they flare into a trapezoidal geometry from their springing points, cross-braced by triangular trusses and ties.

BIHAR MUSEUM

PATNA | 2015

The Bihar Museum was commissioned in order to showcase the long history and rich cultural heritage of Bihar, which boasts such historic personages as Gautama Buddha and Emperor Ashoka. Situated on a 13-acre site, the museum complex has a total built area of 26,000 square metres, with different blocks fulfilling distinct functions: an entryway, three exhibition galleries, an auditorium, a children's space, and office and service blocks.

Interesting structural features in this building include a 20m-deep cantilevered steel truss canopy at the entry, a large double-height elliptical drum of 36m x 32m with an off-centre opening of 18m x 12m for the children's block, and three large-span, long gallery blocks that are connected by a sculptural hexagonal bridge.

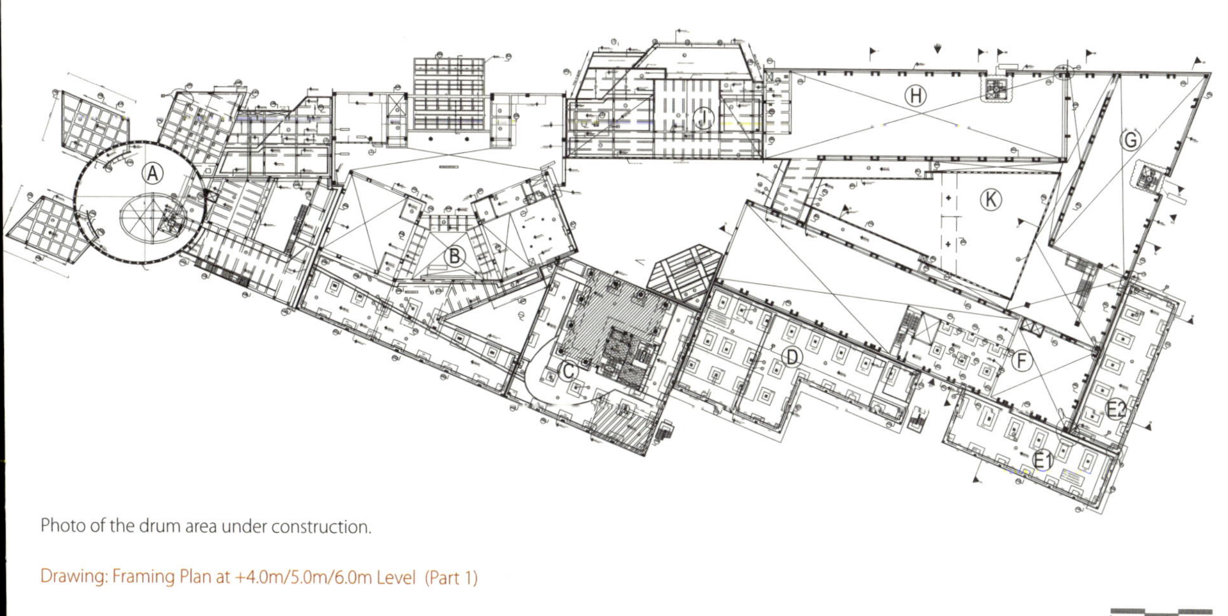

Photo of the drum area under construction.

Drawing: Framing Plan at +4.0m/5.0m/6.0m Level (Part 1)

Previous pages: Aerial view of the complex under construction.

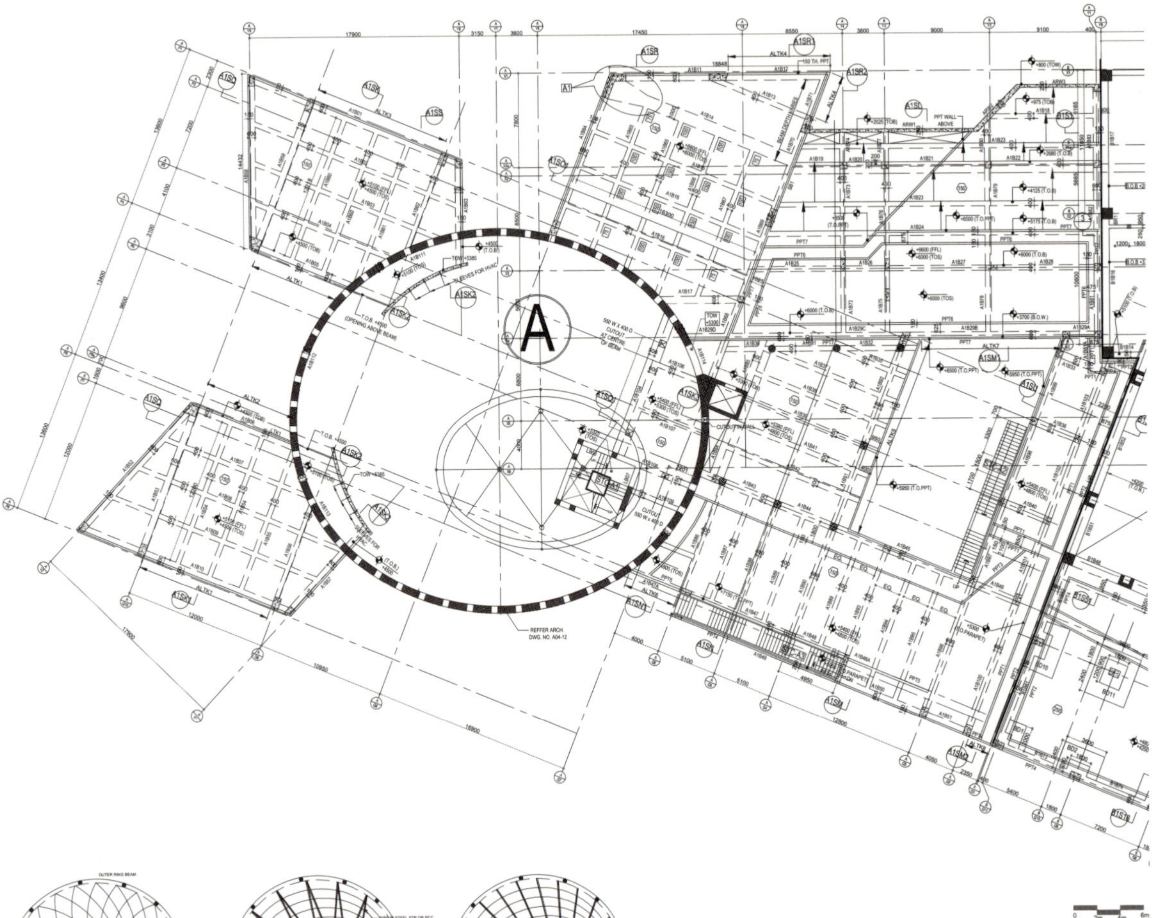

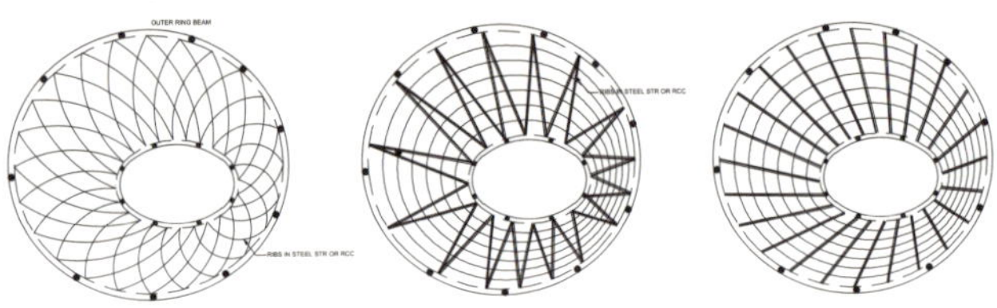

Drawing no. PMP-S-02: Conceptual Sketches of Framing Plan at Roof Level

The three alternative systems of rib, as indicated in the drawings, considered for the catenary roof.

Drawing no. S4-A-MFP-01: Framing Plan at +4.8m/5.3m/6.0m Level (Block A): 15 July 2013

Construction photo of the children's block.

The children's block has a large double-height elliptical drum of 36m x 32m with a 60cm-thick wall. It has a catenary-shaped concrete roof with an eccentrically placed elliptical skylight opening of 18m x 12m. The roof has radial ribs tapering in depth from 80cm to 60cm, supported on an elliptical outer compression ring that is part of the drum, and another elliptical inner tension ring around the eccentric opening. Attached to the periphery of the drum are three rhombic-shaped rooms for children's activities.

Drawing no. S9-A-1FP-TBR-01: Slab Top and Bottom Reinforcement Details at 9.0m Level (Block A): 15 July 2013

Detailed plans of the drum roof at different levels.

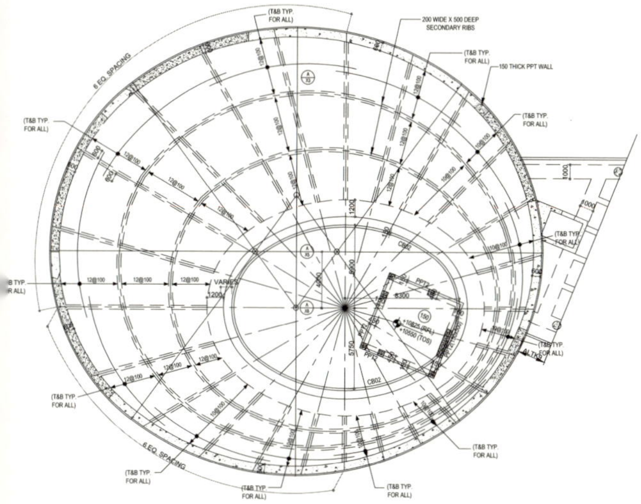

DRUM AREA
SLAB CIRCUMFERENCE REINF. AT ROOFLVL.
SLAB REINF. AT +9.0 M / +10.0 M LVL.

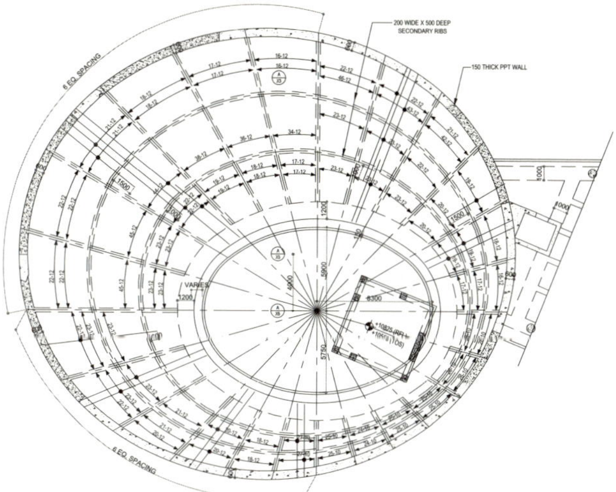

DRUM AREA
SLAB RADIAL REINF. AT ROOF LVL.
SLAB REINF. AT +9.0 M / +10.0 M LVL.

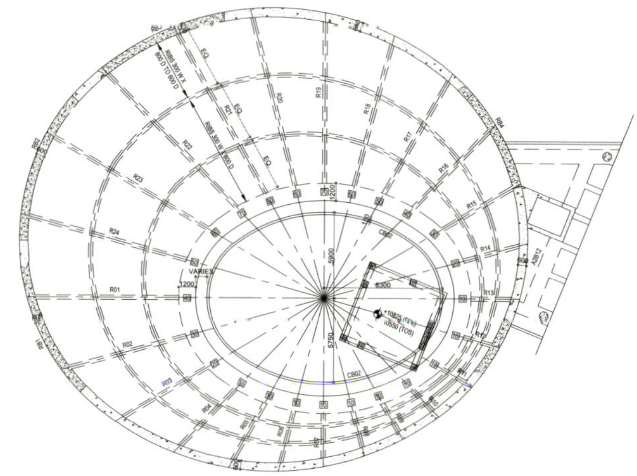

CAMBER TO INNER RING
BEAM//RADIAL&CIRCUMFERENTIAL
RIB IN MM

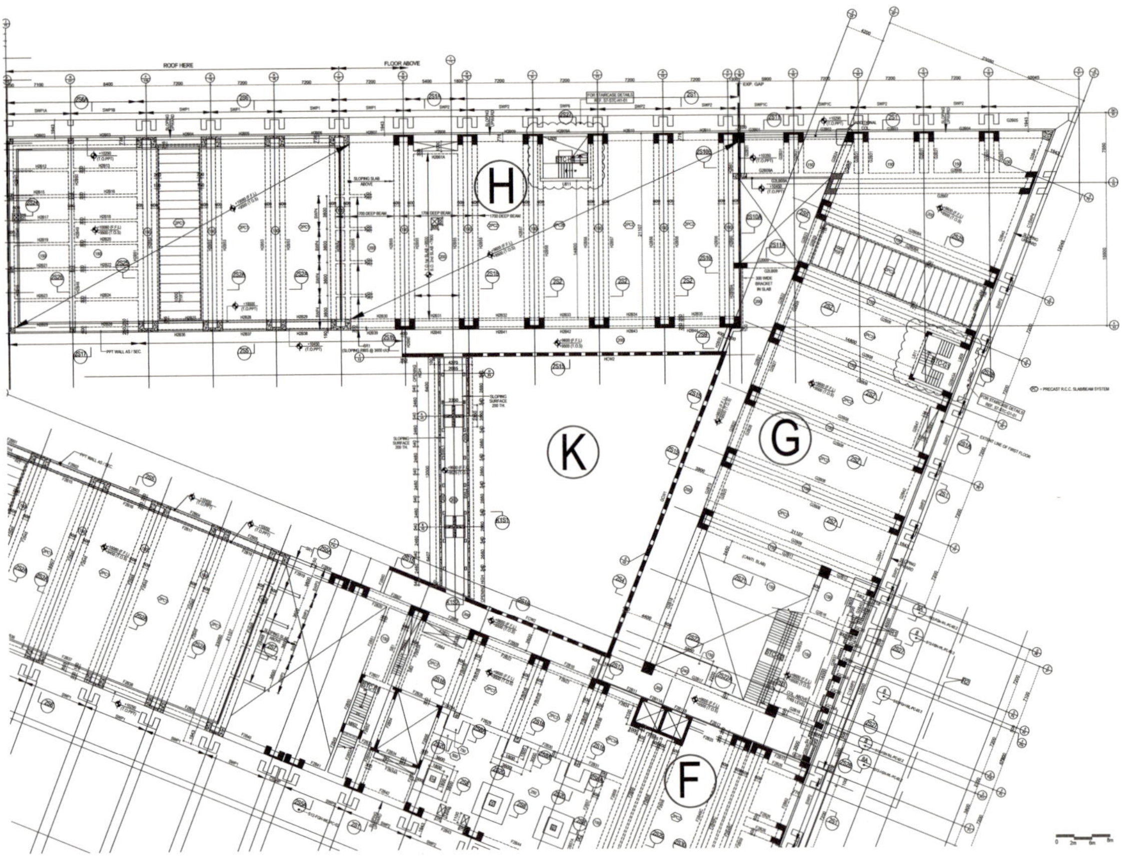

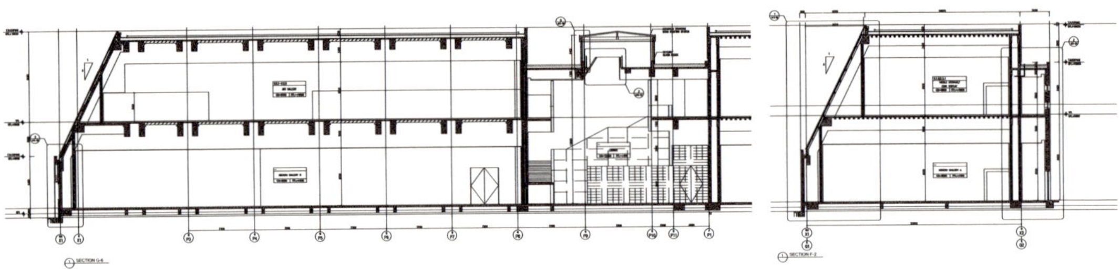

Drawing no. S4-GH-1FP-01: Framing Plan at +9.50m Level (Block G and H): 15 July 2013

(Partial) Drawing: Section–Block G and F

Plan and section of the three exhibition gallery blocks and interconnecting bridge.

Blocks F, G and H are the main gallery spaces, with tapered concrete walls and roofs. These blocks are 28m wide and 103.2m, 77m and 82m long, respectively. The structural system comprises C-shaped cast-in-situ frames spanning 22m, supporting 5m-long pre-cast double Ts.

Construction photos of the gallery blocks: the 20m cantilever canopy with a structure of steel trusses at the entrance; a closed hexagonal bridge with a span of 12m and cantilevers of 6.4m and 7.4m on either sides connecting the galleries; cast-in-situ C-shaped tapering columns; and twin beam frames in the galleries.

Understanding a Practice

Neelkanth Chhaya

Mahendra Raj's career began at a time of momentous change in Indian history. He had graduated as a civil engineer from Punjab College of Engineering and Technology, Lahore, in 1946 and had taken up a position as Assistant Engineer at the Punjab Public Works Department (PWD) in Lahore when the decision to partition India and Pakistan was taken. Raj chose Indian citizenship and was transferred to Shimla, the temporary capital of the state of Punjab in the newly formed India. Despite the erupting violence in Lahore he managed to reach Shimla safely and worked there for a couple of years, first overseeing the maintenance of what he describes as a '200-kilometre mule track' that was the Tibet-Shimla Road. Later, he was given the task of maintaining all government property in Shimla and the road between Shimla and Kalka.

In 1949 plans were made to build Punjab's new capital in Chandigarh at the foothills of the Himalayas and the PWD appointed Raj as the first Resident Engineer of the city. With the internationally famous architect Le Corbusier at its helm, Chandigarh was arguably one of the most important planning and construction projects in post-Independence India, and the position gave the young civil engineer, with just five years of experience, the opportunity to work on a project of international significance. It was also here that he began to learn the art of designing in concrete and about the intricacies of structural design.

Raj's career may be said to have really taken off in 1951 when he was appointed as the Design Engineer for the construction of Chandigarh's High Court, now known as the Palace of Justice. Le Corbusier had conceived of it as a complex building using concrete to create monumental plastic forms, many elements of which were difficult to realise into actual physical structures. A particular challenge was the geometric parasol roof, consisting of shells that presented the engineers with both structural and functional hurdles. Other unconventional features such as the ramp and brise soleil also demanded departures from norms of the time. Working with other engineers, Raj proposed modifications in the design which made the construction of the building possible. One cannot imagine a more difficult test of an engineer's skills, but by the time they had finished the project, Raj had become, in his own words, 'a reasonably good structural engineer.'

In 1953 Raj was promoted to the position of Executive Engineer and given the responsibilities of completing the High Court and designing the Secretariat. Seemingly a simple frame structure, the Secretariat was actually an exercise in the extremely detailed articulation of the building's facade. Raj took the bold step of presenting alternatives that would circumvent some of the problems with the structural requirements of the facade in the original design. Although somewhat resistant to the idea of adapting his designs at first, once he had examined Raj's proposal, Corbusier accepted them.

By mid-1955 Raj had completed both projects. Feeling that despite his experience he needed to learn more about the formal discipline of structural engineering, he headed overseas. He obtained a Master's degree in Structural Design from the University of Minnesota and subsequently moved to New York City to work for Ammann & Whitney, a firm of consulting engineers where structural design formed an important part of the architectural concept. Raj worked on several projects using large-span folded plate as well as arched structures and developed a fluency in visualising the behaviour of forms, which he would use to great advantage in the future. During this period he also honed the ability to work with unconventional forms, for example, in the sculptural framing of architect John Johansen's US Embassy building in Dublin. In 1959 while visiting India to work on an exhibition pavilion for the US Government, Raj was persuaded to consider moving back to set up a practice. Eventually he did make this decision and in early 1960, together with his friend, established the firm 'Raj and Vakil' in Bombay.

The 1960s were an exciting time in post-Independence India, energised by the emergence of a new generation of professionals in every field eager to make their mark. It was an especially important juncture for developments in Indian architecture and engineering, with a whole generation setting up new practices. Many architects, among them AP Kanvinde in Delhi, Balkrishna Doshi in Ahmedabad and Charles Correa in Bombay, who set up their practices in the late 1950s and early 1960s, appointed Raj as their structural consultant. Eventually they would all come to recognise him as an equal collaborator in design.

It was also during this time that the Bombay Presidency was split into two states and Ahmedabad became the capital of the newly formed Gujarat. Though smaller than the major Indian metropolitan cities, it was a centre for a flourishing textile industry and had an established culture of entrepreneurship that encouraged innovation and exploration. Already in the 1940s and 1950s it had invited such famous architects as Claude Batley and Surendranath Kar to build various public and private buildings. With the formation of the new state there was an ambitious development programme in place with proposals for buildings for new educational institutions, research facilities and importantly, public facilities for performing arts and sports. Thus the city of Ahmedabad became the nexus for important new directions in architecture and engineering and all the young aspirants of the field–Kanvinde, Correa, Doshi, Mahendra Raj and Vakil–found a receptive patronage here.

For Raj, Ahmedabad was a time of professional evolution. Even though the city supported development, its business ethos was underscored by an austerity that was wary of waste and unnecessary opulence. The building industry was less technologically advanced as compared to Delhi or Bombay, and many building practices were quite elementary. Remarkably, in the face of such hurdles and limited resources Raj was able to fully realise his creative designs.

Two outstanding projects showcase the approach that Raj developed and fine-tuned while working in Ahmedabad. The first is a cricket stadium that he designed for the Ahmedabad Municipal Corporation with Charles Correa. What is noteworthy about this structure is the delicacy and elegance that it attains as the thin folded plate rises from the ground and forms an integral whole with the spectator-stands and the 20-meter-long cantilever roof. Each part of the structure works with the other in a fine balance. Raj's intuitive grasp of the behaviour of the materials and his keen feel for systemic clarity make this stadium an aesthetic delight even today. The second project is the Tagore Memorial Hall designed by Balkrishna Doshi. This performance hall takes Raj's exploration of folded-plate structures further in the form of a continuous wall-and-roof system for the entire building. An independent under-structure supporting the rear end of the audience-seating on twisted columns designed to follow the lines of stress forms a large cantilever within the foyer space. Set on the banks of the Sabarmati River, the Tagore Hall is a commanding presence that simultaneously exudes tremendous grace and lightness.

In 1964 Ahmedabad created an opportunity for a new voyage of exploration. One of the campuses of the newly established Indian Institute of Management (IIM) was to be constructed in the city. The project was entrusted to the great American architect Louis Kahn with Doshi as the architectural advisor. Completing the team was Raj as the structural advisor.

Kahn's architecture emphasised a faithfulness to the character of his chosen material and the purity of structural concept. He envisioned the entire IIM campus in brick, a material which required a completely different approach to design since the flow of stress is very different from that observed in concrete, with which Raj had more experience. Despite initial reservations, Raj agreed to use brick for the sake of creative vision and soon took to the needs of this material with great interest. The project entailed the use of arches for spanning over openings and Raj's inspired suggestion of using concrete ties to restrain the arches and take care of the outward thrust enlarged the design vocabulary that Kahn developed at IIM. In addition to the design aspect, there was the question of construction technique. One needed to develop details that were robust enough for simple construction in the rough Ahmedabad brick and yet able to lend the sophistication and clarity that Kahn's vision demanded. Here the close collaboration between architectural and structural development was vital, and it resulted in the bold, heroic yet poised forms that came out of the aesthetic of brick.

Mahendra Raj moved his practice to Delhi in 1970. Here at the heart of national life, projects were grander and more challenging. By now his creative approach was very well developed. His mere presence in any design team demonstrably spurred creative spatial and formal exploration as he freed the architect's proposal of constraints and added to the elegance of the structure.

Raj had already worked with Kanvinde on such projects as the Gandhi Memorial Hall and almost immediately after his move to Delhi they collaborated on an industrial assignment to build the Hindon River Mills in Ghaziabad. In order to accommodate the client's requirements for an increased span without compromising the quality of the job, Raj used cast-in-situ, arched bridge-like frames from which were hung concrete air ducts that eventually supported the roof-slab pre-cast elements. Alongside in-situ construction, pre-cast double t-section slab elements were cast on ground and lifted in place with a gantry crane that moved between the arched frames. This was a fine demonstration of Raj's ability to simultaneously manage architectural layout, structural design, construction practice, service integration and the demands of time. Raj and Kanvinde worked together on numerous occasions in the years that followed, including several dairy projects as well as the Sports Stadium in Srinagar.

A completely different and more plastic design sense can be seen in his work with architect Shiv Nath Prasad whose designs, greatly inspired by Le Corbusier, were bold essays in reinforced concrete. For the Shri Ram Centre for Arts and Culture, New Delhi, with its drum-like lower level and two upper floors for practice and teaching areas, Raj made two virtually independent structures: a drum shell for the sculptural lower part and a square upper portion supported on four columns. The services and staircase were skilfully threaded into the ensemble. Another Prasad project, the Akbar Hotel in New Delhi, exhibits Raj's refined sense of structure. The main hotel block required a close-spaced framing configuration that the architect wished to carry into the public areas as well. On Raj's suggestion, however, these areas were reconceived as large column-free spaces achieved with a transfer girder on the transitional floor. A beautifully ribbed slab on the ground floor allowed a large span to be handled with minimal rib depth, conveying a mastery of structural and aesthetic considerations.

Some of Raj's most impressive structures were built in collaboration with the noted architects Kuldip Singh and Raj Rewal whose boldly innovative designs came audaciously alive with his engineering acumen. With Rewal he built the Hall of Nations, a remarkable single exhibition space of 84m x 84m covered with a unique space-frame structure. Normally these structures are made in steel, but since it was scarce and prohibitively expensive, Raj designed it in reinforced concrete instead. This translation is characteristic of the open-minded creativity that he coupled with carefully guided construction practices which were vital given the complexity of the reinforcement at the joints of this space-frame structure. Raj's other prominent work with Rewal was the State Trading Corporation building, where reinforced-concrete Vierendeel trusses were used to create open-plan office spaces. With Kuldip Singh, Raj designed two major office buildings in Delhi–NCDC and NDMC–both of which used the idea of end walls as supports for a form that tapers from the ground up.

The 1980s witnessed a shift in preferences and priorities. Earlier, a bare, 'brutalist' aesthetic had made the structure visible in all its glory, disciplining its engineers to achieve simplicity and clarity of parts and joints. Now, with architects using claddings and finishes to enrich aesthetics, the structure was no longer visible. A lesser or lazier engineer could have relaxed his focus on the finer nuances of practice, but Raj showed remarkable integrity, still working out every single detail of the structure such that it could stand for itself. He continued his zestful explorations of structural clarity while working with younger architects, displaying his playful best in the design for the Alliance Française building in Delhi, with its umbrella hovering over the structure.

Over the years, practicing from Delhi, Raj built some of the most impressive and innovative structures throughout the country, adding that special touch through structural clarity and elegance. One sees in his career a lifetime devoted to the creative exploration of structural design. His close attention to particularities of context, the transformation of possibly negative factors into positive attributes, a great sense of responsibility toward the use of resources, and immense confidence without bravado are all hallmarks of a great engineer. Mahendra Raj can be said to be one of those figures who set the benchmarks for others to aspire to, making his impact on architecture and engineering in contemporary India immeasurable.

External view of one of the Halls of Industries, New Delhi.

Following pages: Double arches and hangers spanning the roof of Hindon River Mills, Ghaziabad.

The Poetics of Engineered Forms

Jaimini Mehta

What is accessible through one or two rare lectures, but mostly via publications of the work of some of India's best architects, is the sense of building forms engineered in exemplary fashion, with materials at hand, low level of technology and with the most unassuming ease. The last is borne out by the fact that when I mention that his career coincides exactly with India's emergence as an independent nation representing herself through the many urban icons that he helped build, he is truly surprised. He limits his explanations to pragmatic issues such as cost of construction, scarcity of means of production and architects' whims as if it was the exigency of the situation and not his own sense of order that determined the outcome.

That is Mahendra Raj for you–generous, gentle and unassuming. A lifetime spent with the uncompromising logic of structural engineering has not dented this essential humanity. Perhaps, the transcendental qualities of the structures that he has given us can be understood more clearly if we realise that only the confluence of the hard logic of technology and human intentionality can distinguish architecture from an ordinary building.

The supposed neutrality of technology prevents us from understanding its true essence. Yet, this very illusion of its instrumental neutrality makes us want to control it. When Martin Heidegger, in his critique of Western technology, with its emphasis on the instrumental, quoted the poet Friedrich Hölderlin's memorable lines,

'Full of merit, yet poetically, man
Dwells on this earth'

he was drawing our attention to that aspect of technology which brought architecture and engineering to the realm of the poetic. Historically, the relationship between architecture and engineering had never been as problematic as it came to be since 1747 when Jean-Rodolphe Perronet established the first school of civil engineering (École des Ponts et Chaussées) in Paris. Prior to this, architects had engineered their own structures keeping the architectural knowledge unfragmented. Until then, the term 'architect' implied both a builder well-versed in the craft of construction and engineering, as well as an artist. The bifurcation of architecture and engineering marked the end of traditional guilds in Paris and gave birth to the new profession of civil engineering. With this seemingly innocuous event, architecture suffered its first crisis of confidence. What had earlier been taken for granted as inseparable–construction of buildings and the creation of a work of art–was now compartmentalised into two distinct domains, each with its own independent practitioner. The representation of architecture and civil engineering as two separate disciplines with their own methods and myths replaced the earlier unified whole. Civil engineering defined for itself and occupied the space of measurable rationality, while architecture, by default, had to contend with its opposite, namely, the immeasurable and indefinable.

The resulting specialisation of the various skills required to make a building created a classificatory grid wherein different professions positioned themselves depending upon the prevailing social priorities. Ironically, by claiming mathematical exactitude and value neutrality in the age of rationalism, engineering was able to establish its credibility only by divorcing itself from the ambit of the poetic, becoming, in the process, an instrumental profession vis-à-vis architecture.

The radical positions assumed by both the professional communities, as well as the general public, compel us to interrogate the complex relationship between architecture and engineering on the one hand, and the social credibility of the professions in their contribution to defining the urban landscape of India as an independent nation, on the other. This may help us understand the uniqueness and the range of Mahendra Raj's work since he began his career as an engineer in 1947, the year India, an ancient civilisation, became a young, independent modern nation.

When one looks at the entire body of work produced by Raj in the six decades since Independence, what is striking is the consistency of the structural narratives in all the projects. Looking for the structural logic and its essence in the architectural form, and not just the utilitarian task of supporting the building concerned, is what has driven this engineer throughout his career. Thus, the roof of Le Corbusier's High Court in Chandigarh, one of his first assignments as a structural engineer, becomes a series of conoidal shell-like structures, though slightly modified to suit the available means of construction in India. Raj confesses to never having learnt about conoidal shells during his training as an engineer, but once the logic of the form became clear, he found a way to engineer it and also build it within the limitations of available technology. Its dramatic yet graceful projection successfully interprets, in technical terms, the intentions of Le Corbusier as expressed in his early sketches.

It is worth mentioning that between his experiences as a civil/structural engineer in Chandigarh and starting his own independent practice in India, Raj spent five years in the United States pursuing focussed studies in structural design to bridge the gaps in his earlier training. This included a four-year association with the firm of Ammann & Whitney founded by the Swiss-born bridge designer Othmar Ammann and the American, Charles Whitney, who was at that time actively developing his plastic theory of reinforced concrete including thin-shell structures through his path-breaking collaboration with Eero Saarinen. The association resulted in iconic buildings such as the Trans World Airlines (TWA) terminal at the John F. Kennedy (JFK) International Airport in New York, the Dulles International Airport in Washington and the Kresge Auditorium at the Massachusetts Institute of Technology (MIT). This constituted significant exposure for the young Indian engineer. It provided him with the necessary intellectual tools to realise his already present sense of structural order.

This search for structural order with a more confident understanding of the usages of concrete is most evident in a few projects of Raj where the initial architectural design did not hint at an obvious structural logic (unlike the High Court in Chandigarh), but had to be painstakingly arrived at. It called for active and mutually supportive collaborations with architects. The stadium in Ahmedabad, designed in collaboration with architect Charles Correa in 1963–65, is a good example of the non-fragmentary relationship between architecture and engineering which much of Raj's work exemplifies. The stadium design posed a twofold problem of designing the stands to support the spectator seating as well as a large cantilevered roof that would keep the stands free of any obstruction. To this, we may add a third issue–the need to evolve an appropriate architectural language for an important public urban marker. It took at least five alternative engineering proposals from Mahendra Raj and several model studies by the architect before the realisation that folded plate offered the clearest structural logic. The plasticity of concrete hinted at the possibility that it could be a self-sustaining structure when moulded into a folded planer plate. The design of the stadium that finally emerged from this collaboration successfully expressed this. The resulting architectural order was indistinguishable from the structural order.

This familiarity and facility with folded plate found expression again in the Tagore Hall, also in Ahmedabad, in collaboration with architect BV Doshi and fifteen years later in collaboration with Kanvinde, Rai & Chowdhury for the sports complex in Srinagar, Jammu and Kashmir. The latter is actually a bold variation of the logic of folded plate. The narrative was of a giant bubble made of huge concrete plates that were folded to make them self-supporting. The plates themselves were too large to be left unmediated. This was resolved by stiffening them with an equilateral triangulated skeleton of concrete bones covered with pre-cast concrete or glass skin. This project illustrates that it is hard to say where architecture ends and engineering begins and vice versa. A building tells the story of its making through a narrative that retains this ambiguity.

The large exhibition halls, Hall of Nations and Halls of Industries, at Pragati Maidan, New Delhi, built in collaboration with architect Raj Rewal preceded the building in Srinagar by almost a decade and, at first impression, seem to have been similarly conceived. But this is not the case. Designed to house national and international exhibitions, these buildings occupy an important position in the consciousness of Delhi citizens as a symbol of modern India aspiring to be an advanced industrial nation. They have the same significance for Delhi as the Crystal Palace did for London in 1851. The architecture of these halls pre-determined their structural order–a space frame–and the method of construction. But here the story takes an interesting turn. Normally, a space frame formed by tetrahedrons is constructed with pre-fabricated tubular steel members assembled in situ around a standard spherical joint element. This technology was not available in India at that time, and it was decided to construct the space frame in cast-in-place concrete. Problems of pouring concrete in a joint that has potentially nine members coming together cannot be underestimated. Yet Mahendra Raj took up the challenge as can be seen in the manually drafted drawings, which testify to his attention to detail and his visual imagination. Seen from up close, the building seems monolithic as a cast-in-place concrete structure would be. However, from a distance, it appears to contain a massive space with an unexpected lightness. It is an effective marker on the urban landscape of Delhi.

There is no ambiguity in the design of Akbar Hotel, New Delhi–its architectural narrative demands priority. The structural order has its own narrative, except that it is apparent only in segments. Transfer of weight from upper floors, which required repeatable small-span spaces, to the large-span public spaces in the lower floors was realised in this building through a bold floor-high transfer girder. When the structure does reveal itself, as it does in the lobby, its lyrical expression of distribution and transfer of weight is an effective counterpoint to the architectonic qualities of the rest of the building.

It is not often that one finds the body of an individual's work mirroring the journey of a nation from birth to maturity. Viewing the work of Raj chronologically makes it abundantly clear that his engineering contribution has been sought after by some of the best architects who have designed most of the iconic structures in the first six decades of Independent India. Together, these buildings indicate the way India has represented herself over this period. The 1950s and 1960s, characterised by the socialistic leanings of a state with a command economy, were represented by large-scale state-sponsored projects. The Capitol Complex in Chandigarh, the stadium and the Tagore Memorial Hall in Ahmedabad, and the Shri Ram Centre and Akbar Hotel in Delhi–built with singular and monolithic exposed concrete–exemplify this period. Architecture stands in splendid Cartesian isolation in the urban landscape of this era.

A gradual opening up of the democratic space and a growing awareness of the city as an expression of local governance is visible from the 1970s onwards. The two projects by BV Doshi in Ahmedabad–the Central Bank building and the Premabhai Hall–show forms that sensitively adjust to the immediate urban context. There is an apparent and implied give and take between the architectural plan and its response to the fabric of the city. The realisation that a building is complete only to the extent that it completes the city makes the engineering of these buildings a complex undertaking.

Even though concrete still predominates, we see more 'fragile' forms emerging in the work of Mahendra Raj in the 1980s. This is reflected in the Conference Complex in Srinagar where the building is an effective counterpoint to the massive hills behind. Elsewhere, brick and stone cladding conjure up visions of craftsmen working with their bare hands and also bring polychromy to the overall experience of architecture. Buildings ought to be experienced not only from afar but also from close quarters. Both, the National Science Centre in Delhi by Kanvinde, Rai & Chowdhury and the SCOPE Complex by Raj Rewal, built in the 1980s, are state-sponsored projects containing massive built space illustrative of the move away from monolithic forms as well as exposed concrete in favour of a more humanistic and comprehensible formal language, and the use of polychromic surfaces.

Throughout the six decades since Independence, though multiple architects have contributed in this endeavour to represent, through architecture, the new emerging India, the one singular and constant presence has been that of the sensitive and unassuming engineer, Mahendra Raj. Architects from across this period have reposed explicit faith in his engineering sensibilities and his sense of order. Two centuries ago, Perronet may have bifurcated the noble profession of architecture into design and engineering but technology has remained a unifying force–technology not in terms of the instrumentality of construction, as a means to an end, but as the unity of the means and the end as symbolised by the word 'design'. Alberty was right when he said in 1485 that architecture consisted of two aspects: design and matter. The cerebral and the corporeal, imagining and making, are inextricably joined together in the same practical system which provides the basis for both, the construction of a building and a work of art. The separation of architecture and engineering ruptured this complex relationship. Fortunately, on both sides of the divide, there still exist professionals who are able to come together to elevate technology beyond the technical.

Trucks are preparing to take away the last exhibition and the empty space of the Hall of Nations silently awaits another exhibition to validate its structure. In this silence, we hear the concrete whisper about India's relentless monsoon that it has to endure. The surfaces are beginning to blacken. Mildew and fungus appear here and there. The windows appear wide open against the bright Indian sun, the glass barely held in place by the rusting steel mullion, made to look slenderer by the bright sky behind. Time and weather, chronology and the atmospheric, are the facts of the materiality of buildings. This 'fragility' of the physical stands in stark contrast to the empty 'room', a slice of the cosmic space, claimed by man. The space, with its dimensionality, and the quality of light, is a miracle and a testimony to the indomitable spirit of man. Only in this emptiness can we sense the complex nature of technology–technology not as technique, but as the idea of 'making assimilated' by both architecture and engineering. Are we then condemned to realise the true essence of technology only when architecture is not inhabited, by hiding all the signs necessary for us to recognise it?

Following pages: Interior view from the mezzanine floor under the upper seating at the Indoor Stadium, Srinagar.

A Conversation
between BV Doshi and Mahendra Raj

Doshi What made you decide to become an engineer?
Raj It was not my decision; it was my father's decision. Because he himself was an engineer, he wanted all his sons to become civil engineers.

Doshi So did all the brothers in the family become civil engineers?
Raj No, my eldest brother, Yograj, didn't. He became an automobile engineer. My next brother, Rishi Raj, became a civil engineer. The brother younger to him, Krishan Raj, also became a civil engineer.

Doshi Who was in dams...
Raj Yes, both of them were in irrigation, and eventually worked on the Bhakra Dam.

Doshi What did you really want to become, if you had been given a choice?
Raj Given a choice I would have pursued mathematics.

Doshi Okay...
Raj I loved mathematics. Actually a couple of times while studying in engineering college I even thought of just leaving engineering studies and moving to mathematics, but it was not so easy at that time to make such changes. I completed my engineering studies and became an engineer, and that was that.

Doshi But if you had become a mathematician, what would have happened? Did you ever think like that? Because I don't see much difference in either...
Raj (Laughs) No! Of course, there's a lot of difference...

Doshi What is the difference?
Raj Mathematics is pursuit of scientific knowledge. Engineering is application of this knowledge. Whether pure or applied, mathematics is all theoretical, and forms the foundation of all engineering concepts. In engineering you apply mathematics to create. But even in mathematics perhaps, if you become a great mathematician, and you discover or evolve theories, you do add to the existing knowledge, but essentially it's still theoretical. As a civil engineer, and particularly with buildings, you apply your knowledge of science and mathematics and create.

Doshi There is also another component ...when you want to develop new things, your thinking is not the same as of an ordinary engineer. Sound knowledge of mathematics gives you an advantage. In addition there is the capacity of visualisation. I presume strong mathematics and a capacity to visualise gives you an advantage as an engineer.
Raj Yes, you see for civil engineering or for any other discipline of engineering you need to be strong in mathematics. Because as I said earlier, it is the foundation of all engineering–mathematics and science. But after that, if you have an artistic mind and a capacity for visualisation, it helps. Now, my mathematics was strong as an engineer, but I couldn't sketch, even though I could visualise...

Doshi No, but since you can visualise, even if you cannot sketch, it helps.
Raj Visualise yes, but it was very important to make drawings too. But visualisation was very important too. This capability helped me considerably in most of the projects but much more so in very complicated details like the joints in the Hall of Nations. It was very tough to visualise and draw joints with nine to thirteen members, and still tougher to deal with the reinforcing bars in all those members, which had to pass through a joint.

Doshi Yes, I can see that. What I'm saying is that, rhythm and relationship is common in every field, more or less. Many fields have that, even poetry. So, what happens is that if your bent was not of mathematics

then your engineering approach would have not been as poetic as it is.
Raj Yes, perhaps...I don't know.

Doshi It must have helped with the abstractions, I mean...
Raj Yes, it did.

Doshi Like Calatrava, became an engineer but he was also an architect you know. So he combined both the disciplines together. And that fluidity of the form came because of his background of architecture as well as engineering.
Raj Yes, sometimes I do have regrets (but it's too late) that when I was studying in America–because in India, of course, it was not possible–some engineers had this twin degree in architecture and engineering which helped quite a bit. Maybe I should have studied that too, but I kept on pursuing post-graduate studies in engineering.

Doshi Yes, I understand, when I went to Calatrava, to see his office, he was busy in a meeting so I sent in a message that I worked with Le Corbusier and I am from Ahmedabad in India. Is it possible to see you? Suddenly, he broke his meeting, came down, he was on the first floor. Then he says, 'Yes I'm so glad, I'm sorry I'm in a meeting.' He was asking his people: Did you show Mr Doshi the model etc.? And then finally, I don't know what happened, but he says, 'Wait, wait, wait!' then he goes up, comes back, no, he goes to the other room down below, comes back and gives me a book of his sketches–all watercolour drawings of human figures only, in different positions, and he says this is what he does in the morning, in the very early mornings. Then he sketched for me on the front cover and gave it to me, signed. But what I liked about him is that he was looking at the human figure as the basis of his dreams. Leonardo did similar things, so they both had this kind of a mathematical and poetic idea of balance, dynamics and how the tensions work together, and even how the joints work. So the whole body was not only a physical body, it was not a body like an artist was looking, but they were looking at it as an engineer and a mathematician. So that is why I am prodding this question.
Raj Yes, but both of them, I mean, they are geniuses...

Doshi No, but that is what I think you have. You remember we did so many buildings and did you not explore all the time something else that was not thought of earlier?
Raj Yes, sure, the attempt was always to do something different, which has not been thought of earlier. Sometimes one succeeded, sometimes one didn't succeed. But every time the attempt was... because whenever a problem came up, we always made an attempt to resolve it in some different way...

Doshi But coming back again to your idea of USA, if you had not come back to India, and were still in USA, what kind of work would you have done?
Raj (Laughs) God knows what kind...!

Doshi No, no but this is how, you know, one looks back, isn't it? Because you went back in between for two years or three years...
Raj Yes, I went back in between, but that was for a different purpose. That was because I had already started my practice here and my friend in New York, Abba Tor, had a small office. At that time in 1968–69 I made an attempt at outsourcing but there were no computers, no mobiles, nothing, and so with the help of telegrams, telex messages and postal mail we made some attempts and did a couple of outsourced jobs. But it meant I had to be present there in USA and it was not very exciting...you know doing only this sort of work...

Doshi But when you were there you were also looking at the building happening in USA and in the world

because it is a different kind of exposure from what you were practicing from India. Like when I went to USA, I was all the time looking at another world... and for me that world was something that gave me more insights, the spirit of asking questions and to do something different. I'm sure it must have happened to you.
Raj Well my case was slightly different. My introduction to structural building was through Corbusier's buildings. Before that I had not even designed a lintel, not even a masonry arch. So I got into it straight. And there were not very many aids available. I think there were some books by...

Doshi ...Nervi?
Raj No, no, Nervi came later. At that time there were some codes by American Concrete Institute, different associations, and booklets etc. You see, when I started work, I was under one engineer who was a year senior to me, Gulzar Singh. He had come back from Canada–did his Master's and came back. So he had some knowledge of these ACI, PCI etc. and when I started working, he gave me 2-3 books that had information. It was on the basis of these books that I started working on structural design. But it was so exciting to work and I got absolutely absorbed in reading... applying... reading... applying.

Doshi So you really were self-taught?
Raj Yes, more or less, in structural designing. And that's how I decided I must go abroad and study more because I realised there's lot more to learn.

Doshi That was a different time indeed. You went twice to USA, once to do this, after Corbusier's project, then in between, when you went again, what happened then?
Raj What had happened was that by the time I went to USA again, I already had a practice, but not a flourishing practice. The first time when I went to USA, I did my Master's in Minnesota and came to New York and started working for Ammann & Whitney. Ammann & Whitney did large span structures.

Doshi Yes, I remember the one in Urbana, that big stadium.
Raj Yes, that big stadium I had contributed quite a bit to its design. I had worked on the roof. It was a folded plate one. It's quite interesting, because it's done by Eero Saarinen, the famous American architect.
At the top there is an opening. Around the opening there is a solid concrete ring and then the folds start from there. When we were analysing it we found that around that opening, the stresses were higher than the strength available in the folds. My senior at the office said, 'Let us create some ribs attached to the folds to absorb heavier stresses.' I thought to myself, 'That is no solution because with ribs, the folds would look very strange. We should try and devise another system of folds so that they are made in such a manner that they can contain the forces.' So I kept on thinking about it, how to reshape the folds, and then a couple of thoughts came to me and I sketched them quickly and showed it to him. First he was like, 'No, no! Let us go ahead with added ribs. The architects won't accept any change.' But I said let us try because this way we don't have to create the ribs. We get strength where we need it. Then both of us went to his senior and he was very excited when he saw my idea. He said, 'Fantastic, lovely idea', and he said, 'Yes, yes why don't we do that, it is much better.' And then they sent it to the architect and he accepted it readily. That gave me a lot of encouragement. Earlier even in Chandigarh, at times, I had faced similar situations and had got my ideas approved. At one time I had to go to Corbusier to see him, about the Secretariat, Sector Four, a facade where the columns were jumping around. He was furious, but later he accepted my suggestions. I had this habit of looking at the structure and seeing if there was any better way to achieve it, which became very useful in America.

Doshi If you remember, when we were doing Tagore Hall, the beams of the seating were becoming heavy and were very deep and you needed a column that was also going parallel to the beams, going down straight. And we discussed that it was not possible. I suggested that we get some other idea by which a column becomes something else rather than just a wall, and you created that twisted column. It was the first time probably in India where somebody had done this. So I think that is always in your mind–to make connections natural, and then they transform into something else.

Raj Yes, I was quite a bit influenced by Nervi because I went especially to Italy to meet him…

Doshi Rome? Turino?
Raj Yes, Rome. I went to his office. I met him. It was quite interesting to meet him…when I met him in Rome. It was the second time that I had twisted a column. The first time, I had done this in New Great Insurance building where we had taken these columns and flared them up, which was also completely my idea, with this Italian architect… Radiconcini…
He had come to work for the Atomic Energy Commission. Atomic Energy had hired his firm to do some buildings in their complex, and he was doing his own practice also. There was this case of a garage somebody had purchased…

Doshi Oh! The multi-storeyed building on the top of the garage, I remember, in Bombay.
Raj Yes, near Churchgate.

Doshi Yes, I remember the building. I came to your office that time.
Raj In that garage there were columns here and there to accommodate the cars etc. and at the top they wanted to add more floors, because additional floor space index was available. The Italian architect Radiconcini was struggling to use those columns but the columns were placed irregularly. There was another Indian architect, Shareef Moloobhai, who was also associated with this project. He was an architect but he was also a businessman, a ship chandler. He had worked with [Frank] Lloyd Wright in America. Somehow, I think, we had met somewhere earlier and we liked each other. So that is how he came to me. He said, there is this building we're trying to do, will you help us do it? The problem is, we have no information about these old columns–steel is not known, foundations are not known, the drawings are not available, but we want to use the additional floor space index, we don't have any idea how to go about it. We have to find out how we can add on to those old columns without any information about their strength and their foundations. I said, 'Let's see what you want to add and how.' So I got the total plan. And then I thought of this: why not create a new structure with a few columns? You see the problem was that for additional columns, additional piles had to be driven within the building. I investigated and found out that one could do piles inside because there was sufficient head room. We also devised certain other ways of doing these piles. And then I thought that within the slabs if we puncture and create new columns, strong columns, just a few columns, and take them above the existing building, and then have transverse girders at the top, we could support the new building. I made some sketches to explain these thoughts and showed them to Radiconcini. He got very excited. He said, 'Yes, Yes!' and the idea was adopted. We constructed the building accordingly. And even though this project was not that well known, but word spread around. So I think I had this flair for finding a different solution than what somebody else was thinking about.

Doshi You were looking at a circumstance as it came and you were thinking afresh. 'This is the circumstance, what can be done.' There is no reference point to that question.
Raj In civil engineering, building engineering, you have to identify the forces and take them to the ground by a path. You know, the path and the forces try to take a direct and easy path–they don't want to jump around, they just want to go to the ground and be supported there. The requirements of the structure, the requirements of the forces, the requirement of the function, and the statics–you try to think of everything and then sometimes the solution suggests itself. It satisfies everything. Because we had no knowledge of the old columns, we had no knowledge of the old foundations, and they were all staggered, a normal office building was not possible, so what I suggested to them was that they create a few regular columns, puncturing the slabs, and support the new building on that.

Doshi So, coming back to when you went to Italy. Did you meet Nervi and see his structures?
Raj I saw all his structures. I went especially to see them. I was very impressed, but some of them had not been maintained properly, that was a disappointment. I met Nervi–a very simple, unassuming guy, a very small office, and a number of models and sketches.

It was quite inspiring and the idiom he used–there are two things that he used the most, one was Ferro-cement.

Doshi Ferro-cement, yes.
Raj The other was folds. He was very fond of folds. And actually it is this idea that I used for the Tagore Memorial Hall on the column. Nervi has tried it in a couple of places, twisting the column, which I found very impressive. So I was looking forward to an opportunity to try that idea and Tagore Memorial Hall presented that first opportunity. I think we've tried this again also, at a couple of other places. It is very interesting to shape the columns to suit the forces. For the Tagore Hall we wanted the column to be aligned with the rib and the rib had a cantilever and that would give it a moment, and we didn't want to take that moment to the foundations, so by twisting it the moment carrying capacity of the column got reduced below, became high above, that is what we wanted, that the moment should disappear by the time it reaches the ground. So I think at that time one approached every problem entirely in a different manner. The Hindon River Mills was another such example. You know it was very interesting how the structure developed.

Doshi Which building is this?
Raj That textile mill with the arches…

Doshi Yes, yes! Which you did with Delhi Cloth Mills [DCM]. It's almost like a bridge, suspended.
Raj It is a bridge structure, essentially. The way it got developed is interesting. Consultants Gherzi Eastern had been designing structures for textile mills in Ahmedabad.

Doshi With the Mafatlals…
Raj They had created a standard structure on a 20m x 24m grid and some on a 24m x 24m grid. Something like that with concrete boxes.

Doshi Yes, carrying air-conditioning…
Raj No, carrying humidified air. Anybody who wanted to do textile mills thought of them and they did big business. But DCM had this one engineer, Mr Kriplani…

Doshi Yes, I remember him
Raj Brilliant, brilliant engineer. He was a very creative person, and he said we have to do something, unlike what everybody else has done. So they had a little sort of a competition, structural competition, which we participated in. So I think we gave some ideas–sketches. There were a couple of other engineers who also gave ideas. During this interaction somehow Kriplani liked whatever we had given and he selected us. But he said we had to increase the span, but we couldn't increase the budget. He said, 'You have this 20m x 24m grid, so why don't you make the span 48m and reduce the 20m and give us one big hall, 48m wide? You can take more columns on either end of this 48m-wide hall.' I said, 'Well let me think about how we can do that.' But the cost had to stay the same more or less–that's a tall order! So I thought over a couple of days. I kept on thinking that just any system will not work. It has to be some arch. But I thought, arch, such a big volume, what will they do, and how to hang the ducts. So the idea came: Why don't we create a tied arch? And the tie carries the humidified air and pre-cast elements. So this idea clicked and then you know, we made a presentation to them and they also liked it, because it became quite a unique sort of a structure–twin arches, the hollow box girder was hung from these and there were openings in the girder...

Doshi I have seen it. I came there, many times on that site.
Raj The twin arches are supported on twin columns for stability.

Doshi Did you ever replicate this? You haven't replicated any of the structures, no?
Raj (Laughs) No.

Doshi You have never repeated any of your ideas?
Raj (Continues laughing) No, no.

Doshi Why?
Raj Why? I don't know because I…

Doshi You want to change; you don't want to repeat.
Raj That is not the thing. When the mill was complete

the clients, they were very excited. They said what a large area, column free, great for textile set-up and so on and so forth. But a few years later, after they were very happy the textile industry started having problems. It somehow started collapsing, then they said, 'Oh! This has been unlucky!' (Laughs) But I said, 'It's the whole textile industry!' But DCM were very happy with it and then after that they gave us another job also in Ropar with Kriplani, but there it was all steel. We used completely steel structures.

Doshi Do you have photographs of all these?
Raj Yes, there are some photographs. In that period as you know every project was a challenge. Whether the consultancy fee was there or not, but it was a challenge, a big challenge. That was when we did most of these challenging structures.

Doshi But when you talk about these structures, did you ever think about Gaudi's work?
Raj Gaudi's work was very interesting–very, very exciting–which I saw much, much later in my life. But Gaudi's was entirely different, you know...

Doshi He combined many things, he changed the rules of brick, or even concrete and then he combined them together–the nature of form became very fluid.
Raj Yes because Gaudi essentially was equivalent to Calatrava, architect and engineer combined and could use the shapes and forms...

Doshi But he was much more spiritually oriented.
Raj Or as his book says, you think of shapes and forms before and then you test them out.

Doshi Yes, catenary models
Raj He used to make models with strings and weights and then he would reverse them. He would take photographs etc. and reverse them. And that is how most of his buildings he did are in the form of reverse catenaries. And he managed to create very exciting buildings and forms.

Doshi What do you think about that kind of working?
Raj When I first saw Gaudi's photographs etc., it was quite late in my career. Nervi, I was exposed to much earlier. Gaudi has exciting forms, but much later I learnt how he had created these. But so far as other engineers are concerned, Nervi was one and then there was this Bengali engineer, Fazlur Khan.

Doshi Yes. He developed those hollow columns...
Raj No, no, he developed systems for tall buildings, you see, bracings and others–tube-in-tube.

Doshi That's right.
Raj But he was a teacher and an engineer.

Doshi But I am more interested in the point that, you know, we have Indian temples and you have seen Indian temples, okay? Then you are doing these structures which are large span, supporting, which carry weight, transfer of load etc. that is why I am referring to Gaudi. Gaudi was another one who does these things in a different way. So didn't it give rise to some questions in your mind? 'What is this?' 'How was it built?' 'How can we do it?'
Raj Of course, it was very exciting to see Gaudi's work...but I saw it very late in my life

Doshi No, but even then, suppose you have to do something having seen Gaudi's work?
Raj His approach is entirely different. I think he was more a problem-solver, you know? His house, for example, which I saw–the villa which he had done, how light comes in the morning sun and the evening, and the afternoon sun, his studies and how he had shaped them were all very impressive. And then his museum in which he had tried different types of structural systems and masonry, they were also very, very exciting.

Doshi He was not a theorist.
Raj He was not a theorist. But it was all very exciting. It didn't sink in enough that I could try something which he had done.

Doshi Why I'm asking is, like Kahn is a purist, but Corbusier was not. He wouldn't mind combining materials, but Kahn, as you know, he would not even allow you to use regular columns. So what I am asking

is that Gaudi is another one who used all the things yet build not only romantic, but very fascinating forms, even as structures within.
Raj Oh, of course, Gaudi's structures...

Doshi He never plastered the buildings.
Raj But I was influenced by, as I said, Nervi and at that time there was Candela too.

Doshi Candela, yes Felix, Mexico
Raj I like their structures.

Doshi Folded plates
Raj And Candela's hyperbolic paraboloids. Then there was Eduardo Torroja...

Doshi Who did the stadium?
Raj He did everything, you know–a tank, a stadium. This and that. So these were very exciting. So far as Candela is concerned, his hyperbolic paraboloid we tried to do a couple of them, and we did some in Delhi also. In the Danfoss factory we did some umbrellas, and in Bombay we did a house. But, I think when we started practice, all these shells etc. were very much in fashion but they went out of fashion very fast. These buildings were daring and I thoroughly enjoyed reading about them. You see all the buildings had different types of challenges. In a certain period of time, even STC that we did had quite a structural challenge.

Doshi STC, I know. You did a lot of work with Vierendeel girders, no?
Raj That was the first time, but STC's challenge was that everything had to be carried on these cores. And there was not much literature available on how we could disperse loads on cores and concrete walls etc.

Doshi Is it a very different thing?
Raj Yeah, it is you know, because...

Doshi Isn't it like point loads?
Raj Well, the heavy concentrated loads, we had to carry them on...

Doshi Girders?
Raj No.

Doshi Stub columns, yes?
Raj On columns. What we required was very heavy. We had created these cores which had elevators. So there were cross walls, front walls, rear walls, and we wanted to engage all of them to carry load because loads were quite large and also because of earthquake. So we had developed a program to understand how the load disperses. Because when applying very heavy concentrated loads you want to see how that heavy, concentrated load disperses and goes down, and if that element through which it passes can take it or not. Then another heavy, concentrated load comes below and then another and so on.

Doshi Yes, it's multi-storeyed.
Raj So you have to see how far the load disperses. And how to provide the foundation, and how to provide the strength in these various elements. So, in STC both things were big challenges.

Doshi Was anything there pre-cast?
Raj No.

Doshi Would it have been better if they were pre-cast, those beams and girders?
Raj We hardly had pre-cast here in India at that time. Hall of Nations we wanted pre-cast, we developed the joint pre-cast, and the type of precision which is required for pre-cast at that time, nobody could achieve. It was all in-situ.

Doshi But if it was pre-cast, would it have made a difference?
Raj No. For this STC building pre-cast was not an apt solution.

Doshi No?
Raj Not for the vertical elements, nor for the horizontal elements...

Doshi I'm talking only about the horizontal elements.
Raj As I said not for horizontal either, because there

were large spans both ways. You had these Vierendeel girders and then, I think, the other set of Vierendeel girders were 45 feet away, so the spans for the ribs were also quite large. Therefore in-situ was more appropriate. For Hall of Nations we were very eager to have pre-cast but nobody was willing.

Doshi Did you ever do any pre-cast building?
Raj The stadium in Srinagar is all pre-cast.

Doshi So how did you get precision there?
Raj In Srinagar we did manage it because the joint was in-situ. Elements were all pre-cast and it was a single-layer structure, not a double-layered structure. It was different–not a space frame–but a space-plated structure. We had created plates with these elements and those plates formed the entire structure. So each plate had a different orientation to carry the loads. It was very exciting. You see when you do structures like that for the first time you realise how the forces will flow. So, many times when you do the analysis and it does not confirm your mental conception, you get very disturbed. This happened in this Stadium: I had realised how the forces had to come to the ground and the engineer who had developed the program, when he analysed them, he gave me the forces and they were behaving in a funny manner. I said, 'My God! What is happening? You did something wrong somewhere! This doesn't appear to be correct, you better check your program or input etc. because what you have shown me are not the type of forces which should develop. Not this magnitude here and that magnitude there.' So he went back and checked, and of course there was some error, which he had committed, and as a result of this error, the output that was coming was entirely different. And then he presented the results to which I said, 'Now, this appears correct.' But then again there was certain behaviour which I had not realised was happening and then I worked quite hard to try to understand it. Why? What was happening? You see because, when you do something for the first time and you have total responsibility and no one is giving you directions, it's a tremendous responsibility on your shoulders. So it was very exciting to work on this, because suddenly in some place the tension became very high and I said, my God!...

Doshi This was actually the question I was going to ask you: when you do this theoretical work then you have to have a test in the laboratory, a mathematical test. Did you have some people who were always cross-checking...you know, as you say?
Raj Unfortunately, no...

Doshi Then that was my question, why is it not developed?
Raj Yes, there was no cross-checking. I think...I don't know, so many times I stuck my neck out, anytime the noose could've tightened. Before I did the Hall of Nations etc. in the space frame, we were doing some project in Iran. And in Iran there were a number of buildings built by a local architect in collaboration with this famous French architect Michel Ecochard...

Doshi Ecochard, yes...
Raj He was also their consultant. So he was also doing some museum and then we were doing a big housing project and a number of other projects. And there, a space frame had to be built somewhere. So those architects said to me, 'Why don't you go to these places, England and Germany, and see the space frames and understand what systems are available because we'll have to import some system.' So I went to see the space frames. In England, they took me to a laboratory where they were testing the joint. Before this I had seen the Mero system in Germany which had these joint nodes etc. and they explained to me how they are tested, the joints and the members, in the laboratory before they are used and so it was quite interesting. I brought all the literature back with me. But the Iran jobs fell through. And so I didn't get to design the space frame. Much later, the Hall of Nations project came, and now we had to do a space frame. Space frame we thought had to be in steel, because of the joint. But there was no time, and the joint had to be worked out. And my mind was thinking back on all the laboratories where I had gone and seen the joints and members being tested. I said there is no time and now I have to take this responsibility and go about it. So, only on the

basis of my knowledge of engineering, I realised that in steel you had a type of joint, with 10 or 11 members. But then we decided that we didn't want to do steel, because with steel only one firm would be able to work, and the steel will not be available. So we decided to do it in concrete. Then I had worked out pre-cast concrete elements, pre-cast joints with plates, and concretes and connections with bolts and all that had been worked out. But where was the time to test it out? So one had to do it and take the risk, and have sufficient confidence that it would work. After that, it was for CPWD, they were the project managers on the project. They had said that they would like to check the design. I said that would be very good! (Laughs) But they had no senior person. There was a junior engineer who just checked the arithmetic of the numbers etc. So the total responsibility was on me and it was tough. Then the question came about using pre-cast. What we did was, because it had a funny type of configuration, we got a model made. At that time there was this pre-cast factory Hindustan Housing Fab. I knew the manager there. He was a good engineer and so he agreed to make a prototype of the joint only–of the joint and the elements–to see how they get connected. So that much we did, but there was no time to test out the structural strength. We had tested out the physical configurations and the feasibility of the connections. But later on, no contractor was willing to do it that way; they wanted to do it in-situ. Here we were: the Hall had to be built, but nobody wanted to do it in pre-cast concrete. Each contractor opted for in-situ construction. We had not thought of in-situ. With in-situ, the joint had different types of problems, I said, 'My God! What to do now?' How to contain these huge forces which come on the joint, you know, the tensions and compressions? So, I had a very tough time, I must say. But then I thought of ways, of transferring heavy tensions and compressions from one member to the other through that small joint. So, it was absolutely, in retrospect, I think foolhardy of me. I devised ways to contain the heavy tensile forces, to contain the heavy compressive forces, to cross each other they had to balance through the joint. But it worked!

Doshi I know it worked, because it is still there.
Raj Nobody in America would have done this nor in England, because they would have said, make prototypes, test them out and...

Doshi So what made it work? Is it a space frame anymore?
Raj It is a space frame, but it is a very ingenious way of achieving it. What I did was actually very simple. What I decided was, at the joint we have tensions and compressions so the compression was easy to contain. Tension, you know, in concrete cracks up. So, heavy tension in concrete members was a problem. Then I thought that I will, in order to contain that, use spiral reinforcement as we use in pre-stressed concrete. With pre-stressed concrete, you have a very large force at the end, at the anchorage where the concrete can burst. In order to prevent that burst (the burst is on account of the tensions created in the other direction), you put spirals. So we thought of that. That the spirals help. So in the member I had put spirals so that the cracking gets solved. But then at the end, how to absorb the tension rods? I thought I will insert all the tension rods into a compression member so that the compression there will hold, it won't allow the rods to be pulled. So, the technique which I used was that all the tension members, because some members were in tension, some in compression, coming from top were embedded into the compression members, below.
So this technique worked.

Doshi Isn't it very much like muscles of our body?
Raj Our muscles?

Doshi You have both you know, you have tension and compression, and you have joints. So, is it not very similar? No, no, I'm only asking, you never think you know, but I'm saying, you were thinking about it subconsciously, about ways, no?
Raj Yes, but we were thinking about the principles, how tension works, how compression works...

Doshi But to think about the spiral? The spiral is really like trying to now say that you're going to use…that's what I'm saying about muscles, because if they are twisted like this, and therefore they release and twist, release and twist. Therefore this happens isn't it?
Raj Here spirals were used to contain the cracks. Because when the steel is in heavy tension and the concrete is less it can crack up.

Doshi Is that the reason why the concrete is still holding all right? Because all over, anywhere you see climate like this, with temperature variations and monsoon and not really having rust-proof bars, we also have this problem of cracks today in the buildings.
Raj That depends on the quality of concrete, if it is properly vibrated and all that...
…But now that's been a lot to take in one conversation…Let's call it a day!

End of Recording, 65:36
Transcript taken from a private audio recording, 02 March, 2012.

APPENDIX

STAFF CREDITS

The projects included in this book, and many others designed by my office listed here as Chronology of Works, would not have been possible without the assistance of the entire staff including the Design Engineers, Draughtsmen and administrative personnel who have worked with me since I started my practice in 1960 as Raj & Vakil, or those who joined me in Engineering Consultants India or Mahendra Raj Consultants Pvt. Ltd. Some worked for a short time, others for many years, but they all made invaluable contributions. I wish to offer my sincere gratitude to each one of them. Without their dedicated contribution all these works would not have been possible.

An attempt has been made to cover all the staff members who worked with me, but if inadvertently someone has been left out, my apologies.

Mahendra Raj

SENIOR DESIGN MEMBERS

MS Jalundhwala, Associate
DV Jain, Senior Design Engineer
DS Parikh, Senior Associate
TT Fatehi, Director
NR Amin, Associate
VV Nori, Associate
Krishna Hegde, Director
GD Juneja, Associate
DB Mali, Senior Design Engineer
Shubhendu Pal, Senior Design Engineer
Dr DD Bhide, Senior Design Engineer
Ashok Nanda, Associate
Sumit Banerjee, Senior Design Engineer
MB Singh, Associate
BB Choudhary, Senior Associate
SR Singh, Senior Design Engineer
MK Gandhi, Senior Design Engineer
CM Sardana, Senior Associate
J Ghose, Senior Associate
RC Jain, Senior Quantity Surveyor
N Saroop, Senior Design Engineer
MS Osan, Senior Design Engineer
Dr KK Ganguly, Senior Design Engineer
Vijay Rewal, Senior Design Engineer
Parminder Singh, Senior Design Engineer
VK Dogra, Associate
Narendra Pratap, Associate
SS Mann, Executive Director
Aditya Sharma, Executive Director
Patil Vijay Khemachandra, Senior Design Engineer
Maya Rao, Executive Director
K Shivakumar, Senior Design Engineer
Suja Alexander, Senior Design Engineer
Ramanjinam, Senior Design Engineer
SK Bhatt, Senior Quantity Surveyor
Dr VP Narayanaswamy, Senior Design Engineer
Dr Alok Madan, Senior Design Engineer

Previous pages: Internal view of the DCM Pavilion, New Delhi.

JUNIOR DESIGN ENGINEERS

AK Sangani, DP Buch, RD Shah, NS Nishawala, CL Mehta, RS Mandrakar, SM Padwal, RD Mathur, PN Mehta, TSB Panicker, RP Garg, SC Sawhney, S Malhotra, RK Hegde, BJ Kini, Ameresh Jane, Hari Om Gupta, VK Jain, SK Saxena, Shivender Singh, MS Narayanan, Ashok Kumar, Ashok Khanna, SC Gupta, PK Laha, P Yusuf, K Gandotra, NK Nagdeva, CV Parthasarathy, Brijendra Singh, Anup Sud, IC Gupta, Ajay Ghosh, Mohan Murthy, Avinash CGupta, CK Bhalla, VY Salpekhar, Alok K Jain, AK Choudhury, AG Mathew, DS Bedarkar, Savin Bhatia, Ajay Kapoor, LS Naik, Mahabeshwar Hegde, SK Gupta, Bharat Rohra, GK Duggal, Sanjay Khanna, GV Saharabhajanas, Tarun Mathur, MG Syal, Kishore Arora, Bhaskar Joshi, SR Roy, RP Mathur, Vijay Chopra, SK Verma, Rakesh K Garg, Sumit K Gupta, Abhinav Gandhi, SC Keshav, TK Aggarwal, Raj Kumar Mehta, YN Dewan, Sunil K Arora, SN Mudgal, Jagdish Muku, TL Sharma, R Dasgupta, RC Sharma, SR Roy, Anil Khanna, AK Mathur, Harish Arora, Vinod Mutneja, TK Nath, AK Jain, DC Batla, Manjit Singh, MR Deshpandey, RK Ahuja, Ajay Kumar, Sudha Saxena, Alok Bharadwaj, AP Kundu, Sanjay Raina, Iqbal Hussain, MA Khan, Nishant Beg, TP Singh, RK Gupta, PRS Krishnan, Avneesh Saksena, Sanjay Gupta, Ashwin Kr Saxena, AK Bedi, Rachana Sood, GD Sharma, Subrata Saha, DK Gupta, Lissa Mariya, S Hubli, Namita Girdhan, M Rastogi, V Kulshrastha, L Shiv Kumar, AR Patankar, R Roy, PC Sheeja, NPM Nambiar, D Rama Devi, Deepak Nautiyal, Anil Prahlad, Gurvinder Singh, Nitin Dhariyal, Sanjeev Jain, Naria Premil, SN Chattopadhya, Deepika Uppal, Sunil K Sharma, Ajay Sahni, Ashwani Kumar Tyagi, Rajiv Sharma, Subrata Paul, Nisith Maji, Ramesh Kumar Dhar, Babita Bordia, Saikatranjan Ghosh, Sirshendu Mondal, Biswajit Roy, Diwakar Bhagat, Bikash Pal, Saibal Dutta, Charan Singh Bhandari, Sushma Agarwal, P Shyam Sunder, Sameer Kumar Jha, HW Keshav Prasad , K Ravi Chandra, Sunita Rohilla, SK Dixit, Naresh Kumar, Dheeraj Sehgal, Amit Kumar Singh, Sarjit S Soni, Parth Harish Thakkar, Maneesh Kumar, Vimal Kumar Suntha, Kunal Tonk, Ravindra Jaiswal, M Raju, Robert B Keisham, V K Murthy, Harmandeep Singh Kamboj, Mahendra Singh Chauhan, Rahul B Sherke, Birendra Kumar Karaiya, Santosh D Wankhade, Deepak Kumar Jain, Devinder Kumar, Kirandeep Bansal, Dania Hussain, Ranjeet Kumar, Mohammad Furqan, Uttpal Bhattacharya, Neeraj Kumar Nagda, Sanjana Jose, Vishal Kumar Raghav, Raj Ali, Omveer Singh, S M Junid Hussain, Dharendra Choudhary, Syed Azam Shafat, Vikas Kamboj, Aman Parkash, Abdul Samad, Sunita Priyadarshani Kumar, Sukumar Bhera, Sameer Ahmad, Abhishek Kumar, Sharada R Hubli, L Shivakumar, Sharanu Ganesh, Prakash T Rao, G Seshagiri Rao, Narayan Rao, Binod Kumar Singh, T Satisha Shenoy, CG Roopa, Aruna Gopal, Manoj A, S Sivakumar, Shailaja R, Sudha Bhatt, Revathi S, Javed Ahmed, Susairaj M, Ganesh Kumar, Lakshmi, Kamna Sinha, Poorna Chandra Hegde, Basavaraj Chennagi

DRAUGHTSMEN

ML Date, SY Dighe, DV Dabir, T Vasi, RJ Patel, Devender Singh, Jagdish Mehndiratta, SK Moorjani, HL Narula, SL Kakar, SK Jain, A Goswami, RK Madan, Naresh Kumar, DK Guha, KN Saxena, KK Ganguly, D Sinha, S Majumder, Ved Pal Singh, Jagmohan, K M Joseph, R P Tuli, J K Vij, RK Vasdev, K Sridhar, AK Aggarwal, SK Gupta, Anil Verma, KS John, Ravinder Julka, Rajesh Bansal, WV Mathews, Jacob Idiaulla, A Majumdar, Ravi K Gupta, Suvinder Marwa, Somenath Bose, Ashok Kr Bose, RS Virdi, Ravinder Julka, Cherain Philps, George Kutty, CO Verghees, Pradip Barua, Dushyant Kumar, Dalbir Singh, Samander Singh, Manjit Singh, Subrata Basu, Pancham Singh, N Mayreinder, PRS. Krishnan, A Saksena, Jishu Bhawmik, RM Kutti, TK Thomas, Niloy Mitra, Jacob, Venu Gopal, Krishna Kumar, Rajbir Singh, S Ganesh, George K Alias, K Krishnan Kutty, M Gopa Kumar, Shaji George, Shaji Philip, Abraham George, Sincy Kurian, VS Rana, Sabu Baby, Abhik Kumar Dhar, Usha, Sreeja M, Annie Celine, MS Bisht, Kamlesh Mishra, Netrapal Singh, Sunita Rohella, MP Muralee Mohanan, J Stanly, Amreesh Thapliyal, Lalit Kumar Sanoria, Dinesh Chandra, Tara Sharma, Prata Singh Thakur, Abraham George, Priti Pathak, Rukhsana, K Shiva Kumar, Revathi S, Sonia Chawla, Ganesh Kumar K, Sangeeta Yadav, Lakshmi SV, Suja Alexander, Javed CM, Vandana, Bhanu Joshi, Jyoti, Soumya MN, Geeta Bisht, Neelam Rawat, Kavita G Prasad, Deepali Jain, Viral R Shah, Sam Chacko, Salma, Atul Kashyap, Amit Gulhare, Ankush Sharma, Anju Rathore, Daljeet Singh, Hema Kumar Bisht, Pooja Bhandari, Ankit Kumar Ray, Janki Nailwal, Pooja Negi, Ruby Saifi, Ravi Dutt Sharma, Pratibha Rawat, Neetu Kumari, Arfa Azmi, Mukesh Prasad, Nand Kishor, Sunita, Rajinder Singh Bisht, Pritam, Seema Singh, Manju Pandey, Bali Ram, Prabhat Anand, Sohan Singh Rawat, Amit Prakash, Neha Yadav

ADMINISTRATIVE STAFF

M Mani, TJ Georgekutty, J Duroga, Atmaram Budha, Chalapam, PV Srinivasan, Hira Singh, RK Chakraborty, Murari Lal, Mohan Mali, Sashi Dua, Veena Sharma, N Purushan, D K Barua, Sarita Sahdev, Pauloose Lall, Ram Prasad, Sunita Shrivastav, Jebi Lall, Arunadey Goswami, Abdul Rehman, Chetana Lehay, Seema Malik, Gopal Tiwari, Krishan Prasad, Chandi Mukherjee, Benu Mehndiratta, Sushmita Chakraborty, Archana Mehrotra, Liza George, Aruna Jain, Lalita Thami, Mahendra Singh, Parkash Chand Vemra, Bhoj Raj, Nirmala Devi G, GK Achariya, Govind Prasad, Chaitali Sinha, Deven Kalra, Satish Pal, Gaya Parsad, Uttpal Pradhan, Rajan, Nitu Chauhan, Veena Hegde, Inderjeet Singh, Shailendra Lal, Naresh Kumar, Nilesh Shrivastava, Sheena Gladstone

CHRONOLOGY OF WORKS

PUBLIC BUILDINGS

EXHIBITION HALLS

1. Hindustan Lever Pavilion
1961, New Delhi
Architect: Charles Correa

2. Hall of Nations and Halls of Industries
1971–72, New Delhi
Architect: Raj Rewal

3. M Pavilion
1971–72, New Delhi
Architect: Kuldip Singh

4. DCM Pavilion
1972, New Delhi
Architect: Jasbir Sawhney

5. Hall of States
1981–82, New Delhi
Architect: Raj Rewal

6. Ford Pavilion, Pragati Maidan
1996, New Delhi

STADIA

1. Municipal Stadium and Sports Complex
1962–65, Ahmedabad
Architect: Charles Correa

2. Talkatora Indoor Stadium
1971–72, New Delhi
Architect: NDMC

3. Gymnasium Complex, Pahlavi University, *Unbuilt*
1975–76, Iran
Architect: Modam

4. Indoor Sports Stadium
1979–82, Srinagar
Architect: Kanvinde, Rai & Chowdhury

5. Indoor Stadium
(Competition: First prize winning entry), *Unbuilt*
1980, New Delhi
Architect: Ram Sharma and Raj Rewal

6. Indoor Stadium
(Competition: Winning entry), *Unbuilt*
1980, New Delhi
Architect: Raj Rewal

7. Sports Hall
1988–89, Dehradun
Architect: Romi Khosla

8. Shivaji Stadium
2010, New Delhi
Architect: Raja Aederi

AUDITORIA & CONCERT HALLS

1. Gandhi Memorial Hall
1962–64, New Delhi
Architect: Kanvinde & Rai

2. Tagore Memorial Theatre
1963–65, Ahmedabad
Architect: BV Doshi

3. Shri Ram Centre for Art and Culture
1966–68, New Delhi
Architect: Shiv Nath Prasad

4. Premabhai Hall
1970–74, Ahmedabad
Architect: BV Doshi

5. Shakuntalam Theatre
1972, New Delhi
Architect: The Design Group

6. Auditorium, National Development Dairy Board (NDDB) Campus
1984–86, Anand, Gujarat
Architect: Kanvinde, Rai & Chowdhury

7. Auditorium, Lal Bahadur Shastri National Academy of Administration
1991–92, Mussoorie
Architect: Kanvinde, Rai & Chowdhury

INSTITUTIONAL BUILDINGS

1. Central Pool
1970, New Delhi
Architect: NDMC

2. Assembly Complex, *Unbuilt*
1972, Kuwait
Architect: BV Doshi

3. NDMC/Palika Kendra, Phase I and Phase II
1973–83; 1995–2000, New Delhi
Architect: Kuldip Singh

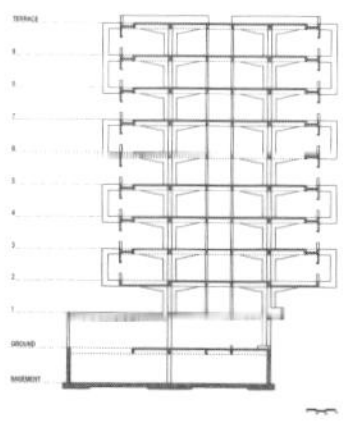

4. Institute of Town Planners
1977–83, New Delhi
Architect: Kanvinde, Rai & Chowdhury

5. Legislative Assembly and High Court
1980–Present, Srinagar
Architect: Kanvinde, Rai & Chowdhury

6. National Institute of Public Finance and Policy
1981–83, New Delhi
Architect: Raj Rewal

7. Vidhan Bhavan
1986–94, Bhopal
Architect: Charles Correa

8. Indian Embassy
1988–89, Kuwait
Architect: Ranjit Sabikhi

9. British Council Library
1989–91, New Delhi
Architect: Charles Correa

10. Permanent Mission of India
1989–92, New York, USA
Architect: Charles Correa

11. Centre for Science and Environment
1992–93, New Delhi
Architect: BV Doshi

12. District Court and Lawyers' Chambers
2000, Rohini, New Delhi
Architect: Jasbir Sawhney

13. Muskaan NGO
2001, New Delhi
Architect: Babbar and Babbar Associates

14. Alliance Française
2004, New Delhi
Architects: SPA Design, ABRD

15. CBI Headquarters
2007, New Delhi
Architect: Raja Aederi

16. National Centre for Biological Sciences, Phase I
2007–10, Bangalore
Architect: ABRD Architects

17 National Centre for Biological Sciences, Phase II
2010–Present, Bangalore
Architect: Kanvinde, Rai & Chowdhury

18. Solar Pavilion
2013, Yarra Village, Madhya Pradesh
Architect: Studio VanRO

EDUCATIONAL & RESEARCH

1. Institute of Catering Technology and Applied Nutrition
1964–66, Bombay
Architect: Charles Correa

2. Indian Institute of Technology Lecture Halls, Core Laboratories and Faculty Buildings
1964–67, Kanpur
Architect: Kanvinde & Rai

3. Institute of Management
(Structural Advisor to NID)
1964, Ahmedabad
Architect: Prof Louis Kahn
Coordinating architects: BV Doshi and A Raje

4. Sir Padampat Research Centre
1974–75, Kota
Architect: Kanvinde & Rai

5. Animal Husbandry and Veterinary Centre
1975–76, Meerut
Architect: Kanvinde & Rai

6. Pahlavi University Dormitories
Unbuilt
1975–76, Iran
Architect: Ranjit Sabikhi

7. International Crop Research Institute for Semi-arid Tropics
1975–78, Hyderabad
Architect: Stein, Doshi & Bhalla

8. Water Research Centre
1977–84, Baghdad, Iraq
Architect: Kanvinde, Rai & Chowdhury

9. University of Agriculture and Technology
1981–83, Faizabad
Architect: Kuldip Singh

10. Indian Institute of Management
1983, Bangalore
Architect: Stein, Doshi & Bhalla

11. SP Jain Advanced Management Research Centre
1984–87, New Delhi
Architect: Ashok Lal

12. Structural Engineers' Research Centre
1985–88, Ghaziabad
Architect: Kanvinde, Rai & Chowdhury

13. SOS Children's Village of India
1987–89, Bhopal
Architect: Ashish Ganju

14. Institute of Carpet Technology
1988–89, Badohi
Architect: Ram Sharma

15. Indian Institute of Health and Medical Research
1988–89, Jaipur
Architect: Ashok Lall

16. TISCO Technical Training Institute
1997–99, Gopalpur
Architect: Charles Correa

17. Indian Institute Of Technology, Lecture Halls and Other Facilities
2012–15, Kanpur
Architect: Kanvinde & Rai

PLACES OF WORSHIP

1. Salvacao Church
1973–77, Bombay
Architect: Charles Correa

2. Mosque
1980–81, Madina, Saudi Arabia
Architect: Sabikhi Choudhury Architects Pvt Ltd

3. Baha'i House Of Worship
(Construction Supervision)
1980–86, New Delhi
Architect: Fariborz Sahba

4. Sai Convention Centre
1999–2000, Bangalore
Architect: Arcop Associates

MUSEUMS & CULTURAL CENTRES

1. Nehru Pavilion
1971–72, New Delhi
Architect: Raj Rewal

2. Museum and Cultural Centre
Unbuilt
1976–78, Karmanshah, Iran
Architects: Modam and Michel Ecochard

3. Crafts Museum
1978–79, New Delhi
Architect: Charles Correa

4. Nehru Science Centre
1979–81, Bombay
Architect: Kanvinde, Rai & Chowdhury

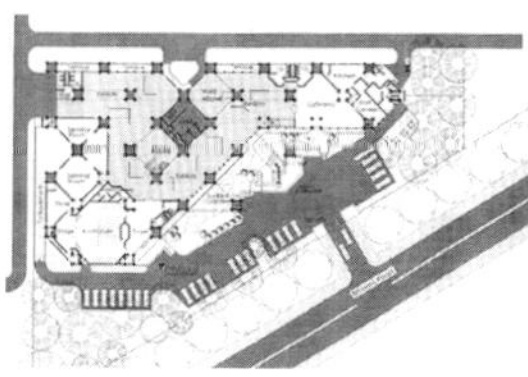

5. National Science Centre
1987–89, New Delhi
Architect: Kanvinde, Rai & Chowdhury

6. Central Gallery of Modern Art
1992–93, Mumbai
Architect: Romi Khosla

7. Salarjung Museum
1995–96, Hyderabad
Architect: Design Team Consultants

8. Rashtriya Manav Sanghralaya
1995–2000, Bhopal
Architect: Ram Sharma

9. Bihar Museum
2015, Patna
Architects: Maki and Associates, Opolis

HOSPITALS & MEDICAL FACILITIES

1. TB Hospital
1964–67, Ahmedabad
Architect: BV Doshi

2. Srinagar Hospital Phase-II
1977–81, Srinagar
Architect: Kanvinde, Rai & Chowdhury

3. Escorts Heart Institute
1999–2000, New Delhi
Architect: Arvind Gupta

4. MCD Primary Health Care
2011–12, New Delhi
Architect: Romi Khosla

5. Indira Gandhi PWD Hospital
Under Construction, New Delhi
Architect: Jasbir Sawhney

COMMERCIAL

OFFICES

1. **New Great Insurance Company of India**
1961–63, Bombay
Architect: Studio Technio

2. **Central Bank**
1966–69, Ahmedabad
Architect: BV Doshi

3. **New Kamani Chambers**
1967–69, Bombay
Architect: Bose Brothers

4. **Bawa Dan Singh Office Building,** *Incomplete*
1971, New Delhi
Architect: Kothari & Associates

5. **Deenar Bhawan**
1974–75, New Delhi
Architect: Vijay Gupta

6. **State Trading Corporation**
1977–88, New Delhi
Architect: Raj Rewal

7. **NCDC Office**
1978–80, New Delhi
Architect: Kuldip Singh

8. **LIC Jeevan Bharti**
1979–85, New Delhi
Architect: Charles Correa

9. **NDDB**
1981–82, New Delhi
Architect: Kanvinde, Rai & Chowdhury

10. **SCOPE**
1982–87, New Delhi
Architect: Raj Rewal

11. **NDDB**
1984–86, Anand, Gujarat
Architect: Kanvinde, Rai & Chowdhury

12. **Mobile Creche**
1987–88, New Delhi
Architect: Benjamin

13. **Computer Maintenance Centre**
1987–91, Hyderabad
Architect: Design Plus

14. **MP State Trading Corporation Office Building**
1989–91, Bhopal
Architect: Charles Correa

15. **MP Urja Nigam Ltd Office Complex**
1989–92, Bhopal
Architect: Sanjay Wadhwa

16. **Hughes Software Systems**
1993–95, Gurgaon
Architect: Arcop Associates

17. **Capital Court, Munirka**
1997–99, New Delhi
Architect: Raja Aederi

18. **Administrative Building for Chambal Fertilisers**
2000, Kota
Architect: William Perry (MRC)

19. **Bharat Diamond Bourse**
1994–98, Bombay
Architect: BV Doshi

20. **Usha India Office Building**
1996–98, New Delhi
Architect: William Perry (MRC)

21. **Birla Corporate Head Office**
1996–2000, New Delhi
Architect: Arcop Associates

22. **First India Place, Gurgaon**
1997–99, Gurgaon
Architect: WS Atkins

23. **God Gift Commercial Complex**
1999–2000, Gurgaon
Architect: RC Gulati

24. **Xansa Software Office**
2001–02, Noida
Architect: Jasbir Sawhney

25. **Bearys Group Research Triangle**
2005–09, Bangalore
Architect: NM Salim & Associates

26. **Triburg Headquarters**
2006, Gurgaon
Architect: SPA Design

27. **Tahiliani Design Headquarters**
2006, Gurgaon
Architect: SPA Design

28. **Artha Infratech Pvt Ltd**
Under Construction
Noida
Architect: Raj Rewal

HOTELS

1. **Akbar Hotel**
1967–69, New Delhi
Architect: Shiv Nath Prasad

6. **Hotel Bhramaputra Ashok**
1982–87, Guwahati
Architect: Bose Brothers

7. **Hotel Kalinga Ashok Phase II**
1984–87, Bhubaneshwar
Architect: Ram Sharma

8. **Hotel Oberoi Novotel**
1991–92, Agra
Architect: William Perry (MRC)

9. **Hotel Residency Manor**
1991–93, Mussoorie
Architect: Shiben Ganju

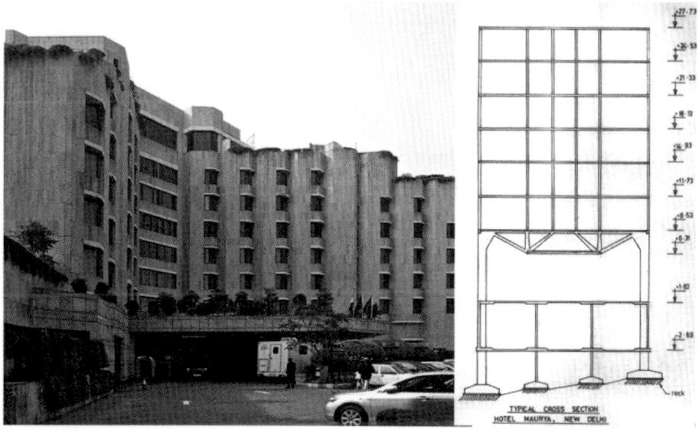

2. **Maurya Hotel and Tower**
1975–83, New Delhi
Architect: Rajinder Kumar

3. **Centaur Lake View Hotel and Conference Complex**
1979–83, Srinagar
Architect: Stein, Doshi & Bhalla

4. **ITDC Motel**
1980, Jammu
Architect: Kanvinde, Rai & Chowdhury

5. **Park Hotel**
1981–86, New Delhi
Architect: Bose Brothers

10. **Jaypee Palace Hotel**
1994–97, Agra
Architect: Arcop Associates

11. **Gokarna Golf Resort**
1998–99, Kathmandu, Nepal
Architect: Romi Khosla

12. **Hyatt Regency**
1999–2000, Goa
Architect: Jasbir Sawhney

13. **Lodhi Hotel**
2003–07, New Delhi
Architect: Kerry Hill Architects

MARKETS

1. **Commercial Centre**
1981–87, Cochin
Architect: Kuldip Singh

2. **Fruit and Vegetable Project**
1985–87, New Delhi
Architect: Design Group

3. **Flower Market**
1987–89, Madras
Architect: Kuldip Singh

4. **Wholesale Fruit and Vegetable Market**
1989–90, Madras
Architect: Kuldip Singh

5. **Crescent Shopping Centre**
2001–04, New Delhi
Architect: Aakar Design Consultants

6. **Sangam Cinema**
2013, New Delhi
Architect: Arcop Associates

INDUSTRIES

1. **Danfoss Factory**
1966–67, Ghaziabad
Architect: Rai & Bali

2. **Silver Cotton Mills**
1971–72, Ahmedabad
Architect: BV Doshi

3. **Hindon River Mills**
1971–73, Ghaziabad
Architect: Kanvinde & Rai

4. **Mother Dairy, NDDB**
1972–75, Bombay, New Delhi, Madras and Calcutta
Architect: Kanvinde & Rai

5. Feeder Dairies and Feeder Balancing Dairies, NDDB
1973–75, Rohtak, Meerut, Patna, Jalgaon, Varanasi, Bikaner, Sangam, Salem, Alwar, Bhilwara, Erode, Sirsa and Ajmer
Architect: Kanvinde & Rai

6. IDC Godowns, NDDB
1974–76, Delhi, Bombay, Calcutta and Madras
Architect: Kanvinde & Rai

11. Soyabean Oil Mill
1984–87, Ujjain
Architect: VK Girdhar

12. Wheels India
1984–85, Rampur
Architect: Kanvinde, Rai & Chowdhury

17. 30 TPD Powder Plants, NDDB
1987–89, Meerut, Ludhiana, Tumkur, Ongole, Erode and Banaskantha
Architect: Kanvinde, Rai & Chowdhury

18. Tannery, *Unbuilt*
1989–90, Kazakhstan
Architect: AAKAR Design Consultants

19. Pepsi Snack & Concentrate
1989–91, Patiala
Architect: Akshay Jain and Rakha Chakraburty (MRC)

20. NDDB Dairies
1989–90; 1994, Lucknow and Sirsa
Architect: Kanvinde, Rai & Chowdhury

21. Taj Bottling Plant
1990–91, Jaunpur
Architect: MRC

22. Press Shop Caparo Maruti
1994–96, Gurgaon
Architect: Charles Correa

23. Delhi Flour Mills
2011–15, Greater Noida
Architect: Studio VanRO

7. DCM Engineering Products
1975–77, Ropar
Architect: Kanvinde & Rai

13. DCM LCV Plant
1984–86, Ghaziabad
Architect: Kanvinde, Rai & Chowdhury

8. Engine Bearing Plant
1977–78, Parwanoo
Architect: Kanvinde & Rai

14. Eicher LCV Plant
1985–86, Indore
Architect: Jasbir Sawhney

15. Shock Absorber Factory, Munjal Showa Ltd
1985–86, Gurgaon
Architect: Jasbir Sawhney

9. Dairies, NDDB
1977–84, Jaipur and Gwalior
Architect: Kanvinde, Rai & Chowdhury

10. Paper Lamination Plant, NDDB
1979–81, Itola
Architect: Kanvinde, Rai & Chowdhury

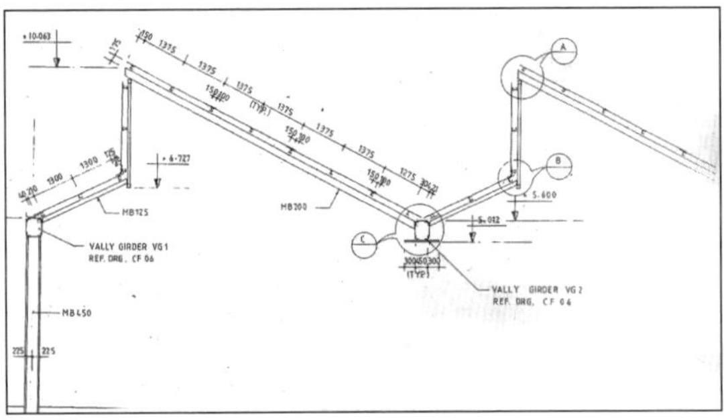

16. Telephone Cable Factory
1987–88, Mohali
Architect: Ram Sharma

INFRASTRUCTURE

AIRPORTS

1. Indira Gandhi International Airport
1982–86, New Delhi
Architect: Charles Correa

2. Hyderabad Airport Integrated Terminal and Expansion
1988–89; 1997; 2003, Hyderabad
Architect: AAI, Kiran Chimalwar

3. Patna Airport, Expansion and Modification
1995–97, Patna
Architect: AAI, Kiran Chimalwar

4. Feasibility Report of Airport at Khurik and Ranikhet
1997–98, Himachal Pradesh

8. Jaisalmer Airport
2008–11, Rajasthan
Architects: AAI, Studio VanRO

5. Leh Airport, *Unbuilt*
2007, Leh
Architect: AAI, Studio VanRO

6. Udaipur Airport Tower
2007, Udaipur
Architect: AAI, Kiran Chimalwar

7. Coimbatore Airport Expansion
2008, Tamil Nadu
Architect: AAI, Studio VanRO

BUS TERMINAL

1. Inter State Bus Terminal
1968–73, Delhi
Architect: NDMC

RAIL DEPOTS

1. DMRC Maintenance and Stabling Depot, Najafgarh
2003–04, New Delhi
Architect: Kuldip Singh

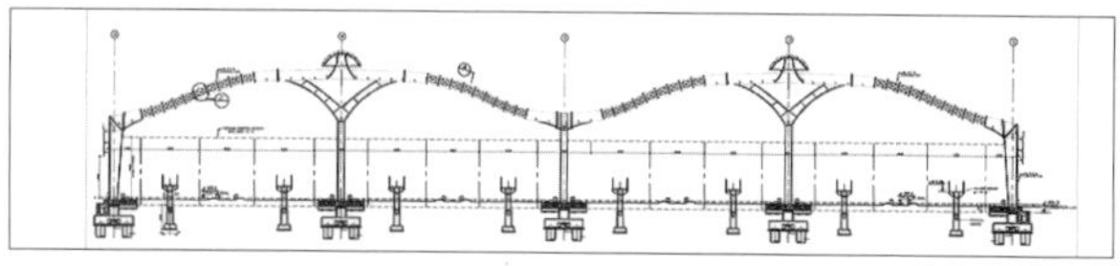

2. DMRC Maintenance and Stabling Depot; Station and Housing
2003–04, New Delhi
Architect: Kuldip Singh

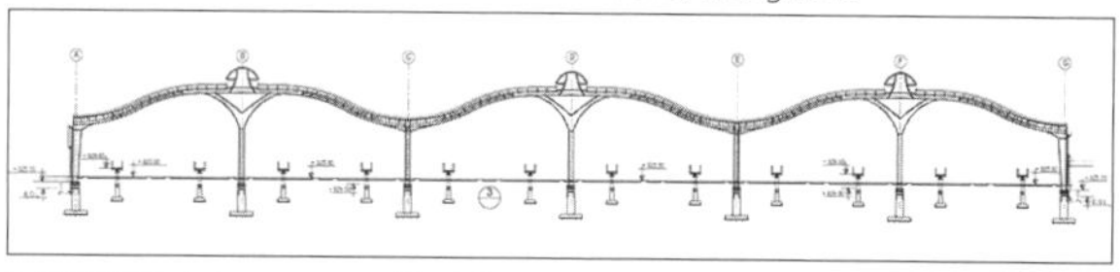

3. BMRC Maintenance and Stabling Depot at Byappanahalli
2007–09, Bangalore
Architect: Kuldip Singh

4. BMRC Maintenance and Stabling Depot at Peenya
2009–10, Bangalore
Architect: Kuldip Singh

5. Rapid Metrorail Gurgaon Ltd
Phase I and II
2010–13, New Delhi
Architect: Jasbir Sawhney

6. DMRC Kalindi Kunj and Ajronda Depot
2012, New Delhi
Architect: Kuldip Singh

7. DMRC Jasola Depot,
Under Construction
New Delhi
Architect: Kuldip Singh

PORTS AND HARBOURS

These projects were designed for piling contractors, Rodio Hazrat, now known as Asia Foundation and Constructions Ltd.

1. Wharf Extension
1966–67, Dabhol

2. Wharf
1970, Mangalore

3. General Cargo Berth
1971, Paradip Port

4. Modifications to Oil Jetty
1971, Kandla

5. Intake Jetty for Desalination Plant
1975, Oman

6. Iron Ore Berth
1978, Mangalore

7. Jetty
1977, Jamnagar

TOWERS

TV towers: 105–115m high towers; the lower 60–75m is a slip-formed tapering concrete shaft and the top 45m is in steel. Each tower has a provision for an extension of about 22.5m

1. Gujarat State Fertilisers Co. Water Tower
1964–67, Baroda
Architect: BV Doshi

2. Signal Station
1977, Visakhapatnam

3. TV Tower
1984–85 Nagpur

4. TV Tower
1984–85, Kasauli

5. TV Tower
1984–85, Panjim

6. ARSR Tower
1993, Bombay

7. ARSR Tower
1993, Bijwasan

RESIDENTIAL

1. Usha Kiran Apartments
1964–67, Bombay
Architect: Karim Noorani & Co

2. Township Gujarat State Fertilisers Co.
1964–67, Baroda
Architect: BV Doshi

3. Madan CGHS Apartments
1965–67, Bombay
Architects: Bhosle, Raja Aederi

4. Pereira House
1966–68, Bombay
Architect: Raja Aederi

5. Adinath CGHS Apartments
1967–69, Ahmedabad
Architect: BV Doshi

6. Residence of Prama Bhandari
1968, New Delhi
Architect: BV Doshi

7. French Embassy Staff Quarters
1969–70, New Delhi
Architect: Raj Rewal

8. Dormitories, Staff College
1973–76, Hyderabad
Architect: BV Doshi

9. Malviya Nagar DDA Housing
1976–79, New Delhi
Architect: Kuldip Singh

10. Cochin Housing
1976–79, Cochin
Architect: Kuldip Singh

11. Arbita Housing, *Unbuilt*
1977–78, Tehran, Iran

12. Social Insurance Building
1977–79, Jeddah, Saudi Arabia
Architect: Ranjit Sabikhi

13. Tara Apartments
1977–80, New Delhi
Architect: Charles Correa

14. LIC Housing
1978–79, New Delhi
Architect: Kuldip Singh

15. Sheikh Sarai DDA Housing
1979–81, New Delhi
Architect: Raj Rewal & Kuldip Singh

16. Zakir Hussain Housing
1980–82, New Delhi
Architect: Raj Rewal

17. AWHO Housing
1983–85, Noida
Architect: Kuldip Singh

18. Residence of Dr Raj Kubha
1983–84, New Delhi
Architect: Ashish Ganju

19. Apartments, Prithviraj Road
1984–86, New Delhi
Architect: Jasbir Sawhney

20. Apartments, South End Lane
1985–87, New Delhi
Architect: CP Sabharwal

21. UP Irrigation Guest House
1986–87, Ghaziabad
Architect: Jasbir Sawhney

22. Residence of Sudhir Kakkar
1987–88, New Delhi
Architect: Ashish Ganju

23. Garden Estate
1987–89, Gurgaon
Architect: Arcop Associates

24. IIT Group Housing
1990–91, New Delhi
Architect: Raj Rewal

25. Laburnum
1994–97, Gurgaon
Architect: Arcop Associates

26. Malibu Towne
1996–98, Gurgaon
Architect: RC Gulati

27. Enron Residential Complex
1996–97, Dabhol
Architect: William Perry (MRC)

28. Hamilton Court Housing
1996–98, Gurgaon

29. IFS Housing
2001–03, Gurgaon
Architect: Abhimanyu Dalal

30. Lakeside Habitat View
2004–05, Bangalore
Architect: NM Salim

31. Orchid Petals
2005–08, Gurgaon
Developer: Orchid Infrastructure

32. The View
2007–10, Mohali
Architect: Arcop Associates

33. Jaypee Greens
2007–11, Noida
Architect: Arcop Associates

34. DMRC Housing,
Unbuilt
2008, Noida
Architect: Studio VanRO

CREDITS

PHOTO CREDITS

Mahendra Raj is grateful to all the photographers, site engineers and architects who have contributed to the enormous collection of photographs over the last sixty years. Any omissions in granting credit are unintentional and, where notified, will be included in later editions. The photographs are listed according to the page number they appear in.

Ariel Huber 2, 6-7, 20-21, 60, 62-63, 64-65 (Top), 65 (Bottom right), 67, 71 (Bottom right), 77, 80-81 (Top), 89, 90, 91, 97, 100-101 (Top), 105, 106, 107, 108-109, 120-121 (Top), 127, 134, 135, 150-151, 157, 161, 179, 192-193 (Top), 200-201, 202-203, 205 (Bottom right), 212-213 (Top), 219, 222-223 (Top), 225, 227, 235, 249 (Bottom right), 254-255, 262-263, 265, 273, 279, 280-281, 285 (Top right), 293 (Bottom), 312-313, 314-315 (Top), 318, 319, 320-321, 330-331, 358-359, 362-363, 364, 365 (Top right and bottom), 391 (Top right), 398-399, 414 (Public Buildings: Stadia: 1, 4), 415 (Auditoria: 3, 4; Institutional Buildings: 5),416 (Educational & Research: 3), 418 (Commercial: Offices: 2, 6, 9, 10, 19), 419 (Hotels: 3; Markets: 3, 4; Industries: 3)

Bärbel Högner 30

Charles Correa Associates 38-39, 40, 41, 42-43, 44, 46, 47, 223 (Bottom right), 224 (Top left, centre left, bottom left, top right, centre right), 351 (Bottom right), 356, 357, 414 (Public Buildings: Exhibition: 1)

Christian Brunner 74, 175, 177 (Bottom), 248, 293 (Top), 300 (Left)

Jasbir Sawhney & Associates 416 (Institutional Buildings: 12)

Madan Mahatta Archives and PHOTOINK 12-13, 16, 54-55, 98-99, 118-119, 121 (Bottom right), 139, 144, 164-165, 170-171, 176, 177 (Top), 178, 180-181, 183 (Bottom), 190-191, 238-239, 242, 251, 253, 259 (Bottom right), 286, 287, 288-289, 290, 291 (Bottom right), 294 (Top right), 301, 302-303, 304, 306-307, 310, 322,329, 392-393, 414 (Public Buildings: Exhibition: 2,4,5), 415 (Auditoria & Concert Halls: 5), 418 (Commercial: Offices: 7, 8), 419 (Hotels: 1), 422 (Residential: 7, 9, 10)

Maki and Associates 380-381, 382-383, 387 (Top right, bottom right)

Opolis 384 (Bottom left), 387 (Bottom left, centre right)

Rohit Raj Mehndiratta 32 (Centre right), 366-367, 368-369, 370, 371, 372-373, 374-375, 376, 377, 378, 379, 387 (Top left, centre left), 413, 421 (Airports: 7, 8)

Stephane Paumier Architects, F Delangle 360-361, 361 (Bottom left), 416 (Institutional Buildings: 14), 418 (Commercial: Offices: 27) (Commercial: 27)

Vastu Shilpa Foundation 5, 213 (Bottom right)

Assembly Hall by Dori at en.wikipedia. Licensed under CC BY 2.0 via Commons - https://commons.wikimedia.org/wiki/File:Assembly_Hall.jpg#/media/File:Assembly_Hall.jpg 32 (Second from top left)

Mahendra Raj Archives 416 (Institutional buildings: 16) ABRD Architects; 224 (Bottom right) Ashwin Mehta; 52 (Top and centre) Bhogilal Thakkar; 220-221, 226 (Left), 305 (Bottom right), 311, 345 (Bottom right), 416 (Institutional Buildings: 10), 417 (Places of Worship: 1), 422 (Residential: 13) Charles Correa Associates;199 (Bottom left and bottom right) Grand Studio; 158, 159, 163, 167, 169, 173, 428 Hellans Industrial and Pictorial Photographers; 128-129, 137 ITDC; 123, 152-153, 154-155, 283 (Bottom right), 285 (Bottom right), 294 (Top left, centre left, bottom left), 315 (Bottom right), 325 (Top right), 414 (Public Buildings: Stadia: 5,6),415 (Auditoria & Concert Halls: 1; Institutional Buildings: 3), 417 (Museums & Cultural Centres: 1), 422 (Residential: 11) Madan Mahatta; 116 (Right), 215 (Bottom left) Parmanand Dalwadi; 32 (Bottom right) The Photo Studios; 182-183 (Top), 185, 189, 199 (Top left, centre), 207, 410-411 Rangoon Studio; Cover, 4 (Bottom left),14-15, 48-49, 51 (Bottom left), 53, 140, 194-195, 196, 209 Stella Snead; 78-79, 96, 115 (Bottom left), 214 (Bottom), 217, 416 (Educational & Research: 10), 422 (Towers: 1; Residential: 2, 6) Vastu Shilpa Foundation; 1, 23 (Bottom right), 24, 25, 27 (Bottom right), 28, 29, 31, 32 (Top left, bottom left), 36, 51 (Bottom right), 52 (Bottom), 66, 68, 71 (Top right), 73, 81 (Bottom right), 85, 87, 95, 110-111 (Top), 111 (Bottom), 112, 113, 115 (Bottom right), 116 (Left), 117, 193 (Bottom left and right), 199 (Top right, centre left, centre right), 205 (Top right), 208, 210-211, 215 (Bottom right), 218, 236, 240-241, 256-257, 260, 261, 267, 269, 275, 277, 308, 309, 327, 333, 340, 342-343, 345 (Top right, centre), 347, 348, 349, 351 (Top right), 353, 355, 409, 414 (Public Buildings: Exhibition: 3, Stadia: 2), 415 (Institutional Buildings: 2, 4, 7, 9), 416 (Institutional Buildings: 13; Educational & Research: 7, 12), 417 (Places of Worship: 4; Museums & Cultural Centres: 3, 4, 5, 6, 7, 8; Hospitals & Medical Facilities: 1, 2), 418 (Commercial: Offices: 1, 11, 16, 17, 22, 24, 26), 419 (Commercial: Hotels: 2, 10, 13; Markets: 5; Industries: 1, 4), 420 (Commercial: Industries: 7, 8, 9, 10, 11, 14, 22, 23), 421 (Infrastructure: Airports: 1, 6; Bus Terminal: 1), 422 (Towers: 2; Residential: 1, 12, 15, 16), 423 (Residential: 23, 25, 26, 27, 28, 29, 30), 425

DRAWING CREDITS

All drawings, calculation sheets and diagrams shown in the book are © Mahendra Raj. The following drawings shown in this book have been donated to Centre Pompidou, Paris.

HALL OF NATIONS & HALLS OF INDUSTRIES, New Delhi, 1972
Drawing no. 101.ITF. 02: Plans, Elevation, Section (Hall of Nations): 16 August 1971
Drawing no. 101.ITF.15B: Details of Pile Caps and Tie Beams (Hall of Nations): 21 October 1971
Drawing no. 101.ITF.59B: (Part) Inner Elevation (Hall of Nations): 16 November 1971
Drawing no. 101.ITF.55C: Reinforcement Layout at Level 1&2 (Hall of Nations): 10 November 1971
Drawing no. 101.ITF.31: Detail of Joints Nos. 79, 65, 52, 40 (Hall of Nations)
Drawing no. 101.ITF.64: Reinf. Details; Typical Joint with Nine Members: Inner Face (Hall of Nations)
Drawing no. 101.ITF.53A: Reinf. Details; Joint Nos. 88, 100, 104, 110 & 113 (Hall of Nations)
Drawing no. 101.ITF.54: Reinf. Details; Joint Nos. 94, 99, 105 & 109 (Hall of Nations)
Drawing no. 101.ITF.52: Reinf. Details; Joint Nos. 115, 116 (Hall of Nations)

SPORTS STADIUM, Srinagar, 1982
Drawing no. 107-SS-S01: Main Structure, Grid System, Members, Plates and their Numbers: 12 May 1980
Drawing no. 107-SS-PS-13: A Proposed Sequence of Construction: 16 August 1979
Drawing no. 107-SS-S01-1: Main Structure—Basic Geometry: 8 May 1980
Drawing no. 107-SS-S08/A: Details of Pedestals (Sheet1) All Pile Caps except PC1: 12 April 1980
Drawing no. 107-SS-S12/A: Main Structure—Reinforcement Details: 29 May 1980

View of a lugget joint during construction,
Indoor Sports Stadium, Srinagar.

Drawing no. 107-SS-S01-2: Main Structure—Joint Details: 25 May 1980
Drawing no. 107-SS-S18/B: Details of Upper Seating: 11 September 1980
Drawing no. 107-SS-S21: Main Structure—Post-Tensioned Members at Levels 2, 3, Details at Anchorage: 28 November 1980

HALL OF STATES, New Delhi, 1982
(Partial) Drawing no. 100SEC-S07.1: Girder Details (Sheet-2): 30 September 1981
Drawing no. 100SEC-S06: Second Floor Framing Plan: 4 April 1981

SCOPE COMPLEX, New Delhi, 1987
Drawing no. 100SC19 A: Details @-1.802m (Upper Basement Level): 4 March 1982

STATE TRADING CORPORATION, New Delhi, 1988
Drawing no. 101ST-S01/A: Layout of Piles: 21 December 1977
Drawing no. 101ST-S19-B: Framing Plan for Typical Floors 4th-13th: 21 August 1983
Drawing no. 101ST-17.6A: Details at 2nd and 3rd Floors CCIC Block: 6 February 1985
Drawing no. 101ST-S19-4: Detail of Girders at 8th to 13th Floor: 8 August 1984

FLOWER MARKET, Madras, 1990
Drawing no. 101FM 34: Plan at +18.59m (showing umbrella roofs) Part C and D: 25 October 1988
Drawing no. 101FM 35: Details of 12.9m Umbrella Roofs: 19 October 1988

CONTRIBUTORS

Abba Tor started his own practice in 1955 in Tel Aviv after graduating from the Israel Institute of Technology (Dip. Eng. 1946) and earning a Master's degree from Columbia University School of Engineering (1953). What started as a significant commission in collaboration with a US firm ended unexpectedly soon after. Tor decided to extend his stay in the US and join the renowned engineering firm of Ammann and Whitney, where he worked for eight years as a senior design engineer and project engineer on significant projects (Eero Saarinen's TWA Terminal, John Deere Corporate Center among others).

In 1966, he accepted a partnership offer from Henry A. Pfisterer, a professor at Yale University School of Architecture. After Prof. Pfisterer's death in 1972, Abba Tor continued to head the firm and its successors until his retirement in 1996. During his career in the US he worked with leading architects such as Eero Saarinen, his successor firm Kevin Roche, John Dinkeloo and Associates, Louis Kahn, Edward Larrabee Barnes, Harrison & Abramowitz, and James Stewart Polshek. Besides his professional work as a structural engineer, Tor has also taught as an adjunct professor at the Columbia University School of Architecture and as guest lecturer and critic at Syracuse University Parsons School of Design, Pratt Institute and Princeton University Schools of Architecture, as well as the Israel Institute of Technology. Abba Tor also served as active member and senior peer reviewer for the American Consulting Engineers Council and is a Life Fellow of the American Society of Civil Engineers (ASCE).

Ariel Huber is an architectural photographer based in Lausanne. He graduated in architecture from ETH Zurich in 1998 and worked for three years as project architect at Studio Daniel Libeskind and Barkow Leibinger Architects in Berlin. In 2001, he returned to Switzerland to start working as an independent photographer focussing on architecture, engineering, art documentation and heritage preservation. Since 2002 he has been travelling to India regularly, photographing projects of Mahendra Raj, Rahul Mehrotra, Studio Mumbai, Opolis and many others.

He has published photographs and writings in various international architecture magazines and publications, and most recently, in the monograph on the Swiss Embassy Building in New Delhi.

During 2004-2005 he was a research assistant with Christian Kerez at ETH Zurich where he explored early steel frame structures in USA and Europe. From 2004 to 2012 he was a lecturer in 3D animation and architectural visualisation at Zurich University of the Arts. Since then he has been giving lectures and conducting workshops on photography and architectural imaging at ETH Zurich, EPFL Lausanne, ZHAW Winterthur and CEPT Ahmedabad, to name a few.

Hans Ulrich Obrist is Co-Director of the Serpentine Galleries, London. Prior to this, he was the curator of the Musée d'Art Moderne de la Ville de Paris. Since his first show *World Soup (The Kitchen Show)* in 1991 he has curated more than 250 shows.

In 2009 Obrist was made Honorary Fellow of the Royal Institute of British Architects (RIBA), and in 2011 received the CCS Bard Award for Curatorial Excellence. Obrist has lectured internationally at academic and art institutions and is contributing editor to several magazines and journals.

Obrist's recent publications include *Lives of the Artists, Lives of the Architects, Ways of Curating, A Brief History of Curating, Do It: The Compendium,* and *The Age of Earthquakes* with Douglas Coupland and Shumon Basar.

Jaimini Mehta is a practicing architect and an independent academic based in Baroda. He studied architecture at MS University of Baroda, and at the University of Pennsylvania in the Louis Kahn Studio. He went on to work in the offices of Louis Kahn and Mitchell/Giurgola Associates in Philadelphia. In 1976 he established his practice under the name of Studio Jaimini Mehta, which has been engaged in designing residential, institutional, commercial and urban design projects. Their work has been published and exhibited in India and abroad.

At present he is the Hon. Director of the Baroda-based Centre for the Study of Urbanism and Architecture, which he instituted in 2006. He was an Adjunct Professor at Rensselaer Polytechnic Institute at Troy, New York, and at CEPT University in Ahmedabad. He also worked as Head of the Schools of Architecture in Baroda and Goa. His published books are *Louis I. Kahn, Architect* co-authored with Romaldo Giurgola (1975), *Rethinking Modernity - Towards Post Rational Architecture* (2011) and *Embodied Vision – Interpreting the Architecture of Fatehpur Sikri* (2014).

Mark Jarzombek, Associate Dean, School of Architecture and Planning, and Professor of the History and Theory of Architecture at Massachusetts Institute of Technology, Cambridge, works on a wide range of historical topics from the 12th century to the modern era with a particular focus on nineteenth and twentieth century aesthetics, and the history and theory of architecture. He is one of the country's leading advocates for global history and has published several books and articles on the topic, including the ground-breaking textbook entitled *A Global History of Architecture* (Wiley Press, 2006) with co-author Vikram Prakash and noted illustrator Francis DK Ching. He is the sole author of *Architecture of First Societies: A Global Perspective* (Wiley Press, 2013), which is a sensitive synthesis of first society architecture through time and includes custom-made drawings, maps and photographs.

He was a CASVA fellow (1985), post-doctoral resident fellow at the J. Paul Getty Center for the History of Humanities and Art, Santa Monica, California (1986), and a fellow at the Institute for Advanced Study, Princeton, NJ (1993), Canadian Center for Architecture (2001) and Sterling and Francine Clark Art Institute (2005) respectively. He serves on the board of several journals and academic institutions including the SSRC and the Buell Foundation, and was a member of the 2011 Deutsche Forschungsgemeinschaft (DFG, German Research Foundation) Excellence Initiative.

Neelkanth H Chhaya is an academic and practicing architect. He has taught at the University of Nairobi, the Institute of Environmental Design at Vallabh Vidyanagar, and was a faculty member at CEPT University, Ahmedabad, from which he retired as Dean of the Faculty of Architecture in 2013. At present he is an adjunct faculty member at Srishti School of Design, Bangalore, and holds the position of Academic Chair at the Goa College of Architecture.

His professional practice has focussed on culturally and environmentally appropriate design and has emphasised innovative application of local skills and materials. Recently he has been closely associated with Hunnarshala Foundation and Khamir Crafts Resource Centre, both of which are organisations working to support the craft and artisanal communities in the Kutch region of Gujarat, India.

Rohit Raj Mehndiratta graduated from School of Architecture CEPT, Ahmedabad, and received an SMArchS degree in Architecture and Urbanism from Massachusetts Institute of Technology (MIT) in 2002. Among other accolades, he received the Merit Recognition Award for the years 2000-01 and 2001-02 from MIT.

His work experience includes an initial training at Frei Otto's Institute for Lightweight Structures in Stuttgart. He went on to work for five years in New York with Davis Brody Bond (DBB) and Elizabeth Kennedy Landscape Architects (EKLA). At DBB he worked on The World Trade Center Memorial, New York, and Nicolas Grimshaw's Experimental Media and Performance Arts Center in RPI, Troy, NY State.

Since 2007 he is the principal of Studio VanRO, an award-winning firm with a body of work that has been widely published and exhibited. Owing to an interest in researching visual urban representations and politics of development he has made presentations at a number of international conferences and published a paper in Sarai Reader 09. He has taught in the Urban Design Program of SPA (School of Planning and Architecture), New Delhi, and is currently a visiting faculty member at the Sushant School of Architecture. He has been a part of expert panel committees on public arts initiatives at the Delhi Urban Art Commission, and UTTIPECís MRTS Corridor Development sub group formed by the Ministry of Urban Development, Delhi Division.

Sanjay Prakash, B. Arch, AIIA, is an architect with a commitment to energy-conscious architecture, eco-friendly design, people's participation in planning, music and production design. Over the years he has integrated all his work with the practice of new urbanism and sustainability in his professional and personal life. His area of practice and research over the last 30 years includes passive and low energy architecture and planning, hybrid air-conditioning, autonomous energy and water systems, bamboo and earth construction, community-based design of common property, and computer-aided design. Under his guidance hundreds of individuals have developed capabilities in performing design, conceptual or management work in these areas.

Vandini Mehta is an architect and independent urban researcher. She received her undergraduate degree from School of Architecture CEPT, Ahmedabad, (1997) and an MS in Architecture from University of California, Berkeley, with a specialisation in Environmental Design in Developing Countries. Her academic and research interests include investigations of urban histories and geographies as revealed in the contemporary city, film and space, especially of current Bollywood cinema. She has presented her research work at several international conferences. She is a visiting faculty member in the Urban Design Program of SPA (School of Planning and Architecture), New Delhi and at Sushant School of Architecture, Gurgaon.

After non-profit community-based development and architectural work early in her career at Seva Mandir, Udaipur, she worked for five years in New York first at McBride & Associates, a small design firm, and later at Magnusson Architecture and Planning, a firm specialising in social housing solutions in partnership with the community and developers, for mixed-use and mixed-income housing and planning projects. Upon her return to India in 2007, she started her own consultancy firm called Studio VanRO in partnership with Rohit Raj Mehndiratta. Since its inception the firm has realised projects of varied scales and programs that have been widely published and exhibited.

Following page: View of the space frame during construction, Hall of Nations, New Delhi.